Lecture Notes in Bioinformatics

Edited by S. Istrail, P. Pevzner, and M. Water

Subseries of Lecture Notes in Computer Science

Ulf Leser Felix Naumann
Barbara Eckman (Eds.)

Data Integration in the Life Sciences

Third International Workshop, DILS 2006
Hinxton, UK, July 20-22, 2006
Proceedings

Series Editors

Sorin Istrail, Brown University, Providence, RI, USA
Pavel Pevzner, University of California, San Diego, CA, USA
Michael Waterman, University of Southern California, Los Angeles, CA, USA

Volume Editors

Ulf Leser
Felix Naumann
Humboldt-Universität zu Berlin, Institut für Informatik
Unter den Linden 6, 10099 Berlin, Germany
E-mail: {leser, naumann}@informatik.hu-berlin.de

Barbara Eckman
IBM Application and Integration Middleware
1475 Phoenixville Pike, West Chester, PA 19380, USA
E-mail: baeckman@us.ibm.com

Library of Congress Control Number: 2006928955

CR Subject Classification (1998): H.2, H.3, H.4, J.3

LNCS Sublibrary: SL 8 – Bioinformatics

ISSN 0302-9743
ISBN-10 3-540-36593-1 Springer Berlin Heidelberg New York
ISBN-13 978-3-540-36593-8 Springer Berlin Heidelberg New York

Springer is a part of Springer Science+Business Media

springer.com

Typesetting: Camera-ready by author, data conversion by Scientific Publishing Services, Chennai, India
Printed on acid-free paper SPIN: 11799511 06/3142 5 4 3 2 1 0

Preface

Data management and data integration are fundamental problems in the life sciences. Advances in molecular biology and molecular medicine are almost universally underpinned by enormous efforts in data management, data integration, automatic data quality assurance, and computational data analysis. Many hot topics in the life sciences, such as systems biology, personalized medicine, and pharmacogenomics, critically depend on integrating data sets and applications produced by different experimental methods, in different research groups, and at different levels of granularity. Despite more than a decade of intensive research in these areas, there remain many unsolved problems. In some respects, these problems are becoming more severe, both due to continuous increases in data volumes and the growing diversity in types of data that need to be managed. And the next big challenge is already upon us: the need to integrate the different "omics" data sets with the vast amounts of clinical data, collected daily in thousands of hospitals and physicians' offices all over the world.

DILS 2006 is the third in an annual workshop series that aims at fostering discussion, exchange, and innovation in research and development in the areas of data integration and data management for the life science. DILS 2004 in Leipzig and DILS 2005 in San Diego each attracted around 100 researchers from all over the world. This year the number of submitted papers again increased. The Program Committee selected 23 papers out of 50 strong full submissions. In an effort to include contributions that do not present a new method but that describe innovative and up-and-running practical systems, we distinguished "research papers" and "systems papers." The seven systems papers can be found in the sections Systems I and Systems II. Among the research papers there are four short papers and 12 full papers.

In addition to the presented papers, DILS 2006 featured two invited talks by Victor M. Markowitz and James H. Kaufmann and a session with updates on projects of world-wide importance: the Taverna eScience project, the BioMoby integration framework, and the BioMart integrated genomics data warehouse. Finally, there was a lively poster session.

The workshop was held at the Wellcome Trust Conference Center on the campus of the European Bioinformatics Institute (EBI) in Hinxton, UK. It was kindly sponsored by Microsoft Research, who also made available their conference management system, IBM Research, metanomics, metanomicshealth, the EBI industry programme, and by Schering AG. We are grateful for the help of Springer in putting together and publishing these proceedings. As Program Co-chairs we thank all authors who submitted their work, and the Program Committee members for their careful (and timely) reviews.

We particularly thank Paul Kersey of the EBI, who served as Local Chair of the workshop, and thus did all the hard work.

June 2006

Ulf Leser
Felix Naumann
Barbara Eckman

Organization

DILS 2006 Co-chairs

Ulf Leser	Humboldt-Universität zu Berlin, Germany
Felix Naumann	Humboldt-Universität zu Berlin, Germany
Barbara Eckman	IBM Healthcare and Life Sciences, USA

Local Chair

Paul Kersey, European Bioinformatics Institute, Hinxton, UK

Program Committee

Emmanuel Barillot	Institut Curie	France
David Benton	GlaxoSmithKline	USA
Laure Berti-Equille	Universitaire de Beaulieu	France
Peter Bunemann	University of Edinburgh	UK
Terence Critchlow	Lawrence Livermore National Laboratory	USA
Jürgen Eils	Deutsches Krebsforschungszentrum DKFZ	Germany
Floris Geerts	University of Edinburgh and Limburgs Universitair Centrum	UK
Amarnath Gupta	San Diego Supercomputer Center	USA
Joachim Hammer	University of Florida	USA
Henning Hermjakob	European Bioinformatics Institute	UK
Mike Hogarth	UC Davis	USA
Stefan Jablonski	Univ. Erlangen-Nuernberg	Germany
H V Jagadish	University of Michigan	USA
Hasan Jamil	Wayne State University	USA
Jacob Köhler	Rothamsted Research	UK
Peter Karp	SRI International	USA
Vipul Kashyap	Partners HealthCare System	USA
Arek Kasprzyk	European Bioinformatics Institute	UK
Anthony Kosky	Axiope Inc	USA
Bertram Ludäscher	UC Davis	USA
Paula Matuszek	GlaxoSmithKline Beecham	USA
Peter Mork	The MITRE Corporation	USA
Jignesh Patel	University of Michigan	USA
Norman Paton	University of Manchester	UK
Christian Piepenbrock	Epigenomics AG	Germany
Erhard Rahm	Universität Leipzig	Germany

Louiqa Raschid	University of Maryland	USA
Otto Ritter	AstraZeneca	USA
Monica Scannapieco	University of Rome "La Sapienza"	Italy
Dennis Paul Wall	Harvard Medical School	USA
Sharon Wang	IBM Healthcare and Life Sciences	USA
Bertram Weiss	Schering AG	Germany
Limsoon Wong	Institute for Infocomm Research	Singapore

Additional Reviewers

Shawn Bowers
Adriane Chapman
Hon Nian Chua
Heiko Dietze
Andreas Doms
Nan Guo
Michael Hartung
Andrew Jones
Toralf Kirsten
Judice Koh
Jörg Lange
Mario Latendresse
Christian Lawerenz
Timothy M. McPhillips
Heiko Müller
Eugene Novikov
Loic Royer
Donny Soh
Silke Trissl
Thomas Wächter

Sponsoring Institutions

Microsoft Research	http://research.microsoft.com/
metanomics	http://www.metanomics.de/
metanomicshealth	http://www.metanomics-health.de/
IBM Research	http://www.research.ibm.com/
EBI Industry Program	http://industry.ebi.ac.uk/
Schering	http://www.schering.de/

Website

For more information please visit the DILS 2006 website at `http://www.informatik.hu-berlin.de/dils2006/`.

Table of Contents

Keynotes

Data Integration

Text Mining

Systems I

Potpourri

Systems II

Short Papers

Workflow

An Application Driven Perspective on Biological Data Integration (Keynote Presentation)

Victor M. Markowitz

Lawrence Berkeley National Lab
Biological Data Management and Technology
1 Cyclotron Road, Berkeley, CA 94720
VMMarkowitz@lbl.gov

Data integration is an important part of biological applications that acquire data generated using evolving technologies and methods or involve data analysis across diverse specialized databases that reflect the expertise of different groups in a specific domain. The increasing number of such databases, the emergence of new types of data that need to be captured, as well as the evolving biological knowledge add to the complexity of already challenging integration problems. Furthermore, devising solutions to these problems requires technical expertise in several areas, such as database management systems, database administration and software engineering, as well as data modeling and analysis.

In practice, biological data integration is less daunting when considered in the context of scientific applications that address specific research questions. Established technologies and methods, such as database management systems, data warehousing tools, and statistical methods, have been employed successfully in developing systems that address such questions. The key challenge is marshaling the scientific and technical expertise required for formulating research questions, determining the integrated data framework for answering them, and addressing the underlying data semantics problems.

Evidence suggests that an iterative strategy based on gradually accumulating domain specific knowledge throughout the integration process is effective in devising solutions for application specific biological data integration problems. This strategy will be discussed in the context of two recently developed integrated genome systems, IMG (http://img.jgi.doe.gov) and IMG/M (http://img.jgi.doe.gov/m).

U. Leser, F. Naumann, and B. Eckman (Eds.): DILS 2006, LNBI 4075, p. 1, 2006.

Towards a National Healthcare Information Infrastructure
(Keynote Presentation)

Sarah Knoop

IBM Healthcare Information Management,
IBM Almaden Research Center,
650 Harry Rd, San Jose, CA 95120
seknoop@us.ibm.com

Many countries around the world have placed an increased focus on the need to modernize their healthcare information infrastructure. This is particularly challenging in the United States. The U.S. healthcare industry is by far the largest in the world in both absolute dollars and in percentage of GDP (>$1.7T - 15% of GDP). It is also quite fragmented and complex. This complexity, coupled with an antiquated infrastructure for the collection of and access to medical data, leads to enormous inefficiencies and sources of error. Driven by consumer, regulatory, and governmental pressure, there is a growing consensus that the time has come to modernize the US Healthcare Information Infrastructure (HII). A modern HII will provide care givers with better and timelier access to data. The launch of a National Health Infrastructure Initiative (NHII) in the US in May 2004 - with the goal of providing an electronic health record for every American within the next decade- will eventually transform the healthcare industry in general...just as I/T has transformed other industries in the past. While such transformation may be disruptive in the short term, it will in the future significantly improve the quality, efficiency, and successful delivery of healthcare while decreasing costs to patients and payers and improving the overall experiences of consumers and providers. The key to this successful outcome will be based on the way we apply I/T to healthcare data and to the services delivered through that I/T. This must be accomplished in a way that protects individuals, allows competition, but gives caregivers reliable and efficient access to the data required to treat patients and to improve the practice of medical science.

In this talk we will describe the IBM Research HII project and our implementation of the standards for interoperability. We will also discuss how the same infrastructure required for interoperable electronic patient records must support the needs of medical science and public health. This can be accomplished by building higher level services upon a National Health Information Network, including discovery services for medical research and data mining and modeling services to protect populations against emerging infectious disease.

U. Leser, F. Naumann, and B. Eckman (Eds.): DILS 2006, LNBI 4075, p. 2, 2006.

Data Access and Integration in the ISPIDER Proteomics Grid

Lucas Zamboulis[1,2], Hao Fan[1,2,*], Khalid Belhajjame[3], Jennifer Siepen[3], Andrew Jones[3], Nigel Martin[1], Alexandra Poulovassilis[1], Simon Hubbard[3], Suzanne M. Embury[4], and Norman W. Paton[4]

[1] School of Computer Science and Information Systems, Birkbeck, Univ. of London
[2] Department of Biochemistry and Molecular Biology, University College London
[3] Faculty of Life Sciences, University of Manchester
[4] School of Computer Science, University of Manchester

Abstract. Grid computing has great potential for supporting the integration of complex, fast changing biological data repositories to enable distributed data analysis. One scenario where Grid computing has such potential is provided by proteomics resources which are rapidly being developed with the emergence of affordable, reliable methods to study the proteome. The protein identifications arising from these methods derive from multiple repositories which need to be integrated to enable uniform access to them. A number of technologies exist which enable these resources to be accessed in a Grid environment, but the independent development of these resources means that significant data integration challenges, such as heterogeneity and schema evolution, have to be met. This paper presents an architecture which supports the combined use of Grid data access (OGSA-DAI), Grid distributed querying (OGSA-DQP) and data integration (AutoMed) software tools to support distributed data analysis. We discuss the application of this architecture for the integration of several autonomous proteomics data resources.

1 Introduction

Grid computing technologies are becoming established which enable distributed computational and data resources to be accessed in a service-based environment. In the life sciences, these technologies offer the possibility of analysis of complex distributed post-genomic resources. To support transparent access, however, such heterogeneous resources need to be integrated rather than simply accessed in a distributed fashion. This paper presents an architecture for such integration and discusses the application of this architecture for the integration of several autonomous proteomics resources.

Proteomics is the study of the protein complement of the genome. It is a rapidly expanding group of technologies adopted by laboratories around the world as it is an essential component of any comprehensive functional genomics

* Currently at International School of Software, Wuhan University, China.

U. Leser, F. Naumann, and B. Eckman (Eds.): DILS 2006, LNBI 4075, pp. 3–18, 2006.

study targeted at the elucidation of biological function. This popularity stems from the increased availability and affordability of reliable methods to study the proteome, as well as the ever growing numbers of tertiary structures and genome sequences emerging from structural genomics and sequencing projects.

The *In Silico Proteome Integrated Data Environment Resource* (ISPIDER) project[1] aims to develop an integrated platform of proteome-related resources, using existing standards from proteomics, bioinformatics and e-Science. The integration of such resources would be extremely beneficial for a number of reasons. First, having access to more data leads to more reliable analyses; for example, performing protein identifications over an integrated resource would reduce the chances of false negatives. Second, bringing together resources containing different but closely related data increases the breadth of information the biologist has access to. Furthermore, the integration of these resources, as opposed to merely providing a common interface for accessing them, enables data from a range of experiments, tissues, or different cell states to be brought together in a form which may be analysed by a biologist in spite of the widely varying coverage and underlying technology of each resource.

In this paper we present an architecture which supports the combined use of Grid data access (OGSA-DAI), Grid distributed querying (OGSA-DQP) and data integration (AutoMed) software tools, together with initial results from the integration of three distributed, autonomous proteomics resources, namely gpmDB[2], Pedro[3] and PepSeeker[4]. The emergence of databases on experimental proteomics, capturing data from experiments on protein separation and identification, is very recent and we know of no previous work that combines data access, distributed querying and data integration of multiple proteomics databases as described here.

Paper outline: Section 2 gives an overview of the OGSA-DAI, OGSA-DQP and AutoMed technologies and introduces the three proteomics resources we have integrated. Section 3 discusses the development of the global schema integrating the proteomics resources within the ISPIDER project, Section 4 presents our new architecture, Section 5 discusses related work and Section 6 gives our conclusions and directions of further work.

2 Background

2.1 OGSA-DAI and OGSA-DQP

OGSA-DAI (Open Grid Services Architecture - Data Access and Integration) is an open-source, extendable middleware product exposing data resources on Grids via web services [2]. OGSA-DAI[5] supports both relational (MySQL, DB2,

[1] See http://www.ispider.man.ac.uk
[2] See http://gpmdb.thegpm.org
[3] See http://pedrodb.man.ac.uk:8080/pedrodb
[4] See http://nwsr.smith.man.ac.uk/pepseeker
[5] See http://www.ogsadai.org.uk/

SQL Server, Oracle, PostgreSQL), XML (Xindice, plans for eXist) and text data sources. It provides a uniform request format for a number of operations on data sources, including querying/updating, data transformation (XSLT), compression (ZIP/GZIP), and data delivery (FTP/SOAP).

OGSA-DQP (Open Grid Services Architecture - Distributed Query Processor) is a service-based distributed query processor [1], offering parallelism to support efficient querying of OGSA-DAI resources available in a grid environment. OGSA-DQP[6] offers two services, the Grid Distributed Query Service (GDQS) or Coordinator, and the Query Evaluation Service (QES) or Evaluator. The Coordinator uses resource metadata and computational resource information to compile, optimise, partition and schedule distributed query execution plans over multiple execution nodes in the Grid. The distributed evaluator services execute query plans generated by the Coordinator. Each Evaluator evaluates a partition of the query execution plan assigned to it by a Coordinator. A set of Evaluators participating in a query form a tree through which data flows from leaf Evaluators which interact with Grid data services, up the tree to reach its destination.

The following steps are needed for a client to set up a connection with OGSA-DQP and execute queries over OGSA-DAI resources. First, the client configures an appropriate GDQS data service resource. As a result of this process, the schemas of the resources are imported and the client is able to access one or more of the databases whose schemas have been referenced within a single query. The client then submits a Perform Document to OGSA-DQP containing an OQL [5] query. The Polar* [21] compiler parses, optimises and schedules the query. The query is partitioned, and each partition is sent to a different Evaluator. The Evaluators then interact with the OGSA-DAI resources and with each other, and send their results back to the GDQS, and, finally, the client.

2.2 AutoMed

AutoMed[7] is a heterogeneous data transformation and integration system which offers the capability to handle virtual, materialised and indeed hybrid data integration across multiple data models. It supports a low-level *hypergraph-based data model (HDM)* and provides facilities for specifying higher-level modelling languages in terms of this HDM. An HDM schema consists of a set of nodes, edges and constraints, and each modelling construct of a higher-level modelling language is specified as some combination of HDM nodes, edges and constraints. For any modelling language $\mathcal{M}$ specified in this way (via the API of AutoMed's Model Definitions Repository), AutoMed provides a set of primitive schema transformations that can be applied to schema constructs expressed in $\mathcal{M}$. In particular, for every construct of $\mathcal{M}$ there is an `add` and a `delete` primitive transformation which add to/delete from a schema an instance of that construct. For those constructs of $\mathcal{M}$ which have textual names, there is also a `rename` primitive transformation.

[6] See http://www.ogsadai.org.uk/about/ogsa-dqp/

[7] See http://www.doc.ic.ac.uk/automed

AutoMed schemas can be incrementally transformed by applying to them a sequence of primitive transformations, each adding, deleting or renaming just one schema construct (thus, in general, AutoMed schemas may contain constructs of more than one modelling language). A sequence of primitive transformations from one schema S_1 to another schema S_2 is termed a *pathway* from S_1 to S_2. All source, intermediate, and integrated schemas, and the pathways between them, are stored in AutoMed's Schemas & Transformations Repository.

Each `add` and `delete` transformation is accompanied by a query specifying the extent of the added or deleted construct in terms of the rest of the constructs in the schema. This query is expressed in a functional query language, IQL, and we will see some examples of IQL queries in Section 4.2. Also available are `extend` and `contract` primitive transformations which behave in the same way as `add` and `delete` except that they state that the extent of the new/removed construct cannot be precisely derived from the other constructs present in the schema. More specifically, each `extend` and `contract` transformation takes a pair of queries that specify a lower and an upper bound on the extent of the construct. The lower bound may be `Void` and the upper bound may be `Any`, which respectively indicate no known information about the lower or upper bound of the extent of the new construct.

The queries supplied with primitive transformations can be used to translate queries or data along a transformation pathway — we refer the reader to [15,14] for details. The queries supplied with primitive transformations also provide the necessary information for these transformations to be automatically *reversible*, in that each `add/extend` transformation is reversed by a `delete/contract` transformation with the same arguments, while each `rename` is reversed by a `rename` with the two arguments swapped.

As discussed in [15], this means that AutoMed is a *both-as-view (BAV)* data integration system: the `add/extend` steps in a transformation pathway correspond to Global-As-View (GAV) rules as they incrementally define target schema constructs in terms of source schema constructs; while the `delete` and `contract` steps correspond to Local-As-View (LAV) rules since they define source schema constructs in terms of target schema constructs. An in-depth comparison of BAV with other data integration approaches can be found in [15,14].

2.3 The Proteomics Resources

Thus far we have integrated three autonomous proteomics resources, all of which contain information on protein/peptide identification:

The Proteome Experimental Data RepOsitory (PEDRo [9]) provides access to a collection of descriptions of experimental data sets in proteomics. PEDRo was one of the first databases used for storing proteomics experimental data. It has also been used as a format for exchanging proteomics data, and in this respect has influenced the standardisation activities of the Proteomics Standards Initiative (PSI[8]).

The Global Proteome Machine Database (gpmDB [6]) is a publicly available database with over 2,200,000 proteins and almost 470,000 unique peptide

[8] See http://psidev.sourceforge.net

identifications. The resource was initially designed to assist in the validation of peptide MS/MS spectra and protein coverage patterns, where patterns in previous assignments could be used to allow some measure of confidence to be assigned to new identifications. Although the gpmDB is restricted to minimal information relating to the protein/peptide identification, it provides access to a wealth of interesting and useful peptide identifications from a range of different laboratories and instruments.

PepSeeker [16] is a database developed as part of the ISPIDER project and is targeted directly at the identification stage of the proteomics pipeline. The database captures the identification allied to the peptide sequence data, coupled to the underlying ion series and as a result it is a comprehensive resource of peptide/protein identifications. The repository currently holds over 50,000 proteins and 50,000 unique peptide identifications.

3 The Proteomics Grid Application

3.1 The ISPIDER Project

Experimental proteomics is an essential component for the elucidation of protein biological functions. It involves the study of a set of proteins produced by an organism with the aim of understanding their behaviour under a variety of experimental conditions and environments. The development of new technologies for protein separation, such as 2D-SDS-PAGE (PolyAcrylamide Gel Electrophoresis), High Performance Liquid Chromatography (HPLC) and Capillary Electrophoresis, together with the availability of public accessible protein sequence databases, has enabled scientists to conduct many interesting proteomics experiments on a daily basis. Also, thanks to techniques such as Multi-Dimensional Protein Identification Technology (MudPIT), a single proteomics experiment may identify hundreds of proteins and, as a result, produce a large amount of valuable biological data.

There is a growing number of resources that offer a range of approaches for the capture, storage and dissemination of proteomic experimental data, reflecting the fact that proteomics has now come of age in the post-genomic era and is delivering large, complex datasets which are rich in information. While the existence of such databases opens up many possibilities for the proteomics community, there is still a need for a support for integrating proteomics data, and tools for constructing proteomics-specific experiments.

The aim of the ISPIDER project is to build on state-of-the-art technologies for e-science and data integration in order to provide an environment for integrating proteomics data, constructing and executing analyses over such data, and a library of proteomics-aware components that can act as building blocks for such analyses. The project is Grid-enabling existing proteomics data resources, creating new resources, producing middleware technologies for the integration of these resources — including tools for data integration, workflows and data analysis — and producing visualisation and other types of client for biologist end users.

3.2 Developing the Global Schema

One of the key questions that arose when we started the integration task, was the scope of the global schema. One choice would be a global schema targeted to answering a specific class of proteomics questions e.g. protein-specific questions.

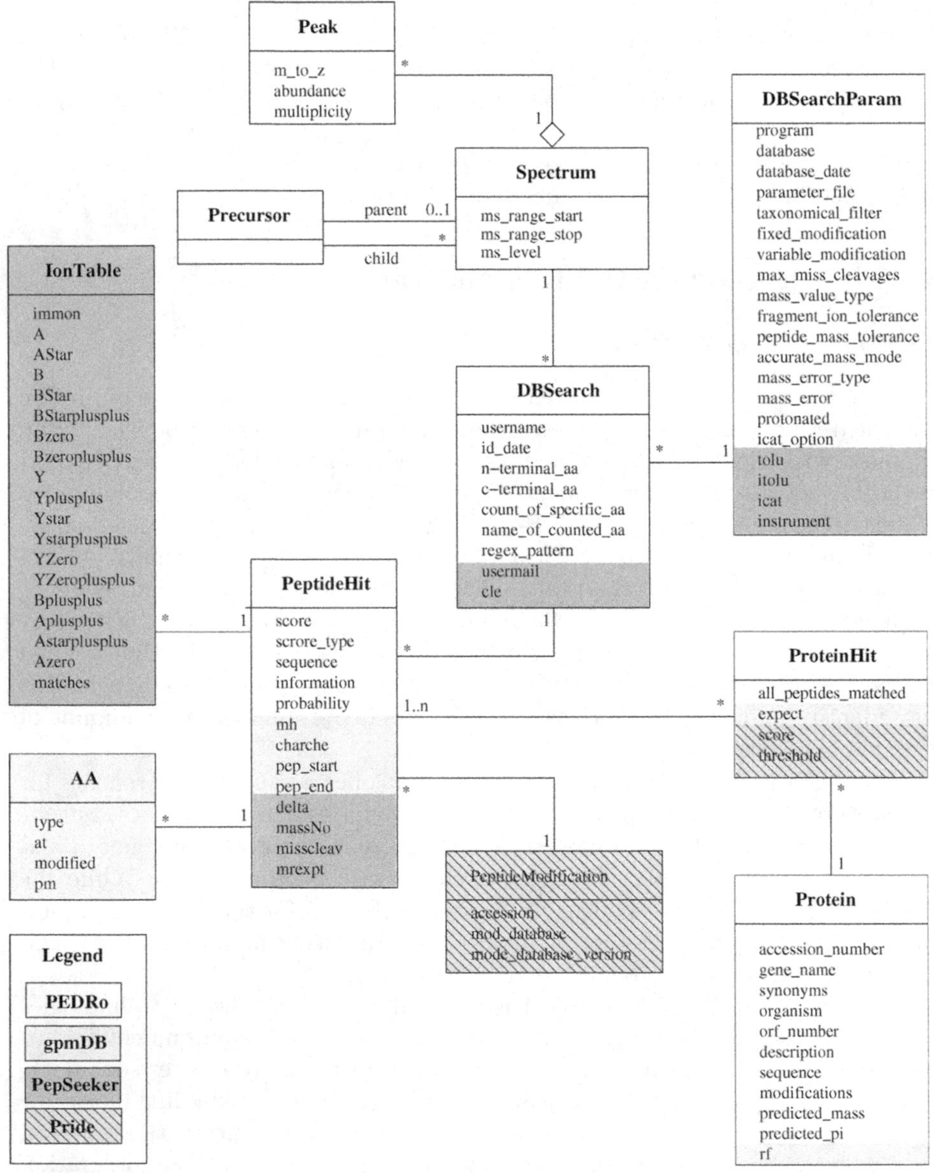

Fig. 1. The Global Schema — colouring denotes the origin of the attributes

A typical query that could be issued in such a case would be *give me the list of proteins that have been identified so far by the proteomics experiments.* Opting for this choice implies a limited usage of the integrated databases. For example, the user will probably not be able to have information on the peptide masses that have been used as inputs to the identification. An alternative choice would be a global schema that is a 'union' schema, integrating the full schemas of the participant databases. Building such a schema together with specifying the mappings between its constructs and the constructs of the participant schemas may, however, turn out to be a complex and lengthy process.

The option we chose is therefore a trade-off between these two alternatives, and the global schema is a subset of the union of the participant schemas. This global schema captures enough information for answering common proteomics questions, particularly queries involving analysis of the results of the proteomics experiments. The scope of this results analysis ranges from the software used for the identification, the peptides produced by the digestion of the proteins, the protein database against which the candidate proteins are compared, to the score and description of the identified proteins. Our choice of results analysis as the scope of the global schema has also been motivated by the fact that it represents the area of overlap between the three databases being integrated, thus allowing the user to pose queries that combine and compare results analysis from the three databases.

Figure 1 gives a UML class diagram of the global schema (the PRIDE[9] data resource mentioned in the figure has not yet been integrated with the other three resources and this is an area of ongoing work). In a protein identification pipeline (a common type of proteomics experiment), a protein is identified using a mass spectrometer which determines the mass-to-charge ratio of the protein ions. The *ms_level* of *Spectrum* describes how many rounds of Mass Spectrometry (MS) have been performed, for example a common and powerful MS technique is tandem MS, in which two rounds of MS are performed. The first round of MS produces a spectra of the precursor ions, predefined selections (determined by *mz_range_start* and *mz_range_end*) of the precursor ions then undergo a second MS round to produce a number of product ion spectra. The relationship between each product spectra and its respective precursor spectra is captured by the *Precursor* association. Individual peaks in each of the precursor spectra are described by the mass-to-charge ratio (*m_to_z*), the peak height (*abundance*) and the isotopic pattern around the main peak (*multiplicity*).

The next step in the protein identification pipeline then involves submission of the ion spectra (described by *Spectrum*) to an identification tool such as Mascot[10] [18] or Imprint[11]. The classes *DBSearch* and *DBSearchParam* capture information about who did the identification, when they did it, what program they used, what database was searched, etc.

---

[9] See http://www.ebi.ac.uk/pride/

[10] See http://www.matrixscience.com/search_form_select.html

[11] Imprint is an in-house software tool for Peptide Mass Fingerprinting (PMF), which involves only a single round of MS.

In tandem MS, several peptide hits are often generated in the identification process. A *PeptideHit* is linked to *IonTable* and *AA*. *IonTable* provides information on ions matching peptide ion fragments. *AA* describes how specific amino acid residues in a Peptide are modified (usually chemical modifications), *modified*, and indicates whether the residue was determined to be a point mutation, *pm*. *ProteinHit* represents the proteins against which all or some of the peptides have been aligned, and links to some information about the protein itself. A *Protein* is characterised by a textual description of the protein, an accession number, the predicted mass of the protein, its amino-acid sequence, any common *in vivo* modifications, the organism in which it is to be found, the open reading frame number, *orf_number*, and the reading frame, *rf*.

To build the above global schema, we adopted an incremental approach. We began with the PEDRo schema, specifically the section of its schema that captures peptide/protein identifications. This was for two reasons. First, the results analysis in the PEDRo schema has significant overlaps with the schemas of the other databases and covers most of our target global schema. Second, the PEDRo schema captures more information compared to the other databases, and thus allows for a more detailed view of the results analysis. For example, in PEDRo, the protein is characterized by the accession number, the synonyms, the organism that was the source of the protein and the sequence of the protein, in addition to other information. In contrast, a protein in PepSeeker, for instance, is simply described by its accession number and name.

Given this initial global schema, we then derived the correspondences between the classes and attributes of gpmDB and PepSeeker with this schema. The limited schema documentation and sometimes cryptic attribute naming of those resources meant interviews with the database providers were needed to identify precisely the meaning of every attribute in the schemas.

The global schema was then incrementally expanded by additional classes and attributes that were captured in those databases and not already in the global schema. This mainly consisted of adding the information about the ions associated to the peptides and the modifications they undergo. For example, from PepSeeker, we added the entity *IonTable* which provides information on the ions matching peptide ion fragments. The schemas of gpmDB and PepSeeker are relatively disjoint, with respect to the set of fields that have been added to the global schema, with few exceptions such as the attributes *pep_start* and *pep_end* of the class *PeptideHit* which exist in both PepSeeker and gpmDB schemas.

To identify the instances of the global schema entities, we chose to use life science identifiers *LSID*s[12]. *LSID* is a Life Sciences Research Uniform Resource Name (URN) specification which provides a standardised naming schema for biological entities in the life sciences domain. The three databases use integers to identify their entity instances, and the usage of LSIDs in the (virtual) global database allowed us to overcome the problem of identifier conflict. For example, the LSID *URN:LSID:ispider.man.ac.uk:pedro.protein:99* refers to the protein

[12] See http://www.omg.org/technology/documents/formal/life_sciences.htm

identified by the number *99* in the Pedro database, where *ispider.man.ac.uk* denotes the authority that issued the LSID[13].

4 System Architecture

While OGSA-DAI supports access of data resources in a Grid and OGSA-DQP supports distributed querying of such resources and location transparency, these technologies do not support schema transformation and schema integration. Thus, if applications require heterogeneous Grid-based data to be transformed and integrated, the onus is on the application to encode the necessary transformation/integration logic. This may impact on the robustness and maintainability of applications, and hence the use of data integration middleware that abstracts out this functionality from applications is advantageous because it enables applications to access resources as one virtual integrated resource, notwithstanding the varying formats and data models used by those autonomous resources. To our knowledge, there is currently no such Grid-enabled middleware, and hence our decision to combine OGSA-DAI/DQP with AutoMed into an architecture that enables both transformation and integration of Grid-based data and distributed query processing over the Grid resources. The main advantage of using AutoMed rather than a LAV or GAV-based data integration system is that it readily supports the evolution of both source and integrated schemas by allowing transformation pathways to be extended — this means that the entire integration process does not have to be repeated, and the schemas and pathways can instead be 'repaired'.

Figure 2 illustrates the architecture we have developed. Data sources are exposed using OGSA-DAI grid services. The AutoMed-DAI wrapper imports schema information from any data source, via OGSA-DAI, into the AutoMed Metadata Repository. Thereafter, AutoMed's schema transformation/integration functionality can be used to create one or more virtual global schemas, together with the transformation pathways between these and the AutoMed representations of the data source schemas. Queries posed on a virtual global schema can be submitted to AutoMed's Query Processor, and this interacts with OGSA-DQP via an AutoMed-DQP wrapper to evaluate these queries. OGSA-DQP itself interacts with the data sources via the OGSA-DAI services.

In the remainder of this section we present the major components of this architecture in greater detail: the mechanisms for enabling data access and integration; how queries posed on a virtual global schema are processed by AutoMed's Query Processor and OGSA-DQP; and the AutoMed-DQP wrapper.

4.1 Data Access and Integration

We assume that each data source is made accessible as a grid data resource using OGSA-DAI's Grid Data Service (GDS). To 'import' a data source schema into AutoMed, we have developed the *AutoMed-DAI Wrapper*. This sends a schema

[13] Note the LSID key attributes are not listed in the UML class diagram in Figure 1.

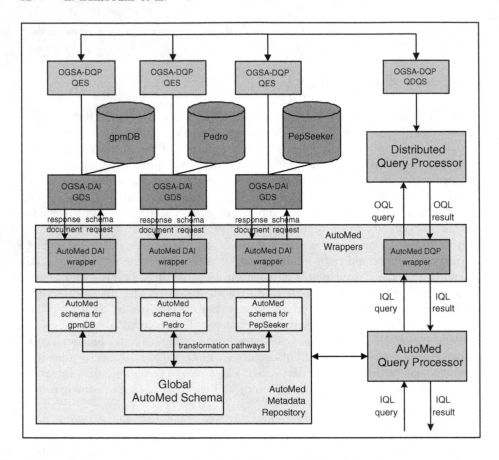

Fig. 2. The AutoMed, OGSA-DAI and OGSA-DQP architecture

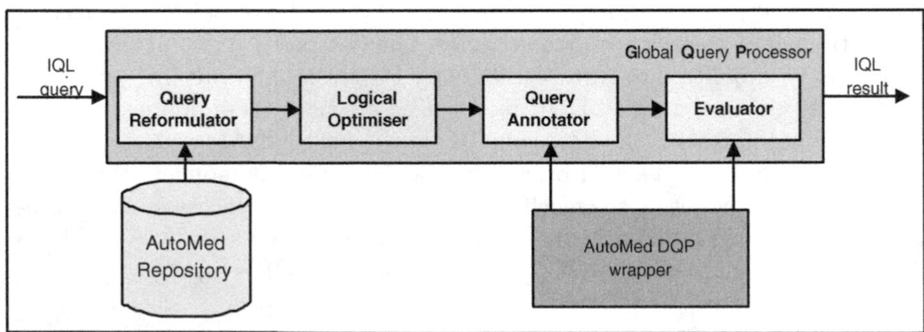

Fig. 3. The AutoMed Query Processor

request document to the GDS, which returns an XML Response Document containing the schema metadata of the data source. The AutoMed-DAI wrapper uses this information to create the corresponding AutoMed schema in the AutoMed Metadata Repository.

These AutoMed data source schemas can now be incrementally transformed and integrated into one or more global virtual schemas, using the API of AutoMed's Schemas & Transformations Repository to issue transformation steps and to create intermediate and final virtual schemas within this repository — we give some examples of transformations below.

4.2 Query Processing

After the integration of the data sources, the user is able to submit to the AutoMed Query Processor (AQP) a query, Q, to be evaluated with respect to a virtual global schema. Q is expressed in AutoMed's own query language, IQL[14].

For example, the following query, Q^1, retrieves all identifications for the protein with accession number `ENSP00000339074`. This query would allow biologists studying this protein to find out more about the kinds of environments in which it has been seen by other scientists. Here, `<<Protein,accession_number>>` denotes the projection of the Protein (virtual) relation onto its primary key attribute (LSID) plus its `accession_number` attribute:

```
[lsid|{lsid,an}<-<<Protein,accession_number>>;an='ENSP00000339074']
```

As a second example, the following query, Q^2, retrieves all protein identifications that match a given peptide. Such a query would allow a scientist working with a protein sequence to ask whether peptide `ATLITFLCDR` has been seen before in other proteomics experiments:

```
[{an,lsid3}|{lsid1,seq}<-<<PeptideHit,sequence>>; seq = 'ATLITFLCDR';
            {lsid2,pr}<-<<ProteinHit,protein>>;
            {lsid3,an}<-<<Protein,accession_number>>; pr = lsid3;
            {pepID,protID}<-<<PeptideHitToProteinHit_mm>>;
            lsid1 = pepID; lsid2 = protID]
```

Figure 3 shows the major components of the AQP and we now consider each of them in turn. Since the initial query, Q, is expressed over a global schema, it references only global schema constructs and needs to be transformed into a query expressed over the data source schemas before it can be evaluated. This is accomplished by the *Query Reformulator* component of the AQP which traverses the schema transformation pathways from the global schema down to the data source AutoMed schemas, and uses the query within each transformation step to incrementally reformulate Q until finally an equivalent query Q_{ref} results which references only schema constructs within the data source schemas.

For example, the following transformation steps within the transformation pathways integrating respectively the PepSeeker, PEDRo and gpmDB schemas are of relevance to reformulating Q^1. Here, `id2lsid` is an IQL function that

[14] IQL is a comprehensions-based language and we refer the reader to [12] for details of its syntax, semantics and implementation. Such languages subsume query languages such as SQL-92 and OQL in expressiveness [4].

generates the global LSID identifiers[15]. In the first step, the relation `Protein` of the global schema is populated from the `proteinhit` relation of PepSeeker; since the latter may contain multiple occurrences of any given protein, the IQL function `distinct` is used to remove duplicates:

```
add(<<Protein,accession_num>>,
    [{id2lsid ['pepseeker.proteinhit:',toString d],x}|
           {d,x}<-(distinct [{k,x}|{k,x}<-<<proteinhit,ProteinID>>])])
...
add(<<Protein,accession_num>>,[{id2lsid ['pedro.protein:',toString d],x}|
                                      {d,x}<-<<protein,accession_num>>])
...
add(<<Protein,accession_num>>,
      [{id2lsid ['gpmdb.proseq:',toString d],x}|{d,x}<-<<proseq,label>>])
```

Q^1 is reformulated to Q^1_{ref} below using the queries appearing within the above transformation steps:

```
[lsid|{lsid,an}<-([{id2lsid ['pepseeker.proteinhit:',toString d],x}|
            {d,x}<-(distinct [{k,x}|{k,x}<-<<proteinhit,ProteinID>>])]
  ++ [{id2lsid ['pedro.protein:',toString d],x}|
                                 {d,x}<-<<protein,accession_num>>]
  ++ [{id2lsid ['gpmdb.proseq:',toString d],x}|{d,x}<-<<proseq,label>>]);
                   an = 'ENSP00000339074']
```

Q^2 makes use of the many-to-many relationship between the `ProteinHit` and `PeptideHit` relations of the global schema for reformulation. The following transformation steps are of relevance to answering Q^2 with respect to the PEDRo schema (we do not list the transformations relevant to the other schemas or the reformulated query Q^2_{ref} itself, due to space limitations)[16]:

```
add(<<PeptideHit,sequence>>,[{id2lsid ['pedro.peptidehit:',toString d],
                             x}|{{d,e},x}<-<<peptidehit,sequence>>])
...
add(<<ProteinHit,protein>>,[{id2lsid ['pedro.proteinhit:',toString d],
                           id2lsid ['pedro.protein:',x]}|
                            {d,x}<-<<proteinhit,protein>>])
...
add(<<Protein,accession_number>>,[{id2lsid ['pedro.protein:',toString d],
                                 x}|{d,x}<-<<protein,accession_num>>])
...
add(<<PeptideHitToProteinHit_mm>>,
     [{k1,k2}|{k1,x}<-[{id2lsid ['pedro.peptidehit:',toString d],x}|
```

[15] It takes as input two string arguments, concatenates them, and prefixes the result by 'URN:LSID:ispider.man.ac.uk:'

[16] In the first step, `peptidehit` has a composite key whose first attribute is used to generate the LSID, the second attribute being a foreign key to another table.

```
                                  {{d,e},x}<-<<peptidehit,db_search>>];
      {k2,y}<-[{id2lsid ['pedro.proteinhit:',toString d],x}|
                                        {d,x}<-<<proteinhit,db_search>>];
      x = y])
```

A reformulated query, Q_{ref}, is next processed by the *Logical Optimiser* component which simplifies Q_{ref} by applying a number of algebraic optimisations. In our context here, one goal of this component is to simplify Q_{ref} is to create the largest possible subqueries that can be pushed down to OGSA-DQP for evaluation, so as to make maximum usage of the data sources' own query capabilities and minimise the resource consumption of AutoMed's Evaluator.

The optimised query, Q_{opt}, is still expressed in IQL and needs to be translated into OQL, the query language supported by OGSA-DQP. We have developed an *AutoMed-DQP Wrapper*, see below, for translating (a subset of) IQL into OQL.

The *Query Annotator* component interacts with the AutoMed-DQP wrapper to identify maximal subqueries translatable by that wrapper and to instantiate wrapper objects within Q_{opt}. The resulting query, Q_{annot}, is finally sent to AutoMed's Evaluator for evaluation. This makes calls to OGSA-DQP to compute the results of the subqueries specified by the Query Annotator, and undertakes any further necessary post-processing of these results.

4.3 The AutoMed-DQP Wrapper

The AutoMed-DQP wrapper undertakes two tasks. First, it needs to inform the AutoMed Query Processor of the subset of IQL it is capable of translating into OQL. As with all other AutoMed wrappers, we have developed a BNF grammar specification from which a parser for the relevant subset of IQL is automatically generated. The AutoMed-DQP wrapper translates IQL comprehensions with one level of nesting (in accordance with the OQL queries supported by OGSA-DQP).

The AutoMed-DQP wrapper is also responsible for making interactions with OGSA-DQP transparent to the remainder of the AutoMed infrastructure. On receiving an IQL query, the wrapper first translates it into the equivalent OQL query. The OQL query is then sent to OGSA-DQP for evaluation. The reply from OGSA-DQP is in the form of an XML Response Document containing the query results. The AutoMed-DQP wrapper translates this document into the IQL type system, and returns the result to AutoMed's Evaluator component.

5 Related Work

The importance of data integration in the life sciences has resulted in diverse technical approaches being followed. In many of these there is little support provided by the infrastructure for resolving schematic heterogeneities. For example, workflow systems enable requests to be formed that both access data resources and invoke analyses on the values retrieved (e.g. [3,17]). However, many bioinformatics web services take and return formatted strings that require custom transformation operations to be developed for converting data between formats.

Perhaps the most widely used data integration system in bioinformatics is SRS [22]. However, like workflow systems, it principally supports storage and access to entries from data resources that started out as formatted textual documents, and the principal mode of access involves navigation between these documents, rather than querying an integrated schema. As such, both of the above approaches make visible the sources from which data is derived and preserve at least some aspects of the source data format.

In approaches building on distributed database technology, there is a tendency to construct views over the underlying data resources, thus hiding schematic heterogeneities from users. In distributed query processing systems that have been designed for or used in bioinformatics, such as DiscoveryLink [11] or Kleisli [7], existing databases or file-based resources are wrapped, and views can be constructed over the wrapped sources using the GAV approach. As such, declarative techniques can be used to provide a more uniform representation of the data in a domain, although with the maintenance challenges widely associated with GAV. There have also been attempts to support querying over domain models expressed as biological ontologies, as in Tambis [10], but again the global schema either directly reflects the structure of the underlying resources or defines the global model using GAV.

Where data is to be subject to intensive integrated analysis, the warehousing approach has also been popular in bioinformatics (e.g. [8,20]). However, the population and maintenance of a centralised warehouse is often laborious, due to inconsistencies between different data sources, naming schemes, etc. However, where data is obtained from databases, these can use queries to populate the warehouse model, as in GAV, or make use of technologies such as AutoMed, as in BioMap [13].

In proteomics, although there are many resources that integrate data about proteins (e.g. [19]), the emergence of databases on experimental proteomics, capturing data from experiments on protein separation and identification, is very recent. As such, we know of no previous work that seeks to support access to multiple proteomics databases, as described here. Furthermore, as the schemas of the databases to be integrated overlap significantly, fine-grained resolution of schema conflicts is crucial to the provision of an effective integration strategy. In essence, in the approach described in this paper, OGSA-DQP provides a query-oriented middleware analogous to that provided by DiscoveryLink or Kleisli, and AutoMed is used to resolve the heterogeneities in the schemas of the sources. We are not aware of a similar approach elsewhere in the life sciences.

6 Conclusions

We have presented an architecture combining Grid data querying (OGSA-DAI/DQP) and data integration (AutoMed) software tools which enables distributed query processing together with the resolution of semantic heterogeneity over autonomous data resources. We have presented results within the ISPIDER project of integrating autonomous resources reflecting various proteomics domains and representations thereof. From a biology viewpoint, the final ISPIDER platform will provide researchers with more information than any of the resources alone, so allowing them to perform analyses that were previously prohibitively difficult or impossible. This integration process both builds on and provides impetus to the development of data standards in the proteomics and related domains.

Additional global schemas may be created as resources holding information relevant to, but disjoint from, the initial global schema are integrated within the ISPIDER platform. To enable querying across such schemas, a global 'super-schema' could then be created. This methodology exemplifies the flexibility and scalability of AutoMed's transformation-based approach which also provides the basis for materialised as well as virtual data integration and tracking data provenance. These facilities too are being pursued within the ISPIDER project.

Beyond data integration, the ISPIDER data sources offer a number of web services to the outside world, performing tasks ranging from simple data retrieval, to significantly more complex operations. We are using Taverna[17], part of the ^{my}Grid[18] middleware, to enable users to construct complex analysis workflows from the available web services. We are currently investigating the interoperation of AutoMed with Taverna for integrating heterogeneous web services.

We are currently evaluating our system in terms of query processing and are considering extensions to the *LogicalOptimiser* of the AQP as well as to the OQL subset supported by OGSA-DQP; this will enable the translation of larger IQL queries into OQL, which we expect will offer a notable performance boost.

Acknowledgements. The work presented in this paper was funded by a grant from the BBSRC. We are also grateful to Steven Lynden for his help with the OGSA-DQP system, Thomas McLaughlin and Julian Selley for their help in defining the correspondences between the database schemas, and Robert Stevens and Carole Goble for their comments and advice.

References

1. M. N. Alpdemir, A. Mukherjee, N.W. Paton, P.Watson, A. A. Fernandes, A. Gounaris, and J. Smith. Service-based distributed querying on the Grid. In *Proc. of the 1st Int. Conf. on Service Oriented Computing*, pages 467–482, 2003.
2. M. Antonioletti et al. The design and implementation of grid database services in OGSA-DAI. *Concurrency - Practice and Experience*, 17(2-4):357–376, 2005.
3. S. Bowers and B. Ludäscher. An ontology-driven framework for data transformation in scientific workflows. In *DILS*, pages 1–16. Springer, 2004.
4. P. Buneman, L. Libkin, D. Suciu, V. Tannen, and L. Wong. Comprehension syntax. *SIGMOD Record*, 23(1):87–96, 1994.
5. R. G. G. Cattell and D. K. Barry. *The Object Database Standard: ODMG 3.0*. Morgan Kaufmann, 2000.
6. R. Craig, J. P. Cortens, and R. C. Beavis. Open source system for analyzing, validating, and storing protein identification data. *Journal of Proteome Research*, 3(6), 2004.

[17] See http://taverna.sourceforge.net

[18] See http://www.mygrid.org.uk

7. S.B. Davidson, C. Overton, V. Tannen, and L. Wong. BioKleisli: A Digital Library for Biomedical Researchers. *Journal of Digital Libraries*, 1(1):36–53, Nov 1997.
8. S. Durinck, Y. Moreau, A. Kasprzyk, S. Davis, B. De Moor, A. Brazma, and W. Huber. Biomart and bioconductor: a powerful link between biological databases and microarray data analysis. *Bioinformatics*, 21(16):3439–3440, 2005.
9. K. Garwood et al. Pedro: A database for storing, searching and disseminating experimental proteomics data. *BMC Genomics*, 5, 2004.
10. C. A. Goble, R. Stevens, G. Ng, S. Bechhofer, N. W. Paton, P. G. Baker, M. Peim, and A. Brass. Transparent access to multiple bioinformatics information sources. *IBM Systems Journal*, 40(2):532–551, 2001.
11. L. M. Haas, P. M. Schwarz, P. Kodali, E. Kotlar, J. E. Rice, and W. C. Swope. Discoverylink: A system for integrated access to life sciences data sources. *IBM Systems Journal*, 40(2):489–511, 2001.
12. E. Jasper, A. Poulovassilis, and L. Zamboulis. Processing IQL queries and migrating data in the AutoMed toolkit. AutoMed Tech. Rep. 20, June 2003.
13. M. Maibaum, L. Zamboulis, G. Rimon, N. Martin, and A. Poulovassilis. Cluster based integration of heterogeneous biological databases using the AutoMed toolkit. In *Proc. Data Integration for the Life Sciences 2005 (DILS'05)*, pages 191–207.
14. P. McBrien and A.Poulovassilis. Defining peer-to-peer data integration using both as view rules. In *Proc. Workshop on Databases, Information Systems and Peer-to-Peer Computing (at VLDB'03), Berlin*, 2003.
15. P. McBrien and A. Poulovassilis. Data integration by bi-directional schema transformation rules. In *Proc. ICDE'03*, pages 227–238, 2003.
16. T. McLaughlin, J. A. Siepen, J. Selley, J. A. Lynch, K. W. Lau, H. Yin, S. J. Gaskell, and S. J. Hubbard. Pepseeker: a database of proteome peptide identifications for investigating fragmentation patterns. *Nucleic Acids Research*, 34, 2006.
17. T. M. Oinn, M. Addis, J. Ferris, D. Marvin, M. Senger, R. M. Greenwood, T. Carver, K. Glover, M. R. Pocock, A. Wipat, and P. Li. Taverna: a tool for the composition and enactment of bioinformatics workflows. *Bioinformatics*, 20(17):3045–3054, 2004.
18. D.N. Perkins, D.J. Pappin, D.M. Creasy, and J.S. Cottrell. Probability-based protein identification by searching sequence databases using mass spectrometry data. *Electrophoresis*, 20(18), 1999.
19. M. Pruess, P. Kersey, and R. Apweiler. The integr8 project - a resource for genomic and proteomic data. *In Silico Biology*, 5, 2004.
20. S.P. Shah, Y. Huang, Y. Xu, M.M.S. Yuen, J. Ling, and B.F.F. Ouellette. Atlas – a data warehouse for integrative bioinformatics. *BMC Bioinformatics*, 6(81), 2005.
21. J. Smith, A. Gounaris, P. Watson, N. W. Paton, A. A. A. Fernandes, and R. Sakellariou. Distributed query processing on the Grid. In *Proc. Grid Computing*, pages 279–290, 2002.
22. E. M. Zdobnov, R. Lopez, R. Apweiler, and T. Etzold. The EBI SRS server-recent developments. *Bioinformatics*, 18(2):368–373, 2002.

A Cell-Cycle Knowledge Integration Framework
Research Paper

Erick Antezana, Elena Tsiporkova, Vladimir Mironov, and Martin Kuiper

Dept. of Plant Systems Biology. Flanders Interuniversity Institute for Biotechnology/Ghent University. Technologiepark 927, B-9052 Ghent Belgium
{erant, eltsi, vlmir, makui}@psb.ugent.be
http://www.psb.ugent.be/cbd/

Abstract. The goal of the EU FP6 project DIAMONDS [1] is to build a computational platform for studying the cell-cycle regulation process in several different (model) organisms (S. cerevisiae, S. pombe, A. thaliana and human). This platform will enable wet-lab biologists to use a systems biology approach encompassing data integration, modeling and simulation, thereby supporting analysis and interpretation of biochemical pathways involved in the cell cycle. To facilitate the computational handling of cell-cycle specific knowledge a detailed cell-cycle ontology is essential. The currently existing cell-cycle branch of the Gene Ontology (GO) provides only a static view and it is not rich enough to support in-depth cell-cycle studies.

In this work, an enhanced Cell-Cycle Ontology (CCO) is proposed as an extension to existing GO. Besides the classical add-ons given by an ontology (data repository, knowledge sharing, validation, annotation, and so on), CCO is intended to further evolve into a knowledge-based system that provides reasoning services oriented to hypotheses evaluation in the context of cell-cycle studies. A data integration pipeline prototype, covering the entire life cycle of the knowledge base, is presented. Concrete problems and initial results related to the implementation of automatic format mappings between ontologies and inconsistency checking issues are discussed in detail.

1 Introduction

The amount of data generated in biological experiments continues to grow exponentially. The shortage of proper approaches or tools for analyzing this information has created a gap between raw data and knowledge. To make matters worse, the lack of a structured documentation of knowledge leaves much of the information extracted from these raw data unused. Moreover, differences in the used technical languages (synonymy and polysemy) have complicated the analysis and interpretation of the data. Currently, there are several efforts for standardizing the used vocabulary. Most importantly, the Gene Ontology (GO) Consortium [9] has been providing a controlled set of terms for gene products whereas the Open

[1] http://www.sbcellcycle.org

U. Leser, F. Naumann, and B. Eckman (Eds.): DILS 2006, LNBI 4075, pp. 19–34, 2006.

Biomedical Ontology(OBO)[2] umbrella has been collecting the most representative ontologies in biological and medical domains. Ontologies clarify scientific discussions providing a shared vocabulary for biologists to communicate their results effectively, explore data and extend scientific investigations. Ontologies also facilitate the implementation of computational approaches and systems to perform data exploration, inference and mining [5].

The goal of the EU FP6 project DIAMONDS is to build and use a systems biology platform of tools to study the cell-cycle process in several different model organisms (S. cerevisiae, S. pombe, A. thaliana and human). Data and information integration and retrieval is essential for studying gene networks, and although several solutions for this already exist (e.g. BioRS[3], SRS[4]; also some ontology-based solutions like TAMBIS [25] and caBIO [8]. A particular challenge is the development of a specific cell-cycle ontology (CCO), as this is relatively poorly developed at present. A rich CCO will be a first step towards more powerful computational approaches to exploit such developed ontology. The process of cell division, or cell cycle, is one of the most fundamental and highly conserved processes in eukaryotic systems. Its cyclical nature makes it a challenging phenomenon for modeling and simulation and a better understanding of it provides significant knowledge for growth in general and human health in particular (cancer related aspects, proliferation disorders issues, prospective therapeutic targets basis and so forth [14], [29]). The available knowledge contained in the cell-cycle literature, however, resides in a format that does not enable straightforward computational processing and consequently, searching and manipulating this information is limited. Moreover, reusing and sharing cell-cycle related data is not facilitated by actual media. Queries within a document are usually limited to simple keyword searches. Therefore, relations between concepts within a document cannot be found unambiguously. For example, two instances, protein X and protein Y can be easily identified by a keyword search. However, unless biologists read at least the text sections comprising those concepts within the document, they will not be able to determine whether these two proteins are related to each other, how this relationship is defined, or in what particular phase of the cell-cycle this relationship is important.

We propose here an ontological paradigm that enables to capture the semantics, temporal aspects and dynamics of the cell cycle regulatory process. Currently, the cell-cycle branch from the bio-ontology GO is too basic to adequately describe the cell-cycle, as it only supports a static view of this process. GO is based on the annotation of gene products (either RNA or proteins). Each of these products may in fact play a role in many molecular processes. Unfortunately, in GO only the prospective activity of a given process is defined without much specification of where or when this process may take place. For particular applications, such as regulatory network modeling and simulation, it is essential to access specific temporal annotations that capture the dynamics of the

[2] http://obo.sourceforge.net/cgi-bin/table.cgi
[3] http://www.biomax.de/products/biors.php
[4] http://www.biowisdom.com/solutions_srs.htm

system. Only two types of relationships are at present considered in GO: subsumption (*is_a*) and partonomic inclusion (*part_of*) (for a formal definition of an ontology structure, refer to the Appendix), which poses a significant limitation for expressing the semantics of a dynamical system. In addition, GO treats its three structured networks as separate ontologies, i.e. no ontological relations are defined among them. Besides, GO suffers of inconsistent treatment of relations such as *is_a*. In spite of these problems, GO has gained a wide appreciation in the life sciences.

The CCO that we propose here belongs to the domain specific ontology type according to the definition given in [12]. As argued in [19], the development of an ontology of a given domain is frequently not a goal in itself, it rather constitutes a skeleton for a set of data that together form a knowledge base. We have set out to build a knowledge-based system founded on CCO, for an in-depth analysis of cell-cycle control mechanisms.

There are several prospective resources that a cell-cycle knowledge base can draw on. Among them, existing ontologies such as GO and some of the ones listed at the OBO repository are key. In addition, databases holding data about gene/protein interactions, such as Reactome [15], BIND [2] and IntAct [13], are also considered. Cell-cycle "slims" from Reactome will provide the first setup. Furthermore, data produced by the DIAMONDS consortium will also feed the repository (E.g. dedicated curation of literature information, annotation information on protein features, protein-protein interaction data).

OWL [21] is a web ontology language that is recommended by the W3C[5] consortium for semantic web applications. OWL comes in three flavors: OWL Full, OWL-DL and OWL Light, ranked in order of their expressivity. For CCO we chose OWL-DL, because of the reasoning capabilities versus computational cost ratio.

Reasoning through a logic approach is best able to deal with the constraints of the gathered knowledge. We have chosen description logics [1] because of its expression power, a well developed theory and consistent semantics. Reasoning packages, such as RACER[6], KAON2[7], Pellet[8] and/or FaCT++[9] are being used for classifying, checking instance consistency and making implicit information explicit. In addition, such reasoning can reveal inconsistencies, hidden dependencies, redundancies and misclassifications. As a result, the CCO becomes more robust.

2 Data Integration Pipeline

A formal specification of a data integration pipeline has been developed (see Figure 1). This specification covers the entire life cycle including the development

[5] http://www.w3.org/
[6] http://www.racer-systems.com/index.phtml
[7] http://kaon2.semanticweb.org/
[8] http://www.mindswap.org/2003/pellet/index.shtml
[9] http://owl.man.ac.uk/factplusplus/

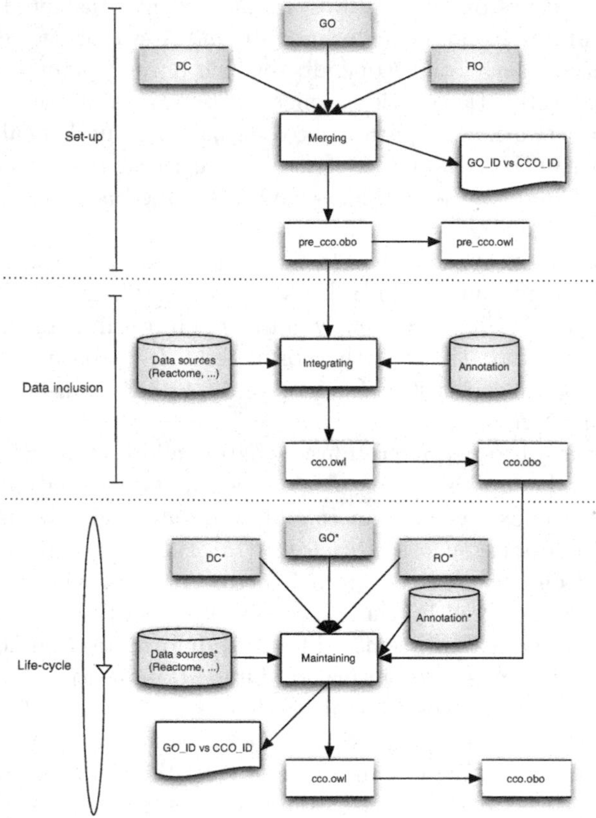

Fig. 1. Data integration pipeline

of the knowledge base. It has three phases: a set-up, a data integration phase, and a maintenance phase. A detailed description of these phases is presented below.

2.1 Set-Up

In the initial phase, the ontology structure and its lexicon (for formal definitions, refer to the Appendix) were engineered. The ontology structure core was based on the GO cell-cycle branch, the Relations Ontology (RO) [24], and the Dublin Core (DC)[10] ontology. The integration pipeline has been implemented in PERL using go-dev[11] and XML::Parser[12]. In order to produce the CCO, ontology pruning and preprocessing has also been done. Whereas this ontology structure has been iteratively refined, changes can still be accommodated until a stable version of

[10] http://www.dublincore.org/

[11] http://www.godatabase.org/dev/doc/go-dev-doc.html

[12] http://search.cpan.org/dist/XML-Parser/

the whole system is reached. The output generated in this phase constitutes the input for the data integration phase.

As the ontology is available in OBO and OWL formats, specific format conversion tools were developed. The OBO format of the CCO is compliant with the version 1.2[13] of the OBO format specification. Since the CCO's ontology structure and lexicon are based on the cell-cycle GO branch, an association table (GO identifier versus CCO identifier) was defined. A CCO entry sample in both OBO and OWL formats showing the correspondence between attributes can be seen in Figure 2.

```
[Term]
id: CCO:P0000016
name: M phase of mitotic cell cycle
def: "Progression through M phase, the part of the mitotic cell cycle during which mitosis and cytokinesis
take place." [GOC:mah, ISBN:0815316194]
xref: GO:0000087
xref: Reactome:68886
relationship: part_of CCO:P0000037
is_a: CCO:P0000038
synonym: "M-phase of mitotic cell cycle" [] {scope="exact"}
```

```
<owl:Class rdf:ID="CCO_P0000016">
      <rdfs:label xml:lang="en">M phase of mitotic cell cycle</rdfs:label>
      <xref rdf:datatype="http://www.w3.org/2001/XMLSchema#string">GO:0000087</xref>
      <xref rdf:datatype="http://www.w3.org/2001/XMLSchema#string">Reactome:68886</xref>
      <rdfs:comment rdf:datatype="http://www.w3.org/2001/XMLSchema#string">Progression through M phase, the
part of the mitotic cell cycle during which mitosis and cytokinesis take place.</rdfs:comment>
      <synonym rdf:datatype="http://www.w3.org/2001/XMLSchema#string">M-phase of mitotic cell cycle</synonym>
      <rdfs:subClassOf>
            <owl:Restriction>
                  <owl:onProperty rdf:resource="#part_of"/>
                  <owl:someValuesFrom rdf:resource="#CCO_P0000037"/>
            </owl:Restriction>
      </rdfs:subClassOf>
      <rdfs:subClassOf rdf:resource="#CCO_P0000038"/>
      <owl:disjointWith rdf:resource="#CCO_P0000270"/>
</owl:Class>
```

Fig. 2. A CCO entry sample

2.2 Data Integration

Data coming from several different sources are being integrated into the knowledge base structure. The main sources are the available gene association files[14], the cell-cycle related data from the Reactome knowledge base and the IntAct database. The output of this phase provides the first release of the system. Besides, cross-references to external public databases, such as UniProt [6] and GenBank [7], will also be supplied in the future.

2.3 Maintenance

The system will support automatic data updates and facilitate targeted manual checking to ensure its consistency. An automatic updating procedure is partially implemented. This will guarantee that the working version of the CCO has the latest data produced by the GO consortium with respect to the cell-cycle data. The cell-cycle knowledge base specifies a given concrete domain state and as

[13] http://www.godatabase.org/dev/doc/obo_format_spec.html

[14] http://www.geneontology.org/GO.current.annotations.shtml

a result, the knowledge base should be constantly updated. In addition, the ontology structure may also be extended and refined. Regarding the update mechanism, the cutting-edge versions of the GO and RO ontologies are fetched by the updating scripts from their version control systems. Then, the checked-out files, in OBO format, are transformed into OWL format. Due to the fact that the accession numbers should remain consistent from update to update, special considerations are taken into account while updating the entries since some terms may have been changed (see section 3.2 for more details). In addition, separate files containing the CCO accession numbers and terms are kept in similar ways as the GO distribution is provided. On the other hand, a version control system has been set up for CCO since the initial results were obtained. The current CCO version as well as previous ones may be retrieved from that repository. In the near future, a web site will provide access to the initial and experimental CCO pre-releases as well as the stable releases.

3 Towards a Formal Cell-Cycle Ontology

Information retrieval and management is enhanced by means of ontologies. The next generation of the web, the semantic web [4], will provide to users and machines a common exchange language avoiding irrelevant search results and increasing the quality of the result searching hits. Such a common and formal language will assist in the maintenance of the daily evolution of the knowledge by checking the incongruities that might arise from periodic updates.

3.1 Motivation and Design Principles

In order to develop an efficient ontology, careful consideration should be given to the purpose that it should serve, and the scientific community that will use it. The main purpose for building the CCO is to capture the semantics of the cell-cycle regulatory process, especially the dynamic aspects of the concepts and their interrelations, and to promote sharing, reuse and enable better computational integration with existing resources. The prospective audience comprises both wet-lab biologists and computational biologists who might have a particular interest in cell-cycle elucidation. Some motivating scenarios or use-cases that have shaped the CCO development are shown in Table 1.

The scope of the ontological concepts ranges from biological processes, cellular components, and molecular functions. Some competency questions that should be answered by the system are summarized in Table 2. Those questions are expressed in natural language and are rather informal. The ontology can be valued by its capability to answer those types of questions and by the extent that it provides support for the motivating scenarios. Those questions correspond to the functional ontology requirements, that is, the system behavior (functions or services).

The following rules, summarized from [23], have been taken into account while engineering the CCO:

Table 1. Some CCO motivating scenarios

1. A **molecular biologist** is interested in knowing the components that interact during the cell-cycle process or in a given event as well as the roles that each component play. He/she is also interested in finding out the prospective components that play a role in a given cell-cycle phase.
2. In the case of a **bioinformatician**, it might be interesting to know the cross-references from one component, that plays a given role in the cell-cycle process, to some external data resources.
3. **General audience** will be interested in finding out all the fundamental components involved in the cell-cycle process. Besides, they will also want to find some synonyms for a given concept.

Table 2. Competency questions

1	What is a X-type **CDK**?
2	What is a Y-type **Cyclin**?
3	In what **events** is **CDK** Z involved?
4	In what **events** is **Rb** involved?
5	Which **CDKs** are involved in the endoreduplication process?
6	Which proteins are **phosphorylated** by kinase X?
7	Which CDK pertains to [G1—S—G2—M] phase?

- Univocity: Terms should have the same meanings on every occasion of use.
- Positivity: Terms should designate genuine classes (do not use for instance `non-inhibitor`).
- Objectivity: Terms should designate biological natural kinds (do not use for instance terms such as `unknown`).
- Single Inheritance: No class in a classification hierarchy should have more than one *is_a* parent on the immediate higher level.
- Intelligibility of Definitions: The terms used in a definition should be simpler than the term to be defined.
- Basis in Reality: The quality of the ontologies depends on the degree to which they represent a certain portion of reality.

Those ontology principles are primarily considered as development guidelines since some rules could limit the representation of real-world situations, e.g. multiple inheritance.

3.2 Development Issues

There are many proposals for ontology development [10]. However, none of them is widely accepted and many of the ontology engineers combine at least some of these methods. Among the most representative methods we can identify the following: Enterprise methodology [27], the TOVE methodology [11], the Unified methodology [28] and METHONTOLOGY [22]. All of them provide guidelines for developing ontologies and most of them consider the following steps:

- choosing an ontology language,
- choosing a development tool,
- acquire domain knowledge,
- reuse ontologies.

In general terms, most of the ontology development methods consider an informal phase, in which some ontology sketches are devised, and a formal phase, in which the ontology is formalized using an ontology language and specific development tools. Capturing knowledge is an expensive and arduous task. Protégé[15] has been used as a developing environment for our CCO. It provides a very user-friendly graphical interface and an extensible architecture based on plug-ins.

GO and the RO have been used as core ontologies for developing the CCO. All the processes from GO under the cell-cycle (GO:0007049) term were taken into account, while RO was completely imported. Thus, 304 terms were adopted from GO and all the 15 relations from RO. The CCO is updated daily and checked using data from GO.

Further enrichment of the CCO is required to merge the relations from RO and the ones provided by Reactome since there is a gap between the top-level relations from RO and the very specific ones from Reactome. For that sake, a mapping and association layer has been implemented. Based on this first skeleton this endeavor has now reached an initial stable state. Additional efforts will consider the integration of instances into a knowledge-based system that will provide a means for hypotheses evaluation.

The CCO is presently available in two formats most widely used by the bio-community: OBO and OWL. The CCO has an average depth of about 3 nodes. Although the mapping between the OBO and OWL (and vice versa) is not totally biunivocal (one-to-one correspondence), all the data has been preserved, i.e., all the terms and relations with their attributes were translated and when necessary some workarounds were implemented for the sake of completeness. The framework of the data integration is shown in Figure 3.

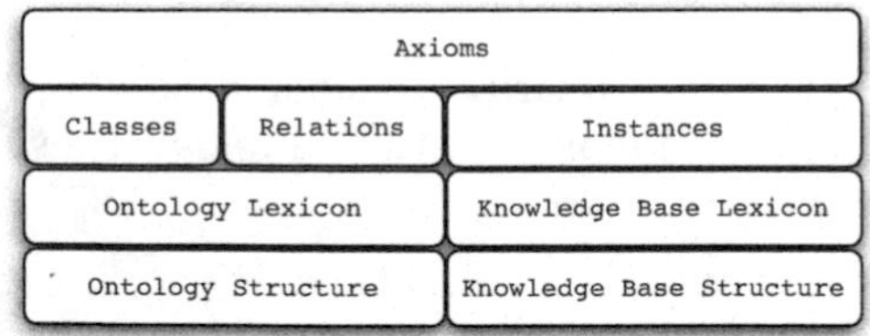

Fig. 3. Data integration framework

[15] http://protege.stanford.edu/

Classes, definitions, their references, and their relations with other classes have been treated as they appeared in their original resources. The DC ontology is being used for coding the terms references. Besides, a number of conventions have been adopted from [26], [3] and GO for identifying the terms within CCO. Each concept has a unique identifier of the form **CCO:cnnnnnnn**, where **CCO** indicates that the concept belongs to the CCO ontology (**CCO** is also known as the ontology namespace), **c** denotes de sub-namespace, and **nnnnnnn** consists of 7 numerical characters as shown in Table 3.

Table 3. CCO accession number. The first character denotes the type of term: **C** stands for cellular component, **F** for molecular function and **P** for biological process.

c	n	n	n	n	n	n	n
[C,F,P]	[0-9]	[0-9]	[0-9]	[0-9]	[0-9]	[0-9]	[0-9]

In order to maintain the accession numbers consistency as much as possible, the following four situations were considered:

1. A totally new added term implies a new accession number.
2. A merge occurs when two or more terms become a new term. The old accession numbers are copied as secondary accession numbers into the new term.
3. A split occurs when one term turns into two or more terms. The original accession numbers are kept in all the derived terms and a new primary accession number is added to each new term.
4. An accession number is dropped only when the data to which it was assigned have been completely removed from the ontology.

3.3 Formats Mapping

This section presents the mapping aspects that were taken into account while engineering the ontology structure. Previous work in this respect has already highlighted[16] many of the problems that we have faced here. The main problem is the insufficiency of information to get an OWL representation from an OBO one. The mapping to OWL has some caveats since currently there are some elements in OWL without any equivalent in OBO. For example, the existencial and universal restrictions cannot explicitly be represented in OBO. Consequently, we assume that all of them are existencial restrictions. Although nowadays no OBO ontology uses either union or intersection constructions, our conversion tools

[16] http://www.godatabase.org/dev/doc/mapping-obo-to-owl.html,
http://b-src.cbrc.jp/source/go-dev/doc/mapping-obo-to-owl.html,
http://gong.man.ac.uk/,
http://www.aiai.ed.ac.uk/resources/go/,
http://bioinfo.unice.fr/equipe/Claude.Pasquier/biowl/index.html

Table 4. Mapping of OBO and OWL terms. NDY stands for not defined yet.

OBO keyword	OWL keyword	OWL element type
[Term]	owl:Class	Class description
id	rdf:ID	Class description
name	rdfs:label	rdf:Property
is_anonymous	NDY	NDY
alt_id	NDY	NDY
def	rdfs:comment	rdf:Property
comment	NDY	NDY
subset	NDY	NDY
synonym	synonym	owl:DataTypeProperty, owl:AnnotationProperty
xref	xref	owl:DataTypeProperty, owl:AnnotationProperty
is_a	rdfs:subClassOf	owl:ObjectProperty
intersection_of	owl:intersectionOf	Class description
union_of	owl:unionOf	Class description
disjoint_of	owl:disjointWith	Class axiom
relationship	NDY	NDY
is_obsolete	owl:DeprecatedClass	Version information
replaced_by	NDY	NDY
consider	owl:equivalentClass	Class axiom

Table 5. Mapping among the OBO and OWL relationships. NDY stands for not defined yet.

OBO keyword	OWL keyword
[Typedef]	owl:ObjectProperty
builtin	NDY
comment	NDY
def	rdfs:comment
exact_synonym	synonym (workaround)
id	rdf:id
inverse_of	owl:inverseOf
is_a	rdfs:subClassOf
is_anti_symmetric	is_anti_symmetric (workaround)
is_reflexive	is_reflexive (workaround)
is_transitive	rdf:type (TransitiveProperty)
NDY	rdf:type (SymmetricProperty)
name	rdfs:label (string)
xref_analog	NDY

support them. Moreover, some terms do not have any definition and in consequence no references (no *dbxref* definition).

A mapping between the OBO specification and the OWL representation of the CCO is shown in Tables 4 and 5. As can be observed, there are some elements that have been mapped in a natural way (e.g. rdfs:label), while some other elements do not have a direct or defined mapping and some non-trivial approaches were taken to bypass this problem. The missing properties in OWL relations are: reflexivity, asymmetry, antisymmetry, intransitivy and partonomic relationships.

The mapping that we introduce is still in an experimental, non-final phase. There are several aspects that will be adapted after some pending decisions are taken. A stable level will be achieved once the OBO specification reaches a sufficient maturity stage. Moreover, because of the OBO metadata, the CCO has adopted OWL Full for its representation. Consequently, an alternative OWL-DL version will additionally become available. The OWL syntax of the OWL generated file is validated automatically by the format conversion script using *vowlidator*[17].

3.4 Handling of Inconsistencies

As stated above, one added value of a description logics approach embedded in the skeleton of the ontology structure is to allow automatic detection and handling of inconsistencies and misclassifications. Thus reasoning environments as RACER can be employed for checking the validity of some of the design principles of CCO, mentioned in Section 3.1. For instance, as already indicated in [23], the way GO uses the *is_a* relation may lead to a violation of the single inheritance principle. After loading the CCO into Protégé (with the Protégé OWL plugin [16]) and adding simple disjointness constraints to some of the CCO classes a certain number of this type of inconsistencies (32 in total which represents 10% of the entire CCO) have been detected by RACER. There are a number of relationships that should have been annotated as *part_of* instead of *is_a* and vice versa. A sample of this analysis is shown in Figure 4 and the corresponding GO cross-references are shown in Table 6. The centriole is an integral part of the centrosome, the microtubule organizing centre of the cell. Accordingly, the term *centriole replication* in GO is linked to its parent term *centrosome duplication* via a *part_of* relationship. Then, in order to be consistent the terms *regulation of centriole replication* and *negative regulation of centriole replication* should be related to their parent terms *regulation of centrosome cycle* and *negative regulation of centrosome cycle* by *part_of* relationships as well. Indeed, the inconsistency problem was solved by replacing the *is_a* with *part_of* relationship for these two pairs of terms in accordance to the True Path Rule[18].

We are presently investigating different approaches to solve these problems so that the divergence against the main ontology source (GO) is minimal. Furthermore, as stated in [23], the *part_of* relation should be specialized using spatial and temporal relations [24] to solve this type of inconsistencies.

---

[17] http://projects.semwebcentral.org/projects/vowlidator/

[18] http://www.geneontology.org/GO.usage.shtml#truePathRule

A major factor for managing the system will be the extent to which it diverges from its main sources. To minimize problems we will seek dialogue with the GO consortium and provide feedback.

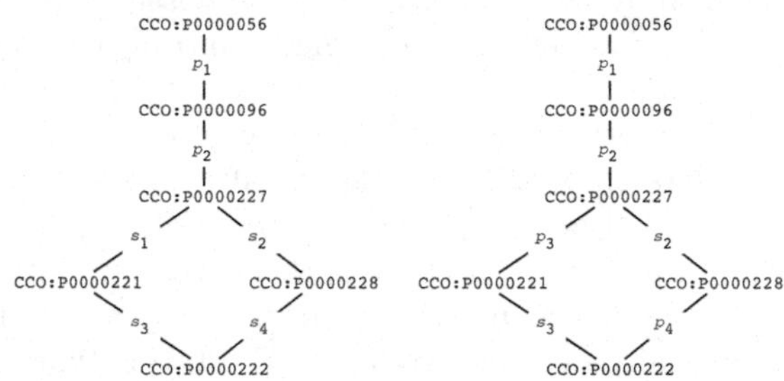

Fig. 4. Comparison of two sample class sub-hierarchies. Let $\mathcal{O}$ be an ontology structure, where $\{s_1, s_2, s_3, s_4\} \subset \mathcal{S}$ and $\{p_1, p_2\} \subset \mathcal{P}$ and $\mathcal{O}'$, where $\{s_2, s_3\} \subset \mathcal{S}'$ and $\{p_1, p_2, p_3, p_4\} \subset \mathcal{P}'$. The ontology $\mathcal{O}$, on the left, has an inconsistent relation s_4. On the right, the ontology $\mathcal{O}'$ is shown with a different, supposedly correct semantics.

Table 6. CCO ID and GO ID for the sample shown in Figure 4

CCO ID	GO ID	Term
CCO:P0000056	GO:0007049	cell cycle
CCO:P0000096	GO:0007098	centrosome cycle
CCO:P0000227	GO:0046605	regulation of centrosome cycle
CCO:P0000221	GO:0046599	regulation of centriole replication
CCO:P0000228	GO:0046606	negative regulation of centrosome cycle
CCO:P0000222	GO:0046600	negative regulation of centriole replication

4 Conclusions and Future Work

The amount of biomedical knowledge is becoming too large for traditional local approaches. Ontologies can increase the likelihood that such knowledge will be found and used by making the data easier to query and transform. A data integration pipeline detailing the issues of creating, updating and maintaining a cell-cycle knowledge base has been introduced. The formalization towards a description logics framework is a multi-staged and iterative process. The principal **contributions** of the knowledge base are:

- facilitate the communication between communities working on the cell-cycle process by providing a *lingua franca* or common terminology;
- saving time and effort by reusing the CCO and integrating it into related applications or systems;

- assist in application tasks such as knowledge acquisition, where semantic representation plays an important role;
- serving as a data repository.

The CCO is expected to be mainly used in the bioinformatics field and naturally, most of the feedback is expected to happen at that level. On the other hand, it is worth mentioning that the role of the CCO is not to compete against the cell-cycle data from GO. Rather, it is intended to complement GO by providing additional structure for the formalization process of the available knowledge in the cell-cycle field. The work so far has confirmed the existing integration obstacles due to the diversity of data formats and lack of formalization approaches as well as the trade-offs that are common in biological sciences.

The knowledge will be weighted or scored according to some defined evidence codes expressing the support media similar to those implemented in GO (experimental, electronically inferred, and so forth). A graphical user interface is also foreseen. The ultimate aim of the project is to support hypothesis evaluation about cell-cycle regulation issues. These hypotheses will be evaluated for consistency against the existing knowledge. The end product intends to include several intermediate milestones:

- An improved cell-cycle ontology, built on the existing ontology from GO and complemented with the temporal/dynamical aspects of the process. The three GO ontologies altogether supply an initial temporal framework for CCO by providing the cellular components (what/where), molecular functions (what) and biological processes (how/when). We are currently investigating approaches for connecting these three ontologies and representing knowledge such as for example `CDK A (what) is located in Cytoplasm (where) during Cytokinesis (when)`.
- A knowledge base holding the CCO as the core structure and data taken from Reactome and some other prospective resources as well as data produced by the DIAMONDS consortium, which is expected to boost the initial evolution of the system by providing data.
- A query [32] system for hypotheses validation, annotation assistance.
- A user interface providing a user-friendly environment for interacting with the system, creating queries and input data, annotation and so forth.

Besides the classical benefits provided by an ontology (data repository, knowledge sharing, validation, annotation, and so on), we aim to build a knowledge-based system that provides reasoning services oriented to hypotheses evaluation in the context of cell-cycle analysis. Consistency checking will further facilitate and improve some tasks done by annotation teams.

Finally, once a stable version of the ontology is released, the cell-cycle community will be invited to contribute to this effort and enhance the system.

Aknowledgements. This work was financially supported by the EU Framework programme for research, contract number LSHG-CT-2004-512143.

References

1. Baader, F., Calvanese, D., McGuinness, D., Nardi, D., Patel-Schneider, P.(eds.): The Description Logic Handbook. Theory, Implementation and Applications. Cambridge University Press (2003)
2. Bader, G.D., Betel, D., Hogue, C.W.V.: BIND: the Biomolecular Interaction Network Database. Nucleic Acids Res, (2003) **31** 1 248-250
3. Bard, J., Rhee, S.Y., Ashburner, M.: An ontology for cell types. Genome Biology, (2005) **6** R21
4. Berners-Lee, T., Hendler, J., Lassila, O.: The Semantic Web. Scientific American (May 2001)
5. Blake, J.: Bio-ontologies-fast and furious. Nature Biotechnology, (2004) **22** 773-774
6. Bairoch, A., Apweiler, R., Wu, C.H., Barker, W.C., Boeckmann, B., Ferro, S., Gasteiger, E., Huang, H., Lopez, R., Magrane, M., Martin, M.J., Natale, D.A., ODonovan, C., Redaschi, N., Yeh, L.S.: The universal protein resource (Uniprot). Nucleic Acids Res., (2005) **33** Database issue D154D159
7. Benson, D.A., Karsch-Mizrachi, I., Lipman, D.J., Ostell, J., Wheeler, D.L.: Genbank. Nucleic Acids Res., (2005) **33** Database issue D34D38
8. Covitz, P.A., Hartel, F., Schaefer, C., De Coronado, S., Fragoso, G., Sahni, H., Gustafson, S., Buetow, K.H. caCORE: A common infrastructure for cancer informatics. Bioinformatics, (2003) **19** 18 2404-2412
9. Gene Ontology Consortium.: The Gene Ontology (GO) database and informatics resource. Nucleic Acids Res., (2004) **32** Database issue D258-D261
10. Gomez-Perez, A., Corcho, O., Fernandez-Lopez, M.: Ontological Engineering : with examples from the areas of Knowledge Management, e-Commerce and the Semantic Web. Springer (2004)
11. Gruninger, M., Fox M.S.: Methodology for the Design and Evaluation of Ontologies. Workshop on Basic Ontological Issues in Knowledge Sharing, IJCAI-95 (Montreal) (1995)
12. Guarino, N.: Formal Ontology In Information Systems. In Proceedings of FOIS '98, Trento, Italy 6-8 June. IOS Press (1998)
13. Hermjakob, H., Montecchi-Palazzi, L., Lewington, C., Mudali, S., Kerrien, S., Orchard, S., Vingron, M., Roechert, B., Roepstorff, P., Valencia, A., Margalit, H., Armstrong, J., Bairoch, A., Cesareni, G., Sherman, D., Apweiler, R.: IntAct: an open source molecular interaction database. Nucleic Acids Res., (2004) **32** Database issue D452-D455
14. Inze, D.: Why should we study the plant cell cycle? J. Exp. Bot., (2003) **54** 385 11251126
15. Joshi-Tope, G., Gillespie, M., Vastrik, I., DEustachio, P., Schmidt, E., de Bono, B., Jassal, B., Gopinath, G.R., Wu, G.R., Matthews, L., Lewis, S., Birney, E., Stein, L.: Reactome: a knowledgebase of biological pathways. Nucleic Acids Res., (2005) **33** Database issue D428D432
16. Knublauch, H., Dameron, O., Mussen, M.A.: Weaving the biomedical semantic web with the Protege OWL plugin (2004)
17. Maedche, A.: Ontology Learning For The Semantic Web. Norwell, Massachusetts, Kluwer Academic Publishers (2003)
18. Maedche, A., Volz R.: The Ontology Extraction & Maintenance Framework Text-To-Onto. In Proceedings of the ICDM-2001 Workshop on the integration of Data Mining and Knowledge Management, San Jose, USA, November, 31 (2001)

19. Noy, N.F., McGuiness, D.L.: Ontology development 101: A guide to creating your first ontology. Technical Report SMI-2001-0880, Stanford University, SMI technical report (2001)
20. Ogden, C., Richards, I.: The Meaning of Meaning: A Study of the Influence of Language upon Thought and of the Science of Symbolism. Routledge & Kegan Paul Ltd., London, 10 edition (1923)
21. McGuinness, D.L., van Harmelen, F.(eds.): OWL Web Ontology Language Overview. http://www.w3.org/TR/2004/REC-owl-features-20040210/
22. Lopez, M.F., Perez, A.G., Juristo, N.: METHONTOLOGY: From Ontological Art Towards Ontological Engineering. Workshop on Ontological Engineering. Spring Symposium Series: Stanford, USA (1997)
23. Smith, B., Kohler, J., Kumar, A.: On the application of formal principles to life science data: A case study in the Gene Ontology. Database Integration in the Life Sciences (DILS), Berlin: Springer (2004)
24. Smith, B., Ceusters, W., Klagges, B., Kohler, J., Kumar, A., Lomax, J., Mungall, C., Neuhaus, F., Rector, A.L., Rosse, C.: Relations in biomedical ontologies. Genome Biology, (2005) **6** R46
25. Stevens, R., Baker, P., Bechhofer, S., Ng, G., Jacoby, A., Paton, NW., Goble, CA., Brass, A. TAMBIS: transparent access to multiple bioinformatics information sources. Bioinformatics, (2000) **16** 2 184-185
26. Thompson, J.D., Holbrook, S.R., Katoh, K., Koehl, K., Moras, D., Westhof, E., Poch, O.: MAO: a multiple alignment ontology for nucleic acid and protein sequences. Nucleic Acids Res., (2005) **33** 13 4164-4171
27. Uschold, M., King, M. Towards Methodology for Building Ontologies. Workshop on Basic Ontological Issues in Knowledge Sharing, held in conjunction with IJCAI-95 (1995)
28. Uschold, M., Gruninger, M. Ontologies: Principles, methods and applications. Knowledge Engineering Review, (1996) **11** 2 93-136
29. Vermeulen, K., Van Bockstaele, D.R., Berneman, Z.N.: The cell cycle: a review of regulation, deregulation and therapeutic targets in cancer. Cell Prolif., (2003) **36** 3 131149
30. Yeh, I., Karp, P.D., Noy, N.F., Altman, R.B.: Knowledge acquisition, consistency checking and concurrency control for Gene Ontology (GO). Bioinformatics, (2003) **19** 2 241-248
31. Yu, A.C.: Methods in biomedical ontology. Journal of Biomedical Informatics, (2005) **78** 315-333
32. Zhang, Z., Miller, J.A.: Ontology query languages for the semantic web. A performance evaluation (2004)

Appendix: Formal Definitions of Ontology and Knowledge Base Structure

The following formal definitions, which are introduced for showing some framework elements, have been adapted from [17], which in its turn has its mainstay in the Ogden-Richards' semiotic triangle [20].

An ontology structure is a 6-tuple: $\mathcal{O} = \{\mathcal{C}, \mathcal{R}, \mathcal{S}^{\mathcal{C}}, \mathcal{P}^{\mathcal{C}}, \rho, \mathcal{A}^{\mathcal{O}}\}$, where:

- $\mathcal{C}$ and $\mathcal{R}$ are two disjoint sets whose elements are called concepts and relations respectively.

- $\mathcal{S}^{\mathcal{C}}$ is a directed relation $\mathcal{S}^{\mathcal{C}} \subseteq \mathcal{C} \times \mathcal{C}$ which is called subsumption. $\mathcal{S}^{\mathcal{C}}(c_1, c_2)$ means that c_1 is a subconcept of c_2. Consequently, c_1 is called the subsumee and c_2 is the subsumer.
- $\mathcal{P}^{\mathcal{C}}$ is a directed relation $\mathcal{P}^{\mathcal{C}} \subseteq \mathcal{C} \times \mathcal{C}$ which is called partonomic inclusion.
- ρ is a function that relates concepts in neither a taxonomical nor a partonomical way: $\rho : \mathcal{R} \rightarrow \mathcal{C} \times \mathcal{C}$.
- $\mathcal{A}^{\mathcal{O}}$ is a set of axioms on $\mathcal{O}$ expressed in a logical language, e.g. a description logic.

The notion of lexicon is also introduced. A lexicon for an ontology structure $\mathcal{O}$ is a 4-tuple $\mathcal{L} = \{\mathcal{L}^{\mathcal{C}}, \mathcal{L}^{\mathcal{R}}, \mathcal{F}, \mathcal{G}\}$, where:

- $\mathcal{L}^{\mathcal{C}}$ and $\mathcal{L}^{\mathcal{R}}$ are two sets whose elements are lexical entries for concepts and relations respectively.
- $\mathcal{F}$ and $\mathcal{G}$ are two relations $\mathcal{F} \subseteq \mathcal{L}^{\mathcal{C}} \times \mathcal{C}$ and $\mathcal{G} \subseteq \mathcal{L}^{\mathcal{R}} \times \mathcal{R}$ called references for concepts and relations respectively such that: $\mathcal{F}(l) = \{c \in \mathcal{C} | (l, c) \in \mathcal{F}\}$ and $\mathcal{F}^{-1}(c) = \left\{l \in \mathcal{L}^{\mathcal{C}} | (l, c) \in \mathcal{F}\right\}$. $\mathcal{G}$ and $\mathcal{G}^{-1}$ are defined analogously.

In [17] only the concept hierarchy ($\mathcal{S}^{\mathcal{C}}$) was hallmarked from the generic function ρ. We have also made evident the partonomic inclusion ($\mathcal{P}^{\mathcal{C}}$) since it plays an important role in our main source ontologies.

In turn, a knowledge base structure is a 4-tuple $\mathcal{KB} = \left\{\mathcal{O}, \mathcal{I}, \iota^{\mathcal{C}}, \iota^{\mathcal{R}}\right\}$, where:

- $\mathcal{O}$ is an ontology.
- $\mathcal{I}$ is a set whose elements are called instances.
- $\iota^{\mathcal{C}} : \mathcal{C} \rightarrow 2^{\mathcal{I}}$ and $\iota^{\mathcal{R}} : \mathcal{R} \rightarrow 2^{\mathcal{I} \times \mathcal{I}}$ are two functions for concept instantiation and relation instantiation respectively.

Again, the notion of lexicon is also introduced. Therefore, a lexicon for a knowledge base structure $\mathcal{KB}$ just is a tuple $\mathcal{L}^{\mathcal{KB}} = \{\mathcal{L}^{\mathcal{I}}, \mathcal{J}\}$, where:

- $\mathcal{L}^{\mathcal{I}}$ is a set whose elements are called lexical entries for instances.
- $\mathcal{J} \subseteq \mathcal{L}^{\mathcal{I}} \times \mathcal{I}$ is a reference relation for instances, such that for any $\mathcal{J}$, let for $l \in \mathcal{L}^{\mathcal{I}}$: $\mathcal{J}(l) = \{i \in \mathcal{I} | (l, i) \in \mathcal{J}\}$ and $\mathcal{J}^{-1}(i) = \left\{l \in \mathcal{L}^{I} | (l, i) \in \mathcal{J}\right\}$.

Link Discovery in Graphs Derived from Biological Databases
(Research Paper)

Petteri Sevon, Lauri Eronen, Petteri Hintsanen, Kimmo Kulovesi, and Hannu Toivonen*

HIIT Basic Research Unit, Department of Computer Science,
P.O. Box 68, FI-00014 University of Helsinki, Finland
{Petteri.Sevon, Lauri.Eronen, Petteri.Hintsanen, Kimmo.Kulovesi, Hannu.Toivonen}@cs.helsinki.fi

Abstract. Public biological databases contain vast amounts of rich data that can also be used to create and evaluate new biological hypothesis. We propose a method for link discovery in biological databases, i.e., for prediction and evaluation of implicit or previously unknown connections between biological entities and concepts. In our framework, information extracted from available databases is represented as a graph, where vertices correspond to entities and concepts, and edges represent known, annotated relationships between vertices. A link, an (implicit and possibly unknown) relation between two entities is manifested as a path or a subgraph connecting the corresponding vertices. We propose measures for link goodness that are based on three factors: edge reliability, relevance, and rarity. We handle these factors with a proper probabilistic interpretation. We give practical methods for finding and evaluating links in large graphs and report experimental results with Alzheimer genes and protein interactions.

1 Introduction

The amount of publically available biological data is growing at a tremendous pace, as new information about genomes, proteomes, interactomes etc. is published daily. Despite the large amount of that information, it is clear that it only represents a tiny fraction of the biological knowledge that potentially will be discovered. For instance, consider the functions of genes: in the Gene Ontology database[1], 29.5% of those gene products that have an annotation for a molecular function, the annotation at the time of writing is "unknown". This example only represents some of the facts we know that we do not know yet.

We present novel computational methods for predicting some of the missing information, with the primary aim of producing and ranking new biological hypothesis for life scientists working on their own specific problems. We assume

* Work done while visiting the University of Freiburg.

[1] http://www.godatabase.org

U. Leser, F. Naumann, and B. Eckman (Eds.): DILS 2006, LNBI 4075, pp. 35–49, 2006.

a fairly simple and generic form for the input data: a graph where biological entities and concepts constitute the set of vertices, and the edges correspond to known and annotated relationships between the vertices. In this framework, a yet undiscovered link between two entities or concepts may be manifested as a path or a subgraph connecting the corresponding vertices. Qualitative hypotheses for the biological mechanisms are generated by discovering such paths or subgraphs. In this paper, we use the term *link* to refer to any connections between two vertices in the graph, potentially output as a hypothesis for a biological relation.

Not all paths represent a biologically meaningful links. Two edges incident on a vertex may constitute a spurious path, or edges may not be completely reliable. To be able to address more interesting questions, such as evaluation of the statistical significance of a link, or ranking a set of vertices in order of strength of linkage to a given vertex, we need a way of quantifying the strength of a link. This will be a central topic of this paper.

In our scenario for the analysis, a life scientist poses queries to a graph database system. In a simple form, such a query can ask if a path exists between two given concepts, and how strong the link is. In a more complex setting, the user may submit sets of vertices and ask the system to find, evaluate and rank subgraphs connecting any pair of given vertices.

As a motivating example, consider gene mapping for a particular phenotype. The mapping may have resulted in a large set of candidate genes. When further expensive analyses are planned for the wet lab, the investigators first compare the candidates in the light of what is known about them in the public databases and literature, hoping to be able to concentrate the efforts and resources on the most promising candidates. Due to the lack of automated methods, the work is mostly done by manually browsing the databases. This is a slow and laborious process, and necessarily limits the extent and coverage of the search. Our methods aim at partial automation of such tasks. As for the specific example, methods for automated discovery and analysis of connections between a candidate gene and a phenotype have only recently started to emerge [1,2].

In this paper, we propose a method for measuring the strength of a link based on the two-terminal network reliability [3] between the end vertices. The main contributions of the paper are a novel application of the network reliability measure, as well as a unique way of assigning probabilities to the edges based on three aspects: reliability, relevance, and rarity. Reliability reflects the confidence to the data source, relevance is a subjective measure of importance, and rarity rewards (informative) edges between nodes with low degrees. We give methods for finding good paths and subgraphs and for evaluating their quality. The applicability of the methods is not restricted to gene–phenotype links; they can be used for analyzing the link between any pair of concepts, and potentially even in completely different application areas.

Related work. Our work can be characterised as link discovery (link mining, see, e.g., [4] for a review)—or, more specifically, as link prediction; we aim at predicting links between pairs of vertices, where none exist in the form of direct edges. We work on the abstract level of graphs. This gives our methods the flexibility to work, in principle, with arbitrary concepts and relations. In contrast, methods for specific prediction and annotation tasks have already been heavily used in bioinformatics, for instance to predict genes from the DNA, to predict protein structures and functions, to analyse metabolic pathways, and so on. Our approach is complementary to these, and characteristically integrates different sources of data on an abstract level. Swanson [5,6] successfully demonstrated that novel, unexpected links can be found between entities that are not directly connected. He was able to find an association between a set of articles on Raynaud's syndrome and another set on fish oil through associations via a third set of articles. Many measures have been proposed for assessing the strength of a link based on overlapping neighborhoods (see, e.g., [7] for a review), i.e., a subgraph consisting of parallel paths of length two. Lin and Chalupsky [8] consider the rarity of path type, in terms of edge types, as a factor of path interestingness. However, little has been published on analysis of connection subgraphs of arbitrary topology. Faloutsos et al. [9] present the idea of using delivered current in resistor networks as a measure for subgraph goodness in (social) networks and give a method for finding a good connection subgraph between two vertices. Asthana et al. [10] use two-terminal network reliability for predicting protein complex memberships from a network of protein interactions. Ramakrishnan et al. [11] assign weights to the edges based on various measures of informativeness, and then extract connection subgraph maximizing a goodness function based on the resistor network model of Faloutsos et al.

Paper organization. The paper is organized as follows. We first describe the data in Section 2. In Section 3, we define measures for the strength of a link for a single path and for a subgraph, and show how to estimate the statistical significance of a link. In Section 4, we report experimental results using a set of known Alzheimer genes and a set of known protein interactions. Finally, in Section 5, we conclude with a discussion.

2 Description of Data

Our graph data model consists of various biological entities and annotated relations between them. Large, annotated biological data sets can be readily acquired from several public databases and imported into our graph model in a straightforward manner. We now describe the databases we use, and then give a formal definition of the data model.

2.1 Biological Databases

NCBI's Entrez[2] is an integrated, text-based search and retrieval system for the major biological databases. We use publically available copies of Entrez databases[3] along with the Gene Ontology Consortium's annotation database

[2] http://www.ncbi.nlm.nih.gov/entrez/query.fcgi

[3] ftp://ftp.ncbi.nih.gov/entrez/links

(GOA) in our own research. The Entrez databases contain several kinds of interlinked entities (e.g. article abstracts, genes, gene clusters and proteins), assembled by NCBI from various source databases such as UniProt and PubMed. The GO annotation database contains information about the biological processes, cellular components, and molecular functions of gene products, and it is linked with Entrez databases. Although many of the Entrez's source databases are themselves available for download, handling Entrez's link files (essentially lists of edges between entities) is far easier than parsing numerous flat data files in each source database's native format. This is our main reason for using the Entrez databases instead of the original databases.

We represent these entities and relationships as vertices and edges in our graph model. As a result, we get a total of 1,968,951 vertices and 7,008,607 edges. The vertex types in our graph database and some statistics are summarized in Table 1. This particular collection of data sets is not meant to be complete, but it certainly is sufficiently large and versatile for real link discovery.

Table 1. Vertex types

Vertex type	Source database	Number of vertices	Mean degree
Article	PubMed	330970	6.92
Biological process	GOA	10744	6.76
Cellular component	GOA	1807	16.21
Conserved domain	Entrez Domains	15727	99.82
Gene	Entrez Gene	395611	6.09
Gene cluster	UniGene	362155	2.36
Homology group	HomoloGene	35478	14.68
Molecular function	GOA	7922	7.28
OMIM entry*	OMIM	15253	34.35
Protein	Entrez Protein	741856	5.36
Structural property	Entrez Structure	26425	3.33

*OMIM entries correspond to phenotype descriptions and gene loci.

2.2 Data Model

Our data model is a directed, labeled and weighted graph $G = (V, E)$. The elements of the vertex set V are biological entities such as genes, proteins and biological processes, as well as more general objects like article abstracts. They are labeled by a type from a set T_v, such as "gene" or "protein". Edge labels (edge types) from set T_e describe the relations between vertices, for example "codes" (e.g., gene codes protein) or "refers to" (e.g., article refers to gene).

For notational convenience, we define the edge set to consist of triplets (u, τ, v), where u and v are vertices from V and $\tau \in T_e$ is the type of the edge between them. Each type τ has a natural inverse, such as "coded by" and "is referred by", which we denote by $\tau^{-1} \in T_e$; in a similar fashion, for each edge $e = (u, \tau, v) \in E$ we define its inverse edge $e^{-1} = (v, \tau^{-1}, u) \in E$ and assume one always exists.

Effectively, the graph could be seen as undirected but with directed labels. We call a directed path $\mathbf{p}$ from s to t an s–t path. Finally, we denote the set of edges incident to any vertex $v \in V$ by $E(v)$ and the set of neighbouring vertices of v by $\Gamma(v) = \{u \in V \mid (v, \tau, u) \in E \text{ for some } \tau \in T_e\}$.

Edges sometimes have natural weights in the source databases. For example, a homology between two proteins could have values denoting the degree of sequence similarity. However, we will use other factors, too, to define the weights of edges. They will be discussed next.

3 Link Goodness and Significance

Our goal is to discover and evaluate links between vertices specified by the user. In order to be able to rank paths, or assess the significance of a connection between two vertices, we need a measure for path goodness. We start by defining edge weights (or probabilities), based on which we define a measure for the quality of a given path, and then outline methods for finding the best paths between a pair of vertices. After that we will address the evaluation of the link as a function of the whole graph, not just the single best path. Finally, we will show how to estimate the statistical significance of links, whether based on the best path, or the graph as a whole.

3.1 Edge Weights

We define edge probabilities (weights) as a function of three aspects:

1. Reliability: how confident are we in the edge? How reliable is the data source, how reliable is the method used to produce or predict the edge, and how strong or probable is the connection estimated to be in the data source?
2. Relevance: how relevant is the edge (type) with respect to the query? We assume that the investigator can give query-specific weights for edge types according to his or her subjective opinions of the importance of each edge type for the query at hand.
3. Rarity: how rare and informative is the edge? As an extreme example, an article that refers to all human genes—and such articles do exist—is not likely to be relevant for a specific gene, whereas an article that only refers to few genes is much more likely to be informative. In our definition, edge rarity will be directly related to the degrees of incident vertices.

We assume that edge relevance is defined by the user, and that edge reliability is defined by the data source and potentially also by the user. We define rarity below, and then combine all aspects to one probability.

Reliability. We envision that the reliabilities of edges are defined using a set of simple rules, such as: if the edge is derived from Swiss-Prot, then its reliability is 0.9, whereas if the edge is derived from the computer-annotated TrEMBL database, then its reliability is 0.5. The interpretation of edge reliability is the degree of belief the investigator has for the edge being correctly annotated.

If there is a value associated with an edge that reflects similarity or confidence, such as a homology score, the value can be transformed into a $[0, 1]$-similarity value. With the interpretation that the similarity of vertices u and v is the probability that any relationship between u and a third vertex t is also true for v and t, the similarity can be multiplied into the reliability of the edge.

Relevance. The relevance of an edge type is the degree of the investigator's belief that edges of that type represents a relevant connection with respect to the query. In a practical system, the investigator has a basic configuration—a set of default relevance values for edge types—and only few adjustments are needed for a typical query.

The relevance values may sometimes be easier to give in terms of vertex types instead of edge types. Then, relevance $q(\tau)$ for a vertex type τ can be decomposed into coefficients for edge types by multiplying all edge types with one end-vertex of type τ by $\sqrt{q(\tau)}$, and edge types with both end-vertices of type τ by $q(\tau)$. As path relevance will be defined as a product of edge relevances, this gives the desired outcome: the relevance of any path visiting a node of type τ is multiplied by $q(\tau)$.

Rarity. We want to give lower scores for paths that visit vertices with high degrees: the higher the degree of vertex v, the less likely it is that any two neighbors of v actually have an interesting connection through v. We define rarity $d(v)$ first for vertices: $d(v)$ is the probability that any two edges incident on v are related to each other and represent a meaningful path.

We propose the *ad hoc* formula $d(v) = (|\Gamma(v)| + 1)^{-\alpha} \in [0, 1]$, with $\alpha > 0$, to determine the penalty for the degree $|\Gamma(v)|$ of vertex v; smaller values mean larger penalty. The parameter α determines how steeply the penalty increases with the degree.

With $\alpha = 1$, rarity $d(v) = 1/(|\Gamma(v)| + 1)$ has a natural probabilistic interpretation. Consider a random walker who, at any vertex, is equally likely to follow any edge, or stop at the vertex. Then, given a path $\mathbf{p}$ through vertices $v_1, v_2, \ldots, v_k$, rarity $d(v_i)$ is the probability that a random walker who has so far traversed nodes $v_1, \ldots, v_i$, will next stay on the path and visit node v_{i+1}. Although lower values of α do not give equally attractive interpretations as random walk probabilities, they can be useful in practice to give relevant penalties for vertex degree that reward parallel paths more than a standard random walker.

The maximum value of $d(v)$ for an non-terminal vertex v of a path is $3^{-\alpha}$. Rarity values of the terminal edges are ignored; they would only add a constant factor to all paths. In principle, the values of α could be set separately for each vertex type, but in this paper we use a single value for all vertices.

As with relevance above, the rarity values are decomposed into edge-specific coefficients by taking the square root of them. Ideally, in the context of analysis of connection subgraphs, the relatedness of edges incident on a vertex should be tested for each pair of edges separately and independently. With the rarity values of vertices decomposed on the incident edges, this is clearly not the case. The approximation is used in order to avoid the quadratic computational cost for each vertex. It has no effect on evaluation of the goodness of a single path.

Total edge weight. Now that we have defined all the components of edge weight, we define edge weight $w(e)$ simply as a product of those factors: $w(e) = r(e)q(e)d(e)$, where $r(e) \in [0,1]$, $q(e) \in [0,1]$, and $d(e) \in [0,1]$ are the reliability, relevance, and rarity of edge e, respectively. Under the assumption that they are probabilities for mutually independent necessary conditions for the edge, the weight $w(e)$ is the probability that edge e exists.

3.2 Discovery of Best Paths

Let us consider random graph model $\mathcal{G}(G, w)$ specified by graph G and edge weights w described above. A realization of the random graph is obtained by independently removing each edge e from G at probability $1 - w(e)$.

We propose the following definition for the goodness $g(\mathbf{p}, w)$ of path $\mathbf{p} = e_1 e_2 \ldots e_k$:

$$g(\mathbf{p}, w) = \prod_{i=1}^{k} w(e_i) \tag{1}$$

With the interpretation that $w(e)$ is the probability that edge e exists, the goodness $g(\mathbf{p}, w)$ is the probability that the whole path exists in a realization of $\mathcal{G}(G, w)$.

The path discovery problem. We now formulate the path discovery task: given two sets S and T of vertices (S and T may overlap), find

1. the k best paths from S to T,
2. all paths whose goodness is at least m, or
3. all paths that consist of at most ℓ edges.

These paths could be shown to the user as most likely hypotheses involving vertices from the given sets or, as will be discussed below, used for further analysis of the link. In any case, before giving final results to the user, it is useful to estimate the statistical significance of the results; this will also be discussed below.

Algorithms. Standard algorithms for finding shortest paths [12,13] can be applied; the probabilities can be transformed into distances required by the standard methods by taking the negative logarithm of the goodness:

$$-\log g(\mathbf{p}, w) = \sum_{i=1}^{k} -\log w(e_i). \tag{2}$$

Any combination of the abovementioned constraints for paths can be easily used.

The number of vertices that can be reached from a single source typically grows exponentially with path length, until it saturates. If the maximum number of edges (or minimum goodness) is set so that the saturation point is not reached at halfway to the maximum number (or minimum goodness), then a bi-directional search starting from both sets will be substantially faster than a standard unidirectional search.

3.3 Evaluation of Graph Connections

The goodness of a single best s–t path is not necessarily a good measure of the strength of the link between vertices s and t. A link consisting of several parallel paths may be considered stronger than a single path, even if all the individual paths are weaker. With a probabilistic interpretation, the quality of a single path reflects the probability that that particular path exists, whereas a more appropriate measure often would be the probability that at least one path exists between s and t.

Graph connection goodness. Based on the probabilistic interpretation, we propose using the two-terminal network reliability [3] as a measure for link goodness $g(G, w, s, t)$ between vertices s and t in graph G. The measure is defined as the probability of a path existing in a realization of the random graph:

$$g(G, w, s, t) = \Pr(\text{“there is an } s\text{–}t \text{ path in a graph generated by } \mathcal{G}(G, w)\text{”}). \tag{3}$$

Algorithms. The two-terminal network problem has been shown to be NP-hard by Valiant [14], but the probability can be estimated using a straightforward Monte Carlo approach: generate a large number of realizations of the random graph, and count the relative frequency of graphs where a path from s to t exists. Monte Carlo estimates that are accurate to within $\pm\varepsilon$ at high probability can be obtained using $O(\varepsilon^{-2})$ iterations. Since we are only interested in cases where $g(G, w, s, t)$ is not very close to zero, we need not worry about the number of iterations required to control relative accuracy. (Reasonable absolute accuracy can be achieved with 100,000–1,000,000 iterations; in practice, our Python-implementation is able to perform 1,000,000 iterations on a graph with 1,000 edges in roughly 1.5 hours on a 3.0 GHz P4 PC.)

A lower bound for $g(G, w, s, t)$ can be computed efficiently by first enumerating all m-good or k best paths from s to t, and then evaluating $g(G', w, s, t)$ in the subgraph G' induced by the set of paths. A graph $G' = (V', E')$ is induced by a set of paths, if V' and E' are the sets of vertices and edges, respectively, occurring in the paths. Since the induced graph is a subgraph of G, it clearly gives a lower bound. Following the terminology of Faloutsos et al. [9], the induced subgraph G' is here called a connection subgraph.

An upper bound for $g(G, w, s, t)$ can be obtained easily when the paths inducing G' are searched unidirectionally starting from, say, s: include all the pruned partial paths in G' and connect them with an edge of probability one to t. This provides the tightest possible upper bound based on G'. With bi-directional search, the upper bound can be obtained in a similar way. Estimation of the upper bound is easily incorporated to the Monte Carlo algorithm, but the procedure is slowed down due to the large number of additional edges from the pruned paths. Our work so far relies on the lower bounds only.

Further efficiency improvements are possible by repeatedly replacing parallel edges by only one edge and by removing vertices (except s or t) with exactly two neighbors as long as there are any. This is a linear-time operation in the size of the graph. For the class of series-parallel graphs, these operations reduce graphs to a single edge and two-terminal network reliability can be computed exactly in linear time.

3.4 Estimation of Link Significance

We eventually want to measure how strongly two given vertices, s and t, are related in graph G. The path probability $g(\mathbf{p}, w)$ (Eq. 1) and the two-terminal network reliability $g(G, w, s, t)$ (Eq. 3) allow ranking of links, but their values may be difficult to put into perspective. Is a probability of, say, 0.4 for the existence of any s–t path high or low? This obviously depends on the data and the specific instances.

Using $\max_{\mathbf{p} \in \mathcal{P}(s,t)} g(\mathbf{p}, w)$, where $\mathcal{P}(s,t)$ is the set of all s–t paths (i.e., goodness of the best s–t path), or $g(G, w, s, t)$ as a test statistic, we can estimate the statistical significance of the link. This tells us how likely it is to obtain, by chance, probability of 0.4 or better. There are a variety of meaningful null hypotheses to be considered:

1. Vertices s and t of types τ_s and τ_t, respectively, are not more strongly connected than randomly chosen vertices s' and t' of types τ_s and τ_t.
2. Vertex s of type τ is not more strongly connected to vertex t than a randomly chosen vertex s' of type τ.
3. Vertices s and t are not more strongly connected in the given graph G than random graph H and edge weights w' generated by model $\mathcal{H}$ similar to the (unknown) model which generated G and w.

The last null hypothesis clearly is the most complicated one, as it is not easy to come up with model $\mathcal{H}$ that generates random graphs that are topologically sufficiently similar to the observed graph. The choice from the first two null hypotheses depends on what we are testing. In a symmetrical case, e.g., testing for significance of connection between two candidate genes, the first null hypothesis is appropriate. If the roles of the vertices are asymmetric, as in testing for the connection from a set of candidate genes to a single phenotype, the second null should be used. In the experiments, we apply the first null hypothesis to assesment of gene–gene link, and the second one to assesment of gene–phenotype link.

Under the null hypotheses 1 and 2, p values can be estimated by sampling vertices s' (Null 1) or pairs (s', t') (Null 2), and computing the test statistic ($g(\mathbf{p}, w)$ of the best s'–t' path $\mathbf{p}$ or $g(G, w, s', t')$) for all (s', t') pairs in the sample. The p value for the connection between s and t is then the proportion of (s', t') pairs giving a test statistic at least as high as the one observed for (s, t). Because vertices of the same type may have wildly varying degrees, we only sample vertices s' and t' that have degrees similar to s and t, respectively.

If a number of hypotheses are to be tested (e.g., several candidate genes), then the resulting p values should be adjusted accordingly to account for multiple testing.

4 Experiments

We demonstrate the use of link goodness by an example in the detection Alzheimer disease genes. We selected a handful of known disease genes, and estimated the significance of the gene–phenotype link for each. We did this this separately for two test different statistics: the probability of the best path between vertices s and t, and the two-terminal reliability computed from the connection subgraph induced by k best paths. In a second experiment, we evaluated the significance of links between genes whose protein products are known to interact. The experiments were performed using the Entrez dataset described in Section 2.1.

Test design is not trivial: for any classified examples, i.e., known disease genes, there are trivial links in the graph (e.g., the OMIM entry for the disease refers directly to the candidate gene). The ideal solution would be to use only edges that are annotated prior to publication of the gene–disease association, but it is difficult to obtain the state of all databases at an earlier date. Instead, we simply removed all trivial paths from the set of k best paths—e.g., paths whose goodness is greater than a given threshold, or paths consisting of at most a given number of edges.

In order to simplify the experimental setting and to avoid introducing a subjective bias, we assume that all edges have the same product rq of reliability and relevance. Consequently, the goodness of a path or subgraph depends only on the topology of the graph and parameters α and rq.

We chose ten known human susceptibility and candidate genes for Alzheimer disease—APP, PSEN1, AD5, AD6, AD9, AD7, COL25A1, APOE, PSEN2, and AD6—obtained by querying the Entrez Gene database with term "Alzheimer". As the vertex representing the phenotype, we used the entry in the OMIM database giving phenotype description of Alzheimer disease. This entry contains trivial links to all known Alzheimer genes, as well as a large number of references to literature on the disease.

For each gene, we sampled 100 genes from the set of all human genes that have similar degree to the tested gene. The goodness values for links between vertices corresponding to these genes and the phenotype constitute our empirical null distribution.

For each gene (candidate or random), we first enumerated the best 100 acyclic paths of at most 6 edges from the gene to the phenotype. For two of the genes, COL25A1 and AD9, no paths to Alzheimer disease were found. Next, in order to eliminate the trivial links, we removed all paths shorter than three edges from this set. Figure 1 shows an induced graph for AD6 (but for clarity only 20 best paths). We used the goodness value of the best of the remaining paths, and the two-terminal network reliability of the graph induced by the remaining paths as test statistics. Two-terminal network reliability was estimated using Monte Carlo algorithm with 100,000 iterations; standard deviation of the estimate is less than 0.0064. Based on these two statistics, we then estimated two p values—one for the best path and another for the connection subgraph—for each candidate gene.

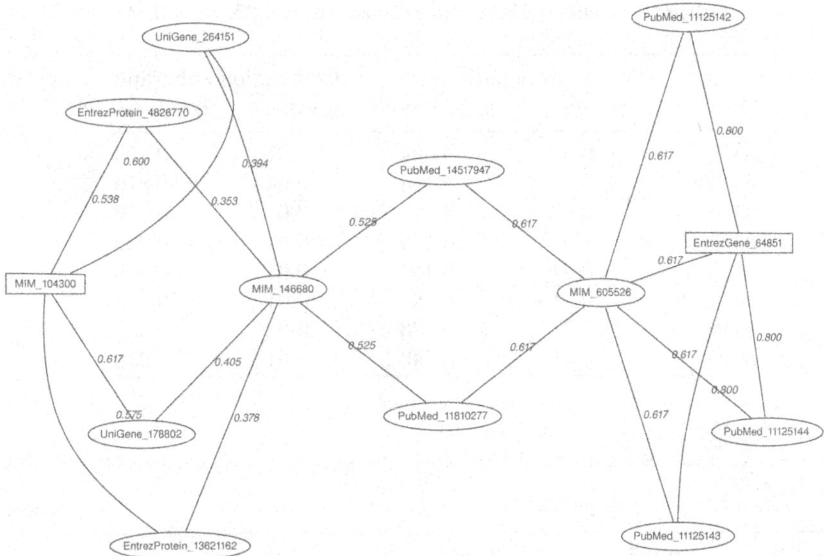

Fig. 1. The graph induced by 20 best paths from gene AD6 to Alzheimer disease. The terminal vertices are rectangular. The edges are labelled with their probabilities ($\alpha = 0.25, rq = 0.8$). Gene AD6 (Entrez Gene entry 64851) is linked to the locus description (OMIM entry 605526) by a direct edge and via three articles. The locus description is in turn linked to another locus description, insulin-degrading enzyme (OMIM entry 146680), via two articles, and, finally, to Alzheimer disease (OMIM entry 104300) via two proteins and two UniGene clusters.

We experimented with the test statistics using parameter values $(\alpha, rq) \in \{0.125, 0.25, 0.5, 1.0\} \times \{0.2, 0.4, 0.6, 0.8, 1.0\}$. For $\alpha = 0.25$ and $rq = 0.8$, the p values and values of the test statistics for each gene are shown in Table 2. The probabilities of best paths and connection subgraphs expectedly vary markedly across genes, and are not alone sufficient indicators of the strength of a link. The estimated p values are more useful here. In this test, they are consistently small; in fact, in many cases none of the 100 randomized data sets produced equally high goodness values. Based on the results, it is difficult to claim that the analysis of connection subgraphs is more powerful than analysis of the best path, but we would expect that to be the case in general.

The goodness values also vary with the values of the two parameters of our test. However, comparable p values were obtained for all combinations of parameter values (except for $\alpha = 1$); mean p values for all combinations are shown in Table 3. This can be seen as an indication of the stability of the measures with respect to the parameters, but it also shows that the links are very strong and rather obvious (as seen in Figure 1), even though all short paths were removed.

Table 2. Results: Alzheimer disease ($\alpha = 0.25, rq = 0.8$)

	Best path		Connection subgraph	
Gene	p value	goodness	p value	goodness
AD7	< 0.01	0.024	0.01	0.153
APOE	< 0.01	0.184	0.01	0.876
APP	0.02	0.123	0.01	0.719
AD8	< 0.01	0.119	< 0.01	0.262
PSEN1	0.04	0.103	0.01	0.963
PSEN2	< 0.01	0.153	< 0.01	0.993
AD6	< 0.01	0.033	< 0.01	0.336
AD5	0.01	0.040	0.01	0.238

Table 3. Mean p values for all combinations of parameter values (best path/connection subgraph)

$rq \setminus \alpha$	0.125	0.250	0.500	1.000
0.2	0.0100/0.0075	0.0175/0.0063	0.0200/0.1325	0.0438/0.3813
0.4	0.0088/0.0075	0.0150/0.0075	0.0163/0.0075	0.0300/0.3813
0.6	0.0088/0.0063	0.0088/0.0075	0.0263/0.0100	0.0063/0.1338
0.8	0.0088/0.0063	0.0088/0.0063	0.0075/0.0075	0.0138/0.0075
1.0	0.0088/0.0200	0.0088/0.0063	0.0088/0.0075	0.0238/0.0088

In a second, more challenging experiment, we evaluated the strength of link between APP and five genes whose protein products interact with the APP protein: HADH2, APBA1, CHRNA7, APOA1, and SHC1. The interactions were obtained from the IntAct-database[4]. The experiments were carried out the same way as with Alzheimer disease, except that we used the first, symmetric null hypothesis (i.e., vertices at both ends were randomized). In the results, two genes show significant linkage to APP (Table 4). The other three genes get non-significant p values despite relatively high values of the test statistics (compared to the Alzheimer experiment), suggesting that pairs of genes are generally

[4] http://www.ebi.ac.uk/intact

strongly connected. A possible remedy is to give higher relevance coefficients for interaction-related edge types. However, it is also possible that simple weighting of edges is not sufficient to distinguish the potential interaction-related paths between the pairs of genes in these cases.

Table 4. Results: interactions with APP ($\alpha = 0.25, rq = 0.8$)

Gene	Best path p value	Best path goodness	Connection subgraph p value	Connection subgraph goodness
HADH7	< 0.01	0.159	0.01	0.917
APBA1	< 0.01	0.137	< 0.01	0.998
CHRNA7	0.17	0.058	0.52	0.359
APOA1	0.56	0.041	0.51	0.530
SHC1	0.15	0.118	0.07	0.937

5 Discussion and Conclusions

In this paper, we have proposed measures and methods for assessing the strength of a link between a pair of vertices in a graph consisting of biological concepts. Such graphs can be easily constructed from many biological databases; due to the simplicity of the data model, integration of data is usually simple and the essential requirement is a referential integrity between the data sources.

We introduced the ideas of assigning probabilities to the edges derived from three factors—reliability, relevance, and rarity. The proposed measures for link strength are based on probabilities of paths that are derived from edge probabilies in a straightforward manner: One is the highest probability of path among all paths connecting the pair of vertices; the other is based on two-terminal network reliability, and approximates (bounds) the probability that at least one path exists between the vertices. We believe that the probabilistic interpretations for link strength are more natural and intuitive for investigators than, e.g., conductance in resistor networks or capacity and maximum network flow.

We demonstrated the link goodness measures for evaluating the strength of gene–phenotype-link using a set of known Alzheimer genes. Both measures gave the known genes low p values, indicating that they would have been successfully identified among the most likely candidates for Alzheimer disease among a random set of genes, except for two genes for which no link was found.

In a second experiment we evaluated the strength of the link between APP and five other genes whose protein products are known to interact with the APP protein. The results suggest that—although two of the genes showed significant linkage to APP—the simplistic experimental setup using a single relevance value for all edge types is not optimal, which was to be expected. We leave the evaluation of expert-specified relevance coefficients as a topic for future research.

Using the goodness of the best path as a test statistic should be less robust than using the two-terminal network reliability. However, in the example case of Alzheimer genes, both methods gave comparable p values. This may be due to several reasons: the function used for rarity, i.e., for penalizing vertex degrees, may be suboptimal, or the test method of removing short paths may still leave some trivial paths that skew the results. Further work is needed to study these issues in detail.

Two of the Alzheimer genes did not have any paths to the disease. This may be due to the limited set of databases we currently use. Several important types of data are missing: protein–protein interactions, tissue specificities, pathways, and Medical Subject Heading annotations of articles, to name a few. Actually, we believe that our probabilistic approach is particularly suitable for analysis of data sets containing uncertain relationships, such as computer annotated interactions or links derived by text mining, as the confidence in the prediction can be easily plugged into the reliability measure.

The use abstract, labeled graphs as a data representation has a number of trade-offs. On one side, it is a generic format, it is easy to convert data into it, and there is a large body of known results and algorithms for graphs. The downside is that information may be lost in the transformation, the vertex or edge types may be too different to be really used in the same graph, and—above all—without built-in knowledge about particular biological concepts, mechanisms, and phenomena, specific discoveries about them cannot be made. It seems obvious to us that several different approaches on different levels of detail and integration are needed, and that they complement rather than compete with each other.

There are several topics for further research. The penalty for vertex degree is now determined for all vertices in a uniform manner, but it might be better to have different rules for different vertex types. The penalty could also be edge type sensitive. For example, consider an article with edges to a large number of genes, one biological process, and one phenotype; we do not want to penalize a path from the biological process to the phenotype from the edges to the genes.

The path queries are now fully specified by the source and target vertex, minimum goodness, maximum length, and edge type relevances. To have more control over the resulting paths, we need a query language that allows an investigator to specify the path types of interest. Earlier suggestions for query languages for paths include regular expressions [15] and context-free grammars [16]. Expressive query languages open possibilities for specifying aspects such as the formulae for degree penalties as background knowledge, or edge relevances, that could be made context sensitive. Another important area for practical applications is visualization of the resulting graphs.

Acknowledgment

This research has been supported by Tekes, Jurilab Ltd., Biocomputing Platforms Ltd., GeneOS Ltd., and Humboldt Foundation.

References

1. Turner, F.S., Clutterbuck, D.R., Semple, C.A.M.: POCUS: Mining genomic sequence annotation to predict disease genes. Genome Biology **4** (2003) R75
2. Perez-Iratxeta, C., Wjst, M., Bork, P., Andrade, M.A.: G2D: A tool for mining genes associated with disease. BMC Genetics **6** (2005) 45
3. Colbourn, C.J.: The Combinatorics of Network Reliability. Oxford University Press (1987)
4. Getoor, L., Diehl, C.P.: Link mining: A survey. SIGKDD Explorations **7** (2005) 3–12
5. Swanson, D.R.: Fish oil, Raynaud's syndrome and undiscovered public knowledge. Perspectives in Biology and Medicine **30** (1986) 7–18
6. Swanson, D.R., Smalheiser, N.R.: An interactive system for finding complementary literatures: A stimulus to scientific discovery. Artificial Intelligence **91** (1997) 183–203
7. Liben-Nowell, D., Kleinberg, J.: The link prediction problem fof social networks. In: Proceedings of the 12th International Conference on Information and Knowledge Management (CIKM'03). (2003) 556–559
8. Lin, S., Chalupsky, H.: Unsupervised link discovery in multi-relational data via rarity analysis. In: Proceedings of the Third IEEE International Conference on Data Mining (ICDM '03). (2003) 171–178
9. Faloutsos, C., McCurley, K.S., Tomkins, A.: Fast discovery of connection subgraphs. In: KDD '04: Proceedings of the tenth ACM SIGKDD international conference on Knowledge discovery and data mining. (2004) 118–127
10. Asthana, S., King, O.D., Gibbons, F.D., Roth, F.P.: Predicting protein complex memebership using probabilistic network reliability. Genome Research **14** (2004) 1170–1175
11. Ramakrishnan, C., Milnor, W.H., Perry, M., Sheth, A.P.: Discovering informative connection subgraphs in multi-relational graphs. SIGKDD Explorations **7** (2005) 56–63
12. Tarjan, R.E.: Data Structures and Network Algorithms. CBMS-NSF Regional Conference Series in Applied Mathematics. SIAM (1983)
13. Eppstein, D.: Finding the k shortest paths. SIAM Journal on Computing **28** (1998) 652–673
14. Valiant, L.G.: The complexity of enumeration and reliability problems. SIAM Journal on Computing **8** (1979) 410–421
15. Lacroix, Z., Raschid, L., Vidal, M.E.: Efficient techniques to explore and rank paths in life science data sources. In: Proceedings of Data Integration in the Life Sciences, First International Workshop (DILS 2004). (2004) 187–202
16. Mork, P., Shaker, R., Halevy, A., Tarczy-Hornoch, P.: PQL: A declarative query language over dynamic biological schemata. In: Proceedings of the American Medical Informatics Association Annual Symposium 2002. (2002) 533–537

Towards an Automated Analysis of Biomedical Abstracts

Barbara Gawronska, Björn Erlendsson, and Björn Olsson

School of Humanities and Informatics, University of Skövde
Box 408
541 28 Skövde, Sweden
barbara.gawronska@his.se,
bjorn.erlendsson@his.se,
bjorn.olsson@his.se

Abstract. An essential part of bioinformatic research concerns the iterative process of validating hypotheses by analyzing facts stored in databases and in published literature. This process can be enhanced by language technology methods, in particular by automatic text understanding. Since it is becoming increasingly difficult to keep up with the vast number of scientific articles being published, there is a need for more easily accessible representations of the current knowledge. The goal of the research described in this paper is to develop a system aimed to support the large-scale research on metabolic and regulatory pathways by extracting relations between biological objects from descriptions found in literature. We present and evaluate the procedures for semantico-syntactic tagging, dividing the text into parts concerning previous research and current research, syntactic parsing, and transformation of syntactic trees into logical representations similar to the pathway graphs utilized in the Kyoto Encyclopaedia of Genes and Genomes.

1 Background and Aim

Text mining has many applications in the area of bioinformatics, where computerized tools are used to analyze data concerning molecular biological objects (genes, proteins, gene regulation pathways, cells, etc) in order to derive new biological insights [1], [2]. The aim of the research described in this paper is to develop a system that applies automated text analysis to support the large-scale analysis of metabolic and regulatory pathways by deriving relevant relations from textual descriptions found in the literature. The need for such a system arises from the fact that molecular biologists today need efficient computer-based tools to navigate the huge amount of knowledge that has been generated over the years and documented in published papers. Since it is becoming increasingly difficult to keep up with the vast number of scientific articles being published, there is a need for more easily accessible representations of the current knowledge. The KEGG pathway database [3] is one example of such an effort to systematically collect the current knowledge on molecular interaction networks in biological processes. Building knowledge bases manually, however, is extremely time-consuming, since each pathway map in KEGG is based on findings from a large number of experiments which have been reported in separate research articles.

U. Leser, F. Naumann, and B. Eckman (Eds.): DILS 2006, LNBI 4075, pp. 50–65, 2006.

Although databases such as KEGG provide easily accessible sources of knowledge for the user, they require enormous amounts of work to build, maintain and keep up-to-date. Therefore, the long-term aim of the research presented here is to provide a semi-automated method of deriving pathway maps using a text corpus as input. As indicated in the overview in Figure 1, we view text analysis as one component in a system that derives pathways from biomedical texts selected from PubMed, using lexical databases and a grammar-based in-depth analysis.

Automated text analysis offers support for the process of structuring knowledge, provided it is conducted using in-depth text comprehension methods. Many of the text mining efforts in bioinformatics, however, have been based on using only statistics regarding co-occurrence of terms [4-11]. As pointed out in [7], this frequent use of simple co-occurrence owes its popularity to the fact that it is easy to implement and allows efficient processing of huge amounts of texts. Such text retrieval and text mining devices can inform the researcher that there seems to be some relation between e.g. a gene and a protein, but in most cases they do not specify what kind of relation it is.

Another line of research is to use pattern- and template-based approaches [12-16]. For example, [17] used a protein name dictionary together with surface clues on word patterns and simple part-of-speech rules to predict protein interactions. In a similar effort, [18] developed a method (BioNLP) based on pattern-matching, which searches for sentences matching a set of rules describing selected functions carried out by proteins. The work in [19] represents a hybrid approach (a stochastic word tagger is combined with rule-based semantic and syntactic analysis). In general, there has been a shift of focus in recent years towards methods which make use of rules and grammars. Examples can be found both in bioinformatics [6], [20] and biomedical information extraction [21], as well as in other domains [22]. As pointed out in [13], a restriction common to most relation extraction models is the lack of ability to extract more than one relation per sentence. Another shortcoming is that relations not expressed by verbs but by, e.g., nouns or participles, are normally omitted.

Among the on-line available information extraction tools, MedScan [6], [20] includes an ambitious attempt to extract positive and negative regulation relations from texts. The developers of the system stress the importance of analyzing subordinated clauses and taking modality into account. We tested the recently released version of MedScan (available at http://www.ariadnegenomics.com/products/medscan.html) on a corpus of 40 biomedical abstracts and found that - although the precision has improved compared to the earlier version - the system is still not reliable enough. As pointed out in [6] the coverage is low. In our corpus of 40 abstracts, only 19 biological relations were found. Out of these 19 relations, it was found upon manual inspection that at most 9 had been correctly extracted. Errors were due mainly to insufficient grammatical analysis. Especially subordinated clauses, ellipsis, appositional constructions, and coordination caused problems. Also long noun sequences caused evident difficulties, and the distribution of the extractions was uneven. Biological relations were extracted from 8 abstracts, while in the remaining 32 no relations were found, although manual inspection revealed that most of these abstracts mentioned several relevant relations. Furthermore, it seems that the system identifies biological objects only if their names are present in the specialized lexicon/ontology it has access to.

In our research, we want to investigate to which extent a system relying mostly on general linguistic knowledge, and to a minimal degree on specialized ontologies can be successful in extracting relations from biological texts. We recently proposed a grammar-based method for extraction of biological relations from scientific texts [23-26]. The method uses an algorithm that searches through the syntactic trees produced by a linguistic parser, identifies relations mentioned in the sentence, and classifies them with respect to their semantic class and epistemic status (facts, counterfactuals, or hypotheses). The semantic categories used in the classification are based on the relation set used in KEGG, so that pathway maps following the same notational convention as KEGG can be automatically generated, and even other relations involving biological objects (coocurrence, part-whole relations) may be extracted. Subsequently, we added several extensions and improvements of the method, such as an improved named entity recognition component and the addition of a method for distinguishing between text describing previous and current work, thereby making it possible to avoid extracting relations from text sections which merely report findings from previous work or common knowledge, rather than new findings [27].

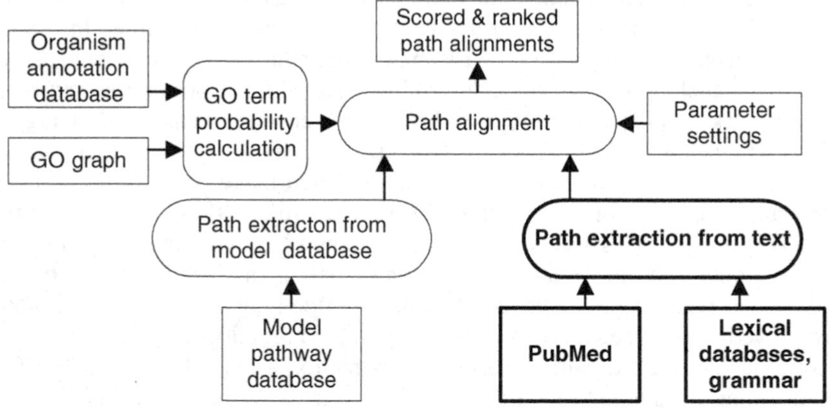

Fig. 1. Overview of the pathway derivation process

In our work, we consider text analysis as a method to extract candidate pathways, which must be further evaluated for biological plausibility. As shown in Figure 1, a set of candidate pathways is extracted from a corpus consisting of a set of selected biomedical articles from PubMed, which allows the user to find all articles containing particular words or combinations of words. The text analysis process uses lexical databases and a grammar to achieve sufficient comprehension of the text for high-precision extraction of pathways corresponding to the textual description. However, even if the candidate pathways extracted by our approach are generally correct with respect to the text (as demonstrated in [24] and [27]), not all of them are necessarily biologically plausible. Therefore, each candidate is evaluated by alignment to and comparison with currently known pathways collected from KEGG and other model databases. As described in detail in [28] the plausibility score is derived by calculating the semantic similarity between the set of gene products that have been paired in the alignment, using the Gene Ontology structured vocabulary of annotation terms. By this technique it is possible to identify similarities between gene products that have annotations that seem to differ at a surface-level comparison, but which actually perform highly similar molecular functions or participate in similar biological processes. Using the derived plausibility scores, the candidate pathways are ranked so that the biologist using the system can select the most biologically relevant pathways that have been derived by the text analysis module. For details on the evaluation of candidate pathways, the reader is referred to [28]. Figure 2 shows the detailed architecture of the text processing system aimed at path extraction.

The syntactic analysis and the algorithm deriving KEGG-like pathways from syntactic trees have been described in [23]. In this paper, we focus on the procedures that prepare the raw text for syntactic and semantic analysis, viz. identification of named entities, tagging, and identification of relevant text parts, but we also mention the latest results of further tests of the syntactic parser and our considerations concerning the design of the final output.

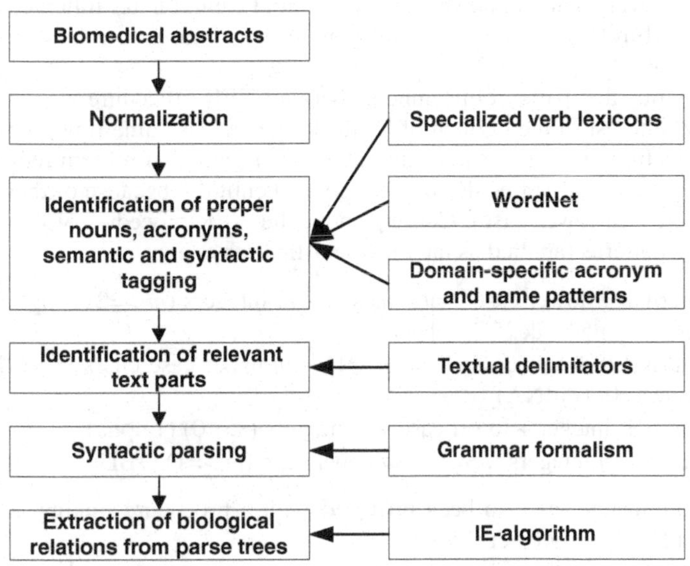

Fig. 2. The information extraction system

2 Named Entity Recognition and Tagging – Domain Adaptation and Testing

Since the precondition for in-depth syntactic and semantic text analysis is appropriate tagging, we started by domain-oriented training and development of the Named Entity Identification (NER) procedure and the semantico-syntactic tagger. The original NER-algorithm and the tagger had been developed for the purpose of processing news reports [29, 30]. The tagger utilized parts of WordNet (version 1.6 [31]), a list of frequent closed-class words and an internally developed lexicon of verbs denoting communication acts (*say*, *report*, *deny*, *suggest*, *indicate* etc). The reason for introducing restricted verb lexicons into the system was the well-known problem of ambiguity in WordNet, especially in its verb part. As a training corpus, we selected a set of biomedical abstracts from PubMed using the search phrases "protein synthesis" and "lymphoma". The corpus consisted of ca 18 000 words (40 abstracts).

The first step of text processing was normalization (also called tokenization): the abstracts were converted to plain text format, illegal characters were removed, headings and bibliographic data marked up. The body text was split into single sentences, and each sentence divided into words. Compound words, where the components were connected by a hyphen, were split into single words.

The NER-procedure, described in detail in [27] identifies proper nouns on the basis of their graphical form, part-of-speech information (e.g. adverbs are not treated as proper noun candidates), internal cues (e.g. presence of indicators like von, de, bin between two strings that both start in capital letters), and external (contextual) cues. In news reports, most frequent external cues are words describing function or semantic category of the named entity (*president*, *minister*, *the city of* ...). Other useful cues are communication verbs: an unknown word with initial capital letter followed by a communication verb refers in an overwhelming majority of cases to a human being or a group of people.

It was obvious that parts of this quite elaborated NER-procedure would be of marginal, if any, interest in the domain of biomedicine. At the same time, the procedure lacked information about most frequent patterns for gene and protein names. Abbreviations like *p53* would not be detected as "named entities" because of their graphical form (first letter in lower case). Consequently, the NER-procedure was enriched by the following patterns (applied as internal identification cues):

- Pattern 1: n lower case chars (n>=1) + m integers (m >=2) + optionally: any character (p53, cdc25C, bcl2)
- Pattern 2: n lower case chars (n>=1) + m upper case chars (m>=1) + k integers (k>=0) (mRNA)
- Pattern 3: integer + lower case + n integers (n>=0) (1alpha)
- Pattern 4: n integers (n>=1) + m upper case (m >=1) (7BL)

The NER-component has also been provided with a procedure linking acronyms to full names of biological objects [27].

For the purpose of Part-of-Speech tagging and semantic tagging, the most frequent verbs denoting relations between biological objects were manually identified in the texts, added to the verb lexicon and classified with respect to the corresponding standard relation set used in the Kyoto Encyclopedia of Genes and Genomes (KEGG, [32]). In this database, most verbs are directly related to KEGG relations (methylation, activation, inhibition, indirect effect). Other relations, introduced after consultation with biologists, are “state change”, “co-occurrence”, “aspectual relation” and “causative relation”. The last two types are of importance for the parsing and extraction procedure, since they indicate that the relevant biological relation is with a high probability degree encoded by the syntactic direct object of the verb (e.g. ...*cause methylation*). In order to be able to interpret this kind of constructions, we added a procedure relating nouns derived from "bioverbs" to the corresponding verbs, and, as a consequence, to the corresponding KEGG-relations.

The training corpus was subsequently re-tagged using the new lexical information. The additions and modifications described above resulted in fairly good performance of the NER-procedure and the tagger: 95% recall and 85% precision on the training corpus. The main shortcoming was the time factor. The extensive search in WordNet is very time-consuming, so it took about 36 minutes to normalize and tag the corpus (i.e. 2 minutes per 1000 input words). In an attempt to reduce the need of WordNet search, the unique tagged words and symbols obtained from the training corpus (about 2500 units) were stored in a database, which will be referred to as the "Tag Memory Database". Each tagged string consists of the following elements:

- the functor "semcat" (semantic class and grammatical category)
- the word form found in the text
- the basic (uninflected) word form
- part of speech information, where the following categories are used:
    - open class words:
        - propername ("named entity", this category includes acronyms)
        - wnn (noun found in WordNet)
        - bionoun (noun derived from "bioverbs", i.e. verbs denoting KEGG-relations, like: *methylation*, *activation*, *inhibition* etc.)
        - wnv (verb found in WordNet)
        - ccv (communication and cognition verb)
        - bioverb (verb denoting a KEGG-relation)
        - adv (adverb)
        - a (adjective)
    - closed class words
        - conj(unction)
        - det(erminer)
        - prep(osition)
        - rel(ative marker)
        - pron(oun, personal)
        - mod(al verb)
        - cop(ula verb)
        - poss(essive pronoun)
        - neg(ation)
        - number
        - punct(uation mark)
        - math(symbol)
- for a subset of verbs and nouns: semantic category
- for verbs and nouns denoting biological relations: corresponding KEGG relation type

A test corpus consisting of about 15 000 words was then selected from PubMed using the protein name *p53* as keyword. This corpus previously unseen by the system) was tagged by a combination of a simple machine learning procedure (matching words against stored tags) and the procedure that had been applied to the training corpus. Figure 3 shows the overall architecture of the process and the knowledge sources utilized at the subsequent stages.

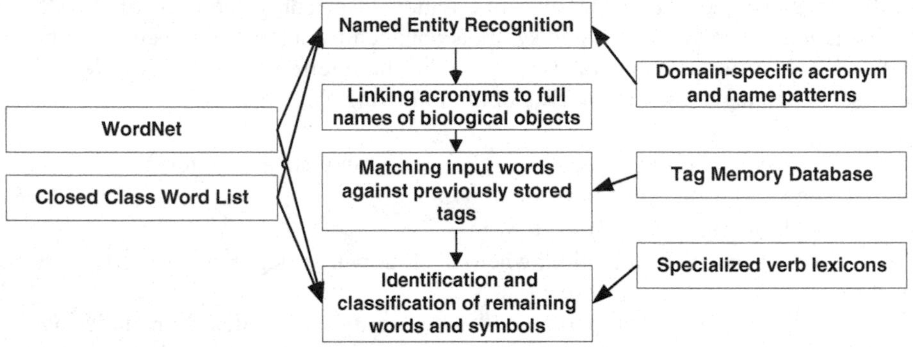

Fig. 3. The modified NER and tagging components and knowledge sources

The tagging process (after NER and acronym linking) is structured as follows:

1. Is the string X part of a string stored in the tagged words database?
   YES → copy the tag and proceed to the next word in the text
   NO → go to 2
2. Is X a closed class word, or an integer, or a mathematical or punctuation symbol?
   YES → provide X with the tag from the closed class database or from rules handling numerical expressions
   NO → go to 3
3. Check if X ends in a suffix typical for verbs or if X is stored in the list of irregular verb forms in WordNet
   YES → create the basic form of X (infinitive) by morphological rules or get the infinitive from the irregular verb list and proceed to 4 with the infinitive as variable value
   NO → go to 4
4. Is X present in some of the internal verb databases (cc_verbs, bioverbs)?
   YES → insert the appropriate tag and proceed to the next word in the text
   NO → go to 5
5. Does X end in a suffix typical for nouns derived from verbs (*-ion*, *-ence*, *-men*, ..)?
   YES → take the stem as the new variable value and go to 4
   NO → go to 6
6. Is X present in the verb part of WordNet?
   YES → insert the verb tag (wnv) and proceed to next word
   NO → go to 7
7. Does X end in a suffix typical for nouns or is X stored in the list of irregular noun forms in WordNet?

YES → create the basic form of X by morphological rules or get the basic form from the irregular noun list and proceed to 8 with the basic form as variable value

NO → go to 9

8. Is X present in the noun part of WordNet?

YES → insert the noun tag (wnn) and proceed to the next word in the text

NO → go to 9

9. Does X end with a suffix typical for adjectives (*-ic*, *-al*,*-ar*, etc.) or adverbs (*-ly*).

YES → insert the appropriate tag and proceed to next word

NO → insert an empty tag and proceed to the next word or, if there are no more words, end.

3 Results of the Tagging Experiments

Tagging of the training corpus resulted in 2500 unique tagged strings (after exclusion of tags identified as numbers). The distribution of morphosyntactic categories among those unique entries is shown in figure 4. It can easily be seen that proper nouns (i.e. names of genes, proteins etc.) and common nouns are the dominating categories. As much as 66% of the unique words (types) found in our corpus were nouns or proper nouns. This fits the observation, made in [23], that biomedical texts are extremely "noun-heavy" (74% of the word occurrences in a 15 000 corpus of biological texts investigated by the authors of [23] belonged to noun phrases).

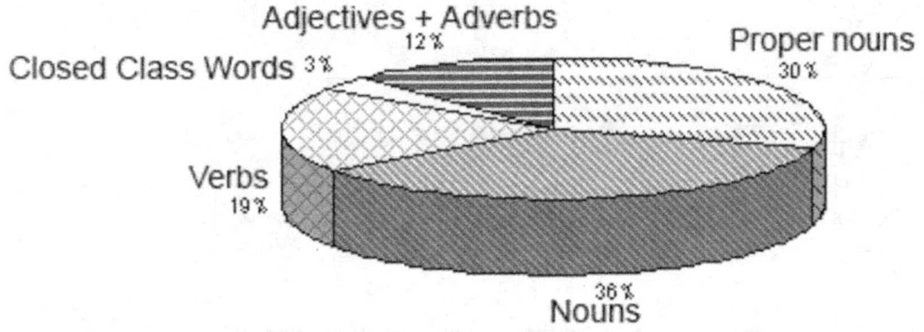

Fig. 4. Unique tagged strings from the training corpus

WordNet was the primary knowledge source responsible for noun identification (95% of the tagged common nouns were found in WordNet). All proper names found in the corpus were identified by the NER-procedure. Verbs, however, were to a considerable degree (38% of all content verb occurrences) identified on the basis of the small internal verb lexicons: the one of cognition and communication verbs (79 lexemes) and the yet smaller lexicon of frequent "bioverbs" (55 lexemes). This shows that the repertoire of verbs in this domain is highly repetitive. The results obtained from the test set confirm this observation.

64% of the words and symbols in the test corpus could be classified using the tags obtained from the training set. The recall values remained unchanged (exactly

95.2%). Figure 5 shows the impact of information learned from the training corpus on different part-of-speech classes. As could be expected, almost all occurrences of closed class words obtained their POS tags from the tag memory database. The use of WordNet was considerably reduced. Only 18.6% of the common nouns in the test material required search in WordNet, compared to 95% of the common nouns in the training set. 87% of all "bioverb" occurrences, 46% of the cognition and communication verbs, and 67% of other content verb occurrences were covered by the Tag Memory Database (TMD). This means that searching WordNet's verb part was needed only for 19% of all verb occurrences in the test corpus, compared to 62% in the training corpus. As expected, the category that was least covered by the TMD was the group of proper nouns. The time needed for preprocessing and tagging decreased by almost 40%.

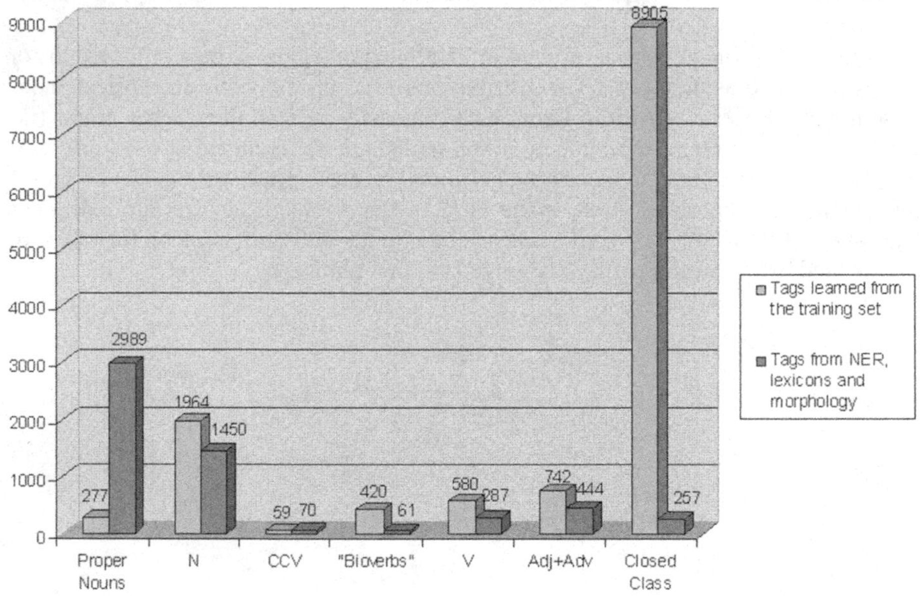

Fig. 5. Tags (occurrences) in the test set in relation to knowledge sources

4 Retrieval of Relevant Text Parts

Since in-depth syntactic parsing may be time consuming, the tagged texts should preferably be divided into parts that should be sent to the syntax module for further analysis, and less relevant parts that can possibly be omitted, as they do not contain any novel information. Scientific texts normally include a description of previous research and common knowledge in the field, in order to put the author's own research into context. For our purpose, it is not desirable to extract relations from sections which concern findings from previous research, since those relations should instead be extracted from the original papers. Otherwise, the same relation would be repeatedly extracted, since it has been reported in many papers referring to the same previous study.

In the training corpus used in this work, we found that in 82.5% of the abstracts the distinction between previous and current research was overtly marked by explicit phrases like *this study* (e.g. *this study utilized/investigated, in this study we investigate(d), this study was intended, we investigate/describe, our aim was*, etc.) In addition, for 5 of the abstracts (7.5%) the distinction could be correctly made by considering the shift of grammatical tense from present to past. Results and common knowledge are usually described in the neutral present tense, whereas switching from present to past frequently indicates attention shift from previous research to the current experiment. On the basis of the analysis of the training corpus, we designed a procedure for distinguishing between the background part and the description of current work [27]. It makes use of lexical cues (as the words and phrases mentioned above), and morphosyntactic cues (tense shift).

As already stated, the idea behind dividing the texts into a background and foreground part was to avoid unnecessary parsing of sentences that should not be sent to the relation extraction procedure. However, immediate deletion of the text portion preceding the foreground 'border' is not to recommend, since the last background sentence may contain antecedents of anaphoric expressions in the first foreground sentence. Thus, the last sentence in the background part is stored and later parsed if the first foreground sentence is found to contain expressions which are clearly anaphoric.

In cases where no background/foreground "border" can be detected, the whole abstract is sent to the parser. The procedure for foreground identification has been implemented recently and is currently being evaluated using the test corpus.

5 Parsing Results and Search for Biological Relations

The syntactic parser utilized for further analysis of the abstracts has been described in [23] and [26]. It is based on a hybrid grammar formalism that combines features of GPSG (Generalized Phrase Structure Grammar) and LFG (Lexical Functional Grammar), and pays special attention to correct co-indexation of elided subjects, subjects of infinitive clauses and elements of relative clauses. In the current experiment, we evaluated the parser on a sample of sentences randomly chosen from the foreground parts of the abstracts in the test corpus. Out of 100 sentences (average length 20 words per sentence), 79 were parsed within reasonable time (1-20 seconds). Sentences that required longer processing time were regarded as unparsed. A close examination of the parse trees revealed that 86% were parsed either fully correctly, or almost correctly. In the latter case, the tree displayed some minor errors, like a wrong attachment of short adverbials (like *rather, yet*, etc.), or that coordination was confused with apposition. These kinds of errors do not have any serious impact on relation identification. If, however, some noun phrases were wrongly delimited, and/or a wrong word marked as the predicate, the tree was judged as incorrect.

The outputs from the parser were converted to slightly simplified XML representations, where all empty nodes (provided for optional adverbials, etc.) were removed. Figure 6 shows a sample result of the conversion.

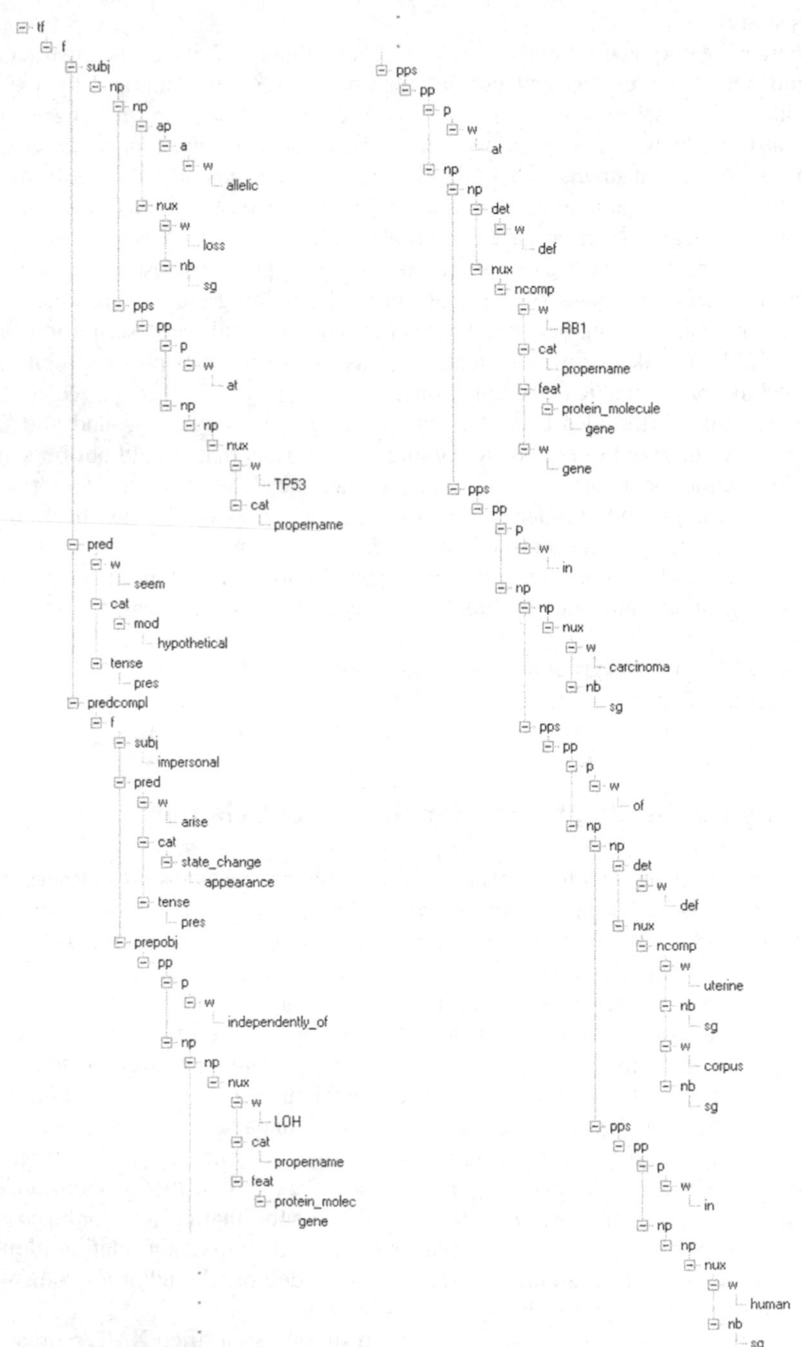

Fig. 6. A sample syntactic tree - the result of parsing the sentence: *Allelic loss at TP53 seems to arise independently of LOH at the RB1 gene in carcinomas of the uterine corpus in humans*

The XML trees were subsequently searched according to an algorithm [23] that has previously been evaluated manually. The main idea behind the search algorithm is to identify predicates and their complements belonging to the same syntactic levels in the tree, and to distinguish predicates that refer to biological relations/processes from those that refer to other activities or express modality. The latter are utilized for marking the extracted relation as true, denied, or hypothetical.

Technically, parts of the search are performed using the Microsoft .Net implementation of the XPath standard version 1.0 recommended by the World Wide Web Consortium (W3C) [33]. The XPath standard provides a possibility to test whether or not a node matches certain patterns and to quickly retrieve a desired fragment of a tree structure. For example, the query *./f/pred[w and tense]* returns the main predicate (the verb placed directly under the main sentence node), while *(./f/predcompl | ./f/obj | ./f/prepobj)/f/pred[w and tense]* finds the embedded predicate (i.e. the main predicate in the subordinated clause). Similar queries may be formulated for finding the complements of a given predicate and for the investigation of their internal structure. The process of tree compression, i.e. reducing the quite elaborated syntactic trees to shorter semantic representations, cannot be performed automatically by XPath and is thus implemented in an additional piece of code (in C#).

In Figure 7, we show the present format of the output from the information extraction procedure. The relation/processes identified in a sentence are presented as a relatively simple graph, where the node "subj"(ect) refers to the biological entity actively involved in the process, nodes "obj"(ect) or" prepobj(ect)" refer to other entities participating in the relations, and "circumstances" contain the information about place, conditions, etc., that may be of relevance. The main relation, including the circumstances, is presented to the user, and other circumstance nodes may be expanded by the user.

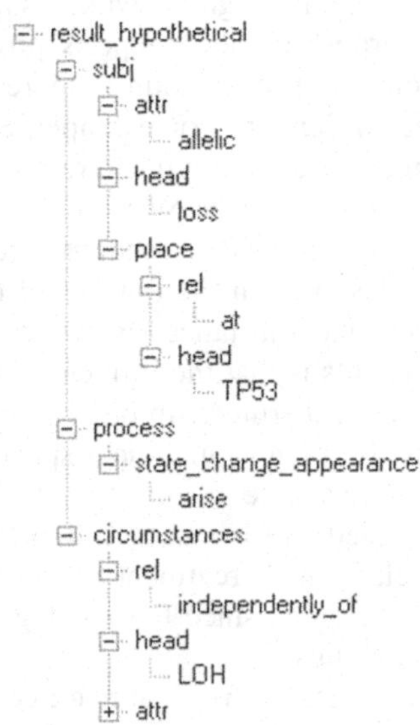

Fig. 7. The current format of the output from the information extraction procedure

The parser employed here is more successful in interpreting syntactic structures than the one utilized in the system MedScan [6,20], as shown in [27]. A new experiment with MedScan and the corpus investigated here confirmed the results of the previous evaluation [27]. MedScan identified biological relations only in 9 of 100 sentences (the sentence shown in Figures 6 and 7 was not present among these 9 sentences). The system has apparent problems with multiple clauses and subordinated sentences, and with multiword predicates like *play a role*, *have a part*, etc. Often, the only relation identified in a sentence is not the central one. For example, in the sentence *In particular, we examined two p53-target genes, p21/WAF1 and p53R2, with a crucial role in p53-induced cell cycle arrest and p53-induced DNA repair respectively* the only relation found by MedScan was: "Regulation (positive): TP53 → DNA nucleotide-excision repair"; which refers to the phrase *p53-induced DNA repair*, whereas the specific roles of the genes p21/WAF1 and p53R2 remain undetected.

6 Conclusions and Implications for Further Work

The results of our tagging experiments confirm the intuitive judgment of biomedical texts as lexically highly repetitive. A very simple memory-based learning procedure reduced the need of searching for word categories in general lexicons from 62% to 19% for verbs and from 95% to less than 19% for common nouns, already after obtaining tags from a rather limited training corpus (18 000 words). This finding suggests that it will be possible to eliminate the need of searching such large lexical databases as WordNet after a relatively short training period.

The most flexible and open word class in the domain consists of named entities, mainly names of biological objects and chemical substances. The NER procedure employed here is quite successful in identifying the most frequent patterns for these names. It does, however, over-generate slightly, which is a heritage from the original procedure, designed for the purpose of news reports processing. In the domain of news reports, the interpretation of a proper noun is more context-dependent than in scientific texts within a restricted domain. For example, Sweden in *Sweden rejected the proposal* should be interpreted as a metonymical reference to human beings, while Sweden in *Stockholm is the largest city in Sweden* should get the category "geographic region/country". The original NER procedure paid a lot of attention to such external cues, which is not as desirable in the domain of biomedicine, where the interpretation of names of objects and substances is less context-dependent. One conclusion to be drawn from the results is that the influence of external cues in the NER component should be restricted. Furthermore, in order to obtain a more detailed classification of biological objects, we plan to include a specialized database (Gene Ontology) as an additional knowledge source.

The relative amount of untagged words was not high (5%), but the results can be further improved by a more elaborated treatment of combinations of mathematical symbols, numbers and measuring units, since these categories constitute most of the untagged, or incorrectly tagged strings.

Another necessary improvement is prefix treatment; compounds like *upregulated, downregulated*, etc., are currently not being recognized when the constituents are spelled as exemplified here (which is quite frequently the case in our corpus). This modification is, however, relatively straightforward to implement.

Although the tagging procedure still suffers from certain shortcomings, it suits the domain of biomedicine better than most freely available taggers for English. The most important difference between the tagger employed here and the widely used Stanford tagger [34-35] concerns the treatment of proper names and acronyms. In our system, full names and acronyms are linked together, and strings of letters and numbers are matched against patterns for names of biological objects. These features facilitate parsing and interpretation of relations mentioned in a sentence. The Stanford tagger (version 2006-01-20) annotates each character string separately. A sequence as *nuclear localization signal (NLS)* gets the following Stanford tags: *nuclear/JJ localization/NN signal/NN -LRB-/-LRB- NLS/NNP -RRB-/-RRB*, while our system provides the information that all words in the sequence refer to the same entity: *semcat('NLS',[propername,acr(['nuclear localization signal'])])*.

Furthermore, each string ending in a number and not beginning in a capital letter is classified as a number by the Stanford tagger, which leads to incorrect classifications of such terms as *cbf2*.

Our continued work will include further development of the Tag Memory Database, testing and development of the procedure for discovering the foreground/background distinction, and the extraction algorithm. The syntactic parser is almost fully developed [24], which is due to the fact that the syntactic patterns in biomedical texts are highly repetitive, similarly to the lexical repertoire. Further development is however needed to improve the recall, where the problems are mostly related to ambiguity. The high frequency of recurrent lexical and syntactic patterns indicates that automatic extraction of pathways from biomedical texts is not an unrealistic task.

Finally, the format used to represent the output from information extraction (see fig. 7) needs further development. At this point it is only a first version to demonstrate one possible representation. A detailed user-study is necessary in order to decide what information must be included in the output and how it should be represented.

References

1. Baxevanis, A.D. and Ouellette, B.F.F.: Bioinformatics: A Practical Guide to the Analysis of Genes and Proteins, 3rd Edition. Wiley-Interscience. (2004)
2. Mount, D.W.: Bioinformatics: Sequence and Genome Analysis. Cold Spring Harbor Press. (2001)
3. Kanehisa, M., Goto, S., Kawashima, S., Okuno, Y., and Hattori, M.,: The KEGG resources for deciphering the genome. Nucleic Acids Res. 32, (2004) 277-D280
4. Becker, K.G., Hosack, D.A., Dennis Jr, G., Lempicki, R.A., Bright, T.J., Cheadle, C. and Engel, J.: PubMatrix: a tool for multiplex literature mining. BMC Bioinformatics 4:61, (2003)
5. Chaussabel, D. and Sher, A.: Mining microarray expression data by literature profiling. Genome Biol. 3(10) (2002) research0055.1–research0055.16.
6. Darasiela, N., Yuryev, A., Egorov, S., Novichkova, S., Nikitin, A. and Mazo. I.: Extracting human protein interactions from MEDLINE using a full-sentence parser. Bioinformatics 20(5) (2004) 604-611.

7. Jelier, R., Jenster, G., Dorssers, L.C.J., van der Eijk, C.C., van Mulligen, E.M., Mons, B. and Kors, J.A.: Co-occurrence based meta-analysis of scientific texts: retrieving biological relationships between genes. Bioinformatics 21(9) (2005) 2049–2058.
8. Jenssen, T.K., Öberg, L.M.K,, Andersson, M.L. and Komorowski, J.: Methods for Large-Scale Mining of Networks of Human Genes In: Proc. of The First SIAM Conference on Datamining, Chicago, April 2001 (2001)
9. Stapley, B., Benoit, G.: Biobibliometrics: Information retrieval and visualization from co-occurrences of gene names in Medline abstracts. In Proceedings of PSB 2000, Hawaii, USA (2000) 529-540.
10. Tanabe, L., Scherf, U., Smith, L.H., Lee, J.K., Hunter, L. and Weinstein, J.N.: MedMiner: an Internet text-mining tool for biomedical information, with application to gene expression profiling. Biotechniques 27(6) (1999) 1210-1217.
11. Wren, J.D., Bekeredjian, R., Stewart, J.A., Shohet, R.V., and Garner H.R.: Knowledge discovery by automated identification and ranking of implicit relationships. Bioinformatics 20 (2004) 389–398.
12. Friedman, C., Kra, P., Yu, H., Krauthammer, M., Rzhetsky, A.: GENIES: A natural-language processing system for the extraction of molecular pathways from journal articles. Bioinformatics 17 (2001)
13. Hahn, U., Romacker, M., Schulz, S.: Creating knowledge repositories from biomedical reports: The MEDSYNDIKATE text mining system. In Pacific Symposium on Biocomputing 2002, Kauai, Hawaii, USA (2002) 338 - 349.
14. Park, J.C. Kim, H.S., Kim, J.J.: Bidirectional incremental parsing for automatic pathway identification with combinatory categorical grammar. In Proceedings of PSB 2001, Hawaii, USA (2001) 396-407.
15. Pustejovsky, J., Castano, J.: Robust relational parsing over biomedical literature: Extracting inhibit relations, Proceedings of PSB 2002, Hawaii, USA (2002) 362-373.
16. Hishiki, T., Collier, N., Nobata, C., Okazaki-Ohta, T.. Ogata, N., Sekimizu,T., Steiner,R., Park, H.S., and Tsuji, J. Developing NLP Tools for Genome Informatics: An Information Extraction Perspective. Proceedings of the 9th Workshop on Genome Informatics (1998) 81-90.
17. Ono, T., Hishigaki, H., Tanigami, A. and Takagi, T.: Automated extraction of information on protein-protein interactions from the biological literature. Bioinformatics 17 (2001) 155-161.
18. Ng, S.-K. and Wong, M.: Toward Routine Automatic Pathway Discovery from On-Line Scientific Text Abstracts. Genome Informatics 10 (1999) 104-112.
19. Rindflesch, T., Tanabe, L., Weinstein, J., Hunter, L.: EDGAR: Extraction of drugs, genes, and relations from biomedical literature. In Proceedings of PSB 2000, Hawaii, USA (2000) 517-528.
20. Novichkova, S., Egorov, S., and Daraselia, N.: MedScan, a natural language processing engine for MEDLINE abstracts. Bioinformatics 19:13 (2003) 1699-1706.
21. Rosario, B. and Hearst, M.A. (2004) Classifying semantic relations in bioscience texts In Proceedings of ACL04, Barcelona, Spain.
22. Roth, D. and Yih, W.: A linear programming formulation for global inference in natural language tasks. In Proc. CoNLL (2004)
23. Gawronska, B. & Erlendsson, B.: Syntactic, Semantic and Referential Patterns in Biomedical Texts: towards in-depth text comprehension for the purpose of bioinformatics. In Sharp, B. (ed.): Natural Language Understanding and Cognitive Science. Proceedings of the 2nd International Workshop on Natural Language Understanding and Cognitive Science NLUCS 2005, Miami, USA, May 2005 (2005) 68-77

24. Gawronska, B., Erlendsson, B. and Olsson, B.: Tracking Biological Relations in Text: A Referent Grammar Approach. Biomedical Ontologies and Text Processing, Workshop held in conjunction with the European Conference on Computational Biology, ECCB 2005, Madrid, Spain, Sept 28, 2005 (2005).
25. Gawronska, B, Olsson, B, de Vin, L.: Natural Language Technology In Multi-Source Information Fusion. In Proceedings of the International IPSI-2004k Conference, Kopaonik, Serbia, April 2004, Published on CD with ISBN 86-7466-117-3 (2004)
26. Olsson, B., Gawronska, B. and Erlendsson, B.: Deriving Pathway Maps from Automated Text Analysis using a Grammar-based Approach. In: Proceedings of the 2nd Moscow Conference on Computational Molecular Biology (MCCMB), Moscow, Russia, July 18-21, 2005 (2005)
27. Olsson, B., Gawronska, B. and Erlendsson, B.: Deriving Pathway Maps from Automated Text Analysis using a Grammar-based Approach. Journal of Bioinformatics and Computational Biology (special issue) (to appear)
28. Gamalielsson, J. and Olsson, B.: Gosap: Gene Ontology Based Semantic Alignment of Biological Pathways (to appear)
29. Gawronska, B., Erlendsson, B. and Duczak, H.: Extracting semantic classes and morpho-syntactic features for English-Polish Machine Translation. Proceedings of the 9th International Conference on Theoretical and Methodological Issues in Machine Translation (TMI-2002), Keihanna, Japan (2002) 63-73.
30. Gawronska, B., Torstensson, N., Erlendsson, B.: Defining and Classifying Space Builders for Information Extraction. In Sharp, B. (ed.): Proceedings of NLUCS- (Natural Language Understanding and Cognitive Science), Porto, Portugal, April 2004 (2004) 15-27
31. Miller, G.A.: WordNet: An on-line lexical database of English. In Communications of ACM 38(11) (1995) 39-41.
32. Kyoto Encyclopaedia of Genes and Genomes. http://www.genome.jp/kegg/, http://www.genome.ad.jp/kegg/document/help_pathway.html (2005)
33. World Wide Web Consortium (W3C). http://www.w3.org/TR/xpath (2005)
34. The Stanford Natural Language Processing Group. http://www-nlp.stanford.edu/software/tagger.shtml (2006)
35. Toutanova, K., Klein D., Manning, C., and Singer, Y. Feature-Rich Part-of-Speech Tagging with a Cyclic Dependency Network. In Proceedings of HLT-NAACL 2003 (2003) 252-259.

Improving Text Mining with Controlled Natural Language: A Case Study for Protein Interactions

Tobias Kuhn[1,2], Loïc Royer[1], Norbert E. Fuchs[2], and Michael Schröder[1]

[1]Biotechnological Center, TU Dresden, Germany
{loic.royer, michael.schroeder}@biotec.tu-dresden.de,
http://www.biotec.tu-dresden.de/schroeder
[2]Department of Informatics, University of Zurich, Switzerland
{tkuhn, fuchs}@ifi.unizh.ch,
http://www.ifi.unizh.ch/attempto

Abstract. Linking the biomedical literature to other data resources is notoriously difficult and requires text mining. Text mining aims to automatically extract facts from literature. Since authors write in natural language, text mining is a great natural language processing challenge, which is far from being solved. We propose an alternative: If authors and editors summarize the main facts in a controlled natural language, text mining will become easier and more powerful. To demonstrate this approach, we use the language Attempto Controlled English (ACE). We define a simple model to capture the main aspects of protein interactions. To evaluate our approach, we collected a dataset of 459 paragraph headings about protein interaction from literature. 56% of these headings can be represented exactly in ACE and another 23% partially. These results indicate that our approach is feasible.

1 Introduction

In this paper we introduce a new paradigm of how to make knowledge of scientific papers accessible by computers. We focus on the fields of life sciences – particular biology – but our approach could be used in other fields as well.

Our approach consists of letting authors express their scientific results in a formal summary that could be an integral part of the papers they publish. We argue that it is more reasonable to let the authors formalize their own results, instead of trying to extract these results from the articles.

This section explains our motivation, introduces the language Attempto Controlled English (ACE) and compares it with other knowledge representation languages. Section 2 shows how ACE is used to build an ontology for protein interactions. In Sect. 3 we use this ontology as foundation for the expression of scientific results and we show how 89 selected articles could have been summarized in ACE. Section 4 shows the benefits of our approach and Sect. 5, finally, gives a short outlook.

U. Leser, F. Naumann, and B. Eckman (Eds.): DILS 2006, LNBI 4075, pp. 66–81, 2006.

1.1 Motivation

Biomedical scientists are challenged by an ever-increasing amount of scientific papers. The indexing service *PubMed*[1] shows the huge quantity of literature that the scientists have to face. It contains at the moment 16 million articles and grows every year by over 600'000 articles. All these biomedical articles are written in natural language. That means that we cannot easily process them with computers. But, facing the quantity of literature, it is clear that we need computational support in order to manage the contained knowledge.

In the last years, *text mining* and *information extraction* – which build both upon natural language processing (NLP) – gained an increasing interest in biomedical sciences. They aim to extract some kind of formal knowledge from natural language texts, which is generally considered a very demanding task. Even the basic problem of *named entity recognition*, that aims to identify named entities (e.g. protein names) in natural texts, is far from being solved. Other major aspects of text mining are the extraction of relationships (e.g. protein interactions), the automatic classification of texts, and the generation of new hypotheses on the basis of the available literature [3]. The *BioCreAtIvE* contest [21] nicely shows, that even sophisticated tools for text mining have a considerable lack of precision and recall: For a simple "named entity recognition"-task the precision ranged up to 86% and the recall was at most 84%. Another attempt is described in [4]: Information about protein-interactions was extracted from a data set of 1.2 million sentences that were taken from biomedical abstracts. They achieved a precision of 91%, but with a poor recall of only 21%. We recommend [3] and [12] for a more comprehensive overview of the "accomplishments and challenges" of text mining.

As a first step towards a better management of biomedical literature, controlled vocabularies like *MeSH*[2] and the *Gene Ontology*[3] have been created. They serve to classify biomedical publications and to link them to other resources. *GoPubMed*[4], for example, is a search engine that connects the abstracts from PubMed with the formal structure of the Gene Ontology. Thus a researcher can exploit the Gene Ontology for the search of relevant literature. Such tools are very valuable for scientists and there has been a notable progress in the last years, but it will never be possible to extract all the information correctly. There is inherent ambiguity and vagueness in natural language that prevents its perfect processing by computers.

For this reason we present an alternative approach: The authors of scientific articles formally summarize their own results. Such formal summaries are added to the articles which makes them processable by computers. This requires a formal language that on the one hand is easy to learn and understand, and on the other hand is expressive enough to represent even complicated scientific results. It is clear that this approach is not applicable for papers that have been

---

[1] `http://www.pubmed.gov`

[2] `http://www.nlm.nih.gov/mesh/meshhome.html`

[3] `http://www.geneontology.org`

[4] See [5] and `http://www.gopubmed.org`

written without the formal summaries, and that means that we still need NLP or manual extraction for such papers. Thus we propose rather a concept for the future than a solution for today's problems. To explore our approach we use Attempto Controlled English as knowledge representation language.

1.2 Formalization of Scientific Results

Since we want to access scientific results by computers, we have to formalize this knowledge at some point. Today researchers write their results in natural language. To extract these results and to formalize them, manual or computer-supported text mining is necessary. Thus the formalization is accomplished by computer-programs or by humans, and in either case it is done without the help of the corresponding researchers. The article is the only source of information. Since such articles are highly domain-specific, they require a lot of background knowledge. Therefore the formalization is a very demanding task, even for humans. Altogether this causes a lot of knowledge to be lost in the vast amount of biomedical literature.

We claim that most of these problems can be solved, if we simply let the authors of scientific articles formalize their own results. The researchers themselves are the most qualified to understand their results, and thus they can give the most precise formal representation. This is not even a big extra-effort for a scientist, since he already has a – more or less – formal model of the domain in his mind, and must write an abstract anyway. He just needs to learn how to express his knowledge in a formal way. This means that we need to provide an intuitive, yet formal language in which a scientist can write his results.

1.3 Attempto Controlled English

Attempto Controlled English (ACE)[5] is a controlled natural language that has been developed by Norbert E. Fuchs and his group at the University of Zurich. ACE is a subset of natural English with a restricted grammar. There are no limitations on the vocabulary, apart from some function words with predefined meanings (e.g. 'every', 'of'). ACE looks like English, but it is in fact a formal language, which means that texts can be translated unambiguously into first-order logic. Some ACE sentences would be ambiguous in natural English, but ACE provides interpretation rules that allow in each case only one reading. The report [13] contains a comprehensive description of the syntax of ACE.

In order to be able to write ACE texts, one has to learn the restrictions on the grammar. Thus, like every formal language, ACE has to be learned. However, since it looks like natural English, everyone is able to understand ACE texts with almost no training. This is a big advantage over other formal languages.

The Attempto parser APE[6] translates ACE texts into Discourse Representation Structures [6]. Such structures are equivalent to expressions in first-order logic, and thus every ACE sentence has a logical representation. Furthermore,

[5] See [7], [8], and `http://www.ifi.unizh.ch/attempto/`

[6] `http://www.ifi.unizh.ch/attempto/tools/`

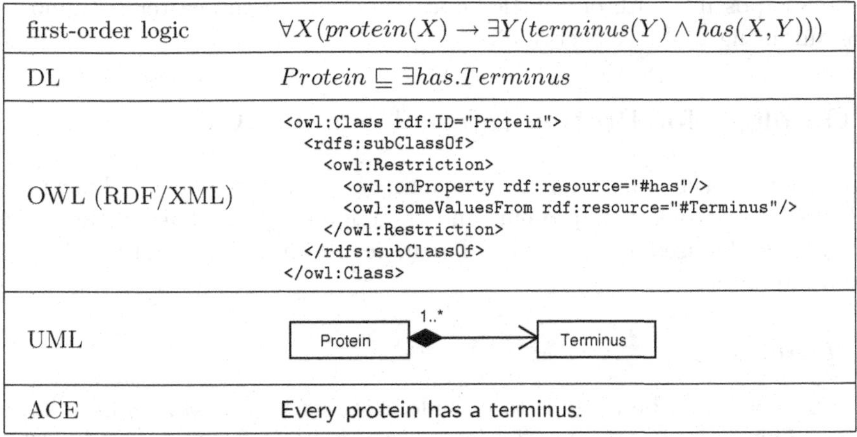

first-order logic	$\forall X(protein(X) \rightarrow \exists Y(terminus(Y) \wedge has(X,Y)))$
DL	$Protein \sqsubseteq \exists has.Terminus$

OWL (RDF/XML)	`<owl:Class rdf:ID="Protein">`
	`<rdfs:subClassOf>`
	`<owl:Restriction>`
	`<owl:onProperty rdf:resource="#has"/>`
	`<owl:someValuesFrom rdf:resource="#Terminus"/>`
	`</owl:Restriction>`
	`</rdfs:subClassOf>`
	`</owl:Class>`

UML	Protein 1..* Terminus
ACE	Every protein has a terminus.

Fig. 1. Comparison of first-order logic, DL, OWL, UML, and ACE

APE creates a paraphrase that shows the interpretation of an ACE text. If a writer is not familiar with the ACE interpretation rules, then he can check the paraphrase for the validation of his ACE text.

1.4 Comparison of Knowledge Representation Languages

In order to show the benefits of ACE, we compare it with four other knowledge representation languages: first-order logic [9], Description Logics (DL) [15], Web Ontology Language (OWL) with its RDF/XML-syntax [14], and Unified Modeling Language (UML) [2].

We have to state that these four languages are not independent. DL and ACE build upon first-order logic, and DL are the basis for OWL. While first-order logic and DL focus on the theoretical concepts of knowledge, OWL, UML, and ACE concentrate on the implementation and application of knowledge representation. Nevertheless we dare to give a direct comparison between these five languages.

Figure 1 shows how the fact 'everything that is a protein has a terminus' is expressed in the five different languages. The OWL representation (using the RDF/XML syntax) is the most verbose and – from the human perspective – the least readable one. The representations in first-order logic and DL are more concise, but they are still not understandable for people who are not familiar with formal notations. The graphical notation of UML looks nice, but for a non-specialist it is hard to guess the meaning of all the shapes and arrows. The ACE representation, in contrast, should be immediately understandable for any English speaking person. It looks perfectly like natural English and thus the reader might not even recognize that it is a formal language.

We can state that controlled natural languages like ACE minimize the gap between machines and humans. A reader is able to understand such languages with almost no training. Furthermore, writing sentences in a controlled natural language is possible with only little effort, especially if the writer is supported by an authoring tool (see Sect. 3.3).

2 Ontology for Protein Interactions in ACE

In order to have a clear basis for the formal representation of scientific knowledge, we defined an ontology for proteins and their interactions. This section shows how ACE can be used as an ontology language, and introduces our ontology for protein interactions.

2.1 Ontologies

The main goal of an ontology is to provide a *shared understanding* of a certain domain. This shared understanding can serve as basis for the communication between people, for the interoperability between systems, for the improvement of reusability and reliability of software systems, and for the specification of software [20]. Furthermore ontologies are an excellent basis for the formal representation of knowledge [11].

Ontologies are not yet broadly established in science, but they are expected to gain a very important role in the future, especially in life sciences. The *Gene Ontology* is the most famous example, although it is actually more a controlled vocabulary than a real ontology.

2.2 Ontology Elements

In order to provide basic structures for ontologies in ACE, we adopt the elements of DL – individuals, concepts, and roles – and we call them *ontology elements*. Furthermore we introduce an additional structure: context information.

Individuals. Individuals stand for single objects of the domain. They are represented in ACE as *proper names* like 'Bub1' (that stands for a protein) or 'Alzheimer' (that stands for a disease).

Concepts. Concepts stand for sets of objects, and there are two possibilities to express them in ACE. *Common nouns* are the most straight-forward way. The noun 'protein', for example, can stand for the concept of all proteins. As a second possibility we can use *adjectives* (in their positive form). The adjective 'organic', for example, can be used for the concept of all organic substances.

Roles. Roles stand for binary relations between objects, and they can be expressed in four different ways. First of all, we can use *transitive verbs* for expressing roles. For example, we can use 'interacts-with' to express a relationship between proteins. Next, we can combine transitive verbs with *adverbs*. For example, we can use the adverb 'directly' together with the transitive verb 'interacts-with' to express the role 'directly interacts-with'. As a third possibility we can use *of-constructs* like 'is a part of'. Due to the syntax of ACE, 'of' is the only allowed preposition for nouns. Finally, we

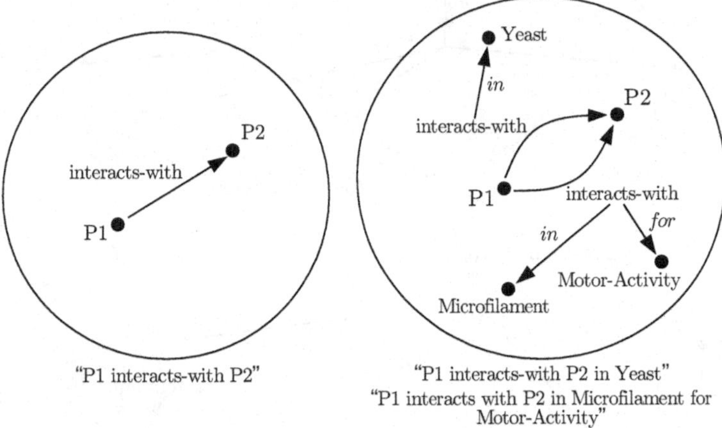

Fig. 2. Context information

can use *constructs with comparative forms of adjectives* like 'is larger than'. Such constructs typically represent transitive relationships.

Context Information. The examination of the results of scientific papers on protein interactions showed that normal roles are often not sufficient to express the needed information. We can express simple statements like 'P1 interacts-with P2', but we cannot express statements with contextual information like 'P1 interacts-with P2 in Yeast' or 'P1 interacts-with P2 in Microfilament for Motor-Activity'. In order to be able to express such results, we want to allow roles to have such additional information. In natural English we usually express such information with prepositional phrases, and this is exactly the way we will do it in ACE. Figure 2 illustrates the examples without and with context information.

Using these ontology elements, we can state for example

P1 is a protein and directly interacts-with P2 in Yeast.

where 'P1', 'P2', and 'Yeast' are individuals, 'protein' stands for a concept, and 'directly interacts-with' stands for a role. The phrase 'is a' is used to assign the individual 'P1' to the concept 'protein'. The conjunction 'and' connects the statements flanking left and right. The preposition 'in', finally, connects to the context 'Yeast'.

2.3 Ontology for Protein Interactions

Since we found no existing ontology that fits our needs, we had to create our own ontology for protein interactions. First, we defined concepts that allow us to make statements about the structure of proteins and protein-complexes. For the sake of a clear structure, we introduced the concept *protein-unit*, which is

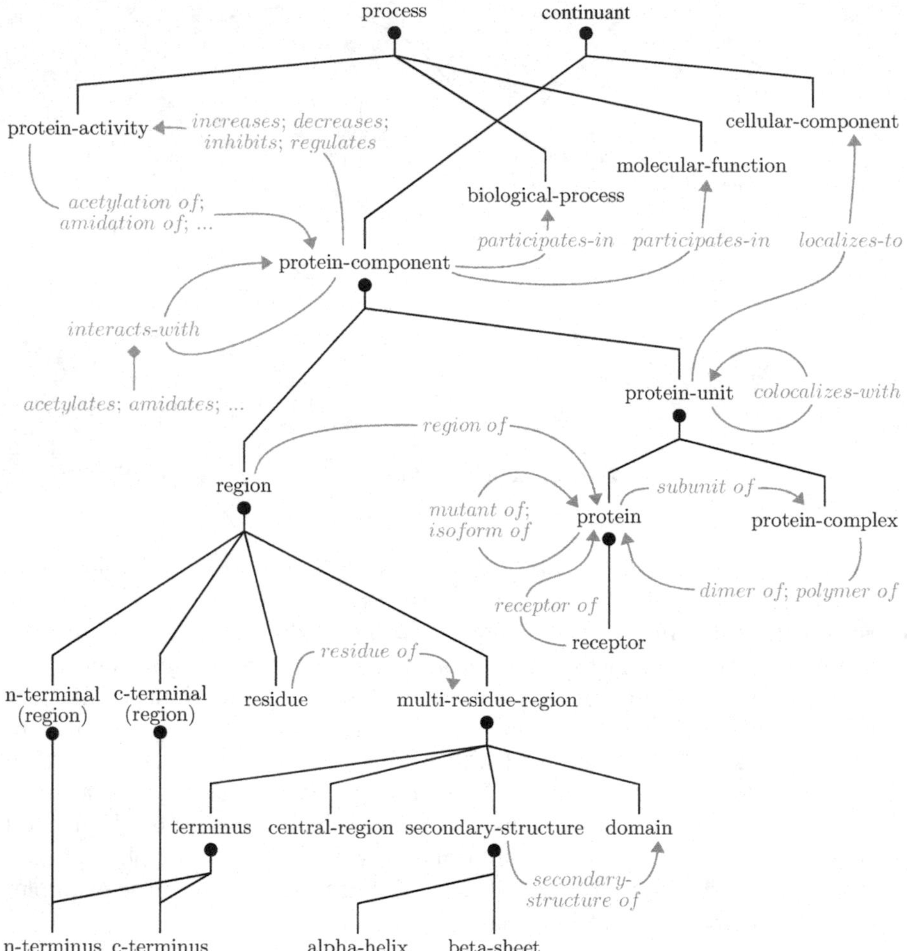

Fig. 3. The structure of the ontology for protein interactions

either a protein or a protein-complex, and *protein-component*, which is either a protein-unit or a region of a protein. In order to describe the structure of such regions, we defined concepts like 'residue', 'secondary-structure', and 'domain'.

Next, we defined the roles for the description of interactions between proteins like 'interacts-with' or 'binds'. We can also express more complicated interactions like 'increases the phosphorylation of'.

Furthermore, we defined some concepts for expressing additional information about proteins, like the localization to a certain cellular component or the participation in a certain process. The big picture of this ontology is shown in Fig. 3.

3 ACE Summaries

Our goal is to show how scientists could write formal summaries of their results. There are some questions that naturally arise: What are these results about? How complex is it to formulate them in a formal language? In the following we present an empirical study of the feasibility of our approach.

3.1 ACE Summaries for 89 Selected Articles

Since we want to show how results of papers about protein interactions could have been written in ACE in the first place, we picked 89 *Elsevier*-articles that concern protein interactions. Such articles mostly have a section called "Results" which is subdivided into subsections. The headings of these subsections are short descriptions of the corresponding results. It turned out that these headings are highly suitable for a manual translation into ACE. Please note that the intended methodology is *not* to express the results first in natural language and then to translate them into ACE. We do this just to demonstrate the feasibility of our approach.

The 89 articles contain 457 such headings. 184 of them are ignored, because they are not formulated as facts (e.g. "Functional characterization of Pellino2"[7]) or because they contain information that is not about protein interactions.

total:		457 *(100%)*
ignored:	(not a fact)	87 *(19%)*
	(off-topic)	97 *(21%)*
used:		273 *(60%)*

We then tried to translate the 273 remaining headings into ACE. For 154 of them there is a perfect match, which means that the complete information can be expressed in ACE; e.g. the heading "Interaction of Act1 with TRAF6"[8] can be rephrased perfectly as "Act1 interacts-with TRAF6". For another 62 headings only a part of the information is expressed; e.g. the heading "The mtFabD protein is part of the core of the FAS-II complex"[9] can only partially be rephrased as "MtFabD is a subunit of FAS-II". For the remaining 57 headings there is no translation at all.

used:		273 *(100%)*
matched:	(perfect)	154 *(56%)*
	(partial)	62 *(23%)*
unmatched:		57 *(21%)*

Let us take a closer look at the reason, why 119 headings cannot be rephrased in ACE at all, or only partially. 56 of them could not be rephrased because their content is not covered by our model, but they could be expressed with an extended model. Another 21 headings describe relations of relations, like the heading "Kal-GEF1 activation of Pak does not require GEF activity"[10]. In this

[7] See article *PMID 12860405*.

[8] See article *PMID 12459498*.

[9] See article *PMID 16213523*.

[10] See article *PMID 15950621*.

case, there is a relation between two objects ("Pak activates Kal-GEF1") and this relation itself stands in another relation ("... does-not-require GEF-activity"). At the moment, we cannot express such structures in ACE in a satisfying way. But there are attempts to extend the language ACE, and we hope that we will be able to express such statements in the future. Furthermore there are 11 headings with fuzzy statements (e.g. "ANKRD contains potential CASQ2 binding sequences ..."[11]) and 31 headings that we could not understand without reading the whole article.

not perfectly matched:	119 *(100%)*
not covered by our model:	56 *(47%)*
relations of relations:	21 *(18%)*
fuzzy:	11 *(9%)*
not understood:	31 *(26%)*

Thus, altogether we could rephrase 79% of the relevant headings, either partially or perfectly. This makes us confident that our approach is feasible for practical use. The reason, why 119 headings are not rephrased perfectly, is mostly our simple model and our lack of understanding. If we used a more detailed model, and if we let the scientists themselves express their own results in ACE, then we expect to be able to express much more than 79% of the results.

3.2 ACE Summary as an Integral Part of an Article

Since ACE looks like natural English, every reader of a scientific article is able to understand ACE texts. Thus the ACE summary of the results could be an integral part of the article. Figure 4 shows how an article with an ACE summary could look like[12]. Figure 5 shows the corresponding logical representation as a Discourse Representation Structure (consult [6] for details). As we see, the natural looking ACE summary can be translated automatically into a formal representation which is processable by computers.

Together with the abstract and a keyword list, the ACE summary gives a concise insight into the content. In contrast to the abstract, the ACE summary is readable by both, humans and machines; and in contrast to the keyword list, the ACE summary does not only mention the objects of interest, but describes the relations among them. Thus, every published article could be a contribution to a constantly growing knowledge base.

3.3 Authoring Tool

Now we sketch a tool that would help writing ACE texts. It would guide the user step by step and would need almost no training. Similar systems are the look-ahead editor *ECOLE* [17,18], the natural language interface *LingoLogic* [19], and the *Ginseng*-system [1]. Our tool would solve several problems:

[11] See article *PMID 15698842*.

[12] Article *PMID 12419313* is used for this example.

The β2-adaptin clathrin adaptor interacts with the mitotic checkpoint kinase BubR1

Corinne Cayrol, Céline Cougoule, Michel Wright

Abstract

The adaptor AP2 is a heterotetrameric complex that associates with clathrin and regulatory proteins to mediate rapid endocytosis from the plasma membrane. Here, we report the identification of ...

Keywords: Protein interactions; Two-hybrid; Vesicular traffic; Adaptor protein; Protein kinase; Mitotic checkpoint.

ACE Summary: Beta2-Adaptin binds BubR1 in Yeast-Two-Hybrid. A trunk-domain of Beta2-Adaptin interacts-with BubR1. Bub1 interacts-with the trunk-domain of Beta2-Adaptin. Bub1 interacts-with every beta-sheet of AP and BubR1 interacts-with every beta-sheet of AP.

Fig. 4. Article with ACE summary

```
A B C D E F G H I
object(A,atomic,named_entity,object,cardinality,count_unit,eq,1), named(A,'Beta2-Adaptin')
object(B,atomic,named_entity,object,cardinality,count_unit,eq,1), named(B,'BubR1')
object(C,atomic,named_entity,object,cardinality,count_unit,eq,1), named(C,'Yeast-Two-Hybrid')
object(D,atomic,named_entity,object,cardinality,count_unit,eq,1), named(D,'Bub1')
object(E,atomic,named_entity,object,cardinality,count_unit,eq,1), named(E,'AP')
predicate(F,unspecified,bind,A,B), modifier(F,unspecified,in,C)
object(G,atomic,'trunk-domain',object,cardinality,count_unit,eq,1)
relation(G,'trunk-domain',of,A)
predicate(H,unspecified,interact_with,G,B)
predicate(I,unspecified,interact_with,D,G)

  J                                             K
  object(J,atomic,'beta-sheet',object,     =>   predicate(K,unspecified,interact_with,D,J)
     cardinality,count_unit,eq,1)
  relation(J,'beta-sheet',of,E)

  L                                             M
  object(L,atomic,'beta-sheet',object,     =>   predicate(M,unspecified,interact_with,B,L)
     cardinality,count_unit,eq,1)
  relation(L,'beta-sheet',of,E)
```

Fig. 5. DRS-representation of the ACE summary

- The tool would help the user to comply with the standard nomenclature. The user would only be allowed to use the defined words. It would also prevent typing errors.
- It would make sure that the created sentences comply with the ACE syntax. At every stage, the tool would allow to proceed only in a way that leads to a correct ACE sentence. Thus the user would not need to know about the syntax of ACE.
- The tool would be aware of the structure of the ontology. In this way it would make sure, for example, that the domains and ranges of roles are respected.

We give now an example how this tool could be used. Suppose that an author of the article that is shown in Fig. 4 wants to write down the fact that the protein *Bub1* interacts with the protein *β2-Adaptin* via its *trunk domain.*

The sentences are created step by step by a simple menu. At the beginning there is just an empty sentence that might look like this:

« ... »

The quotes indicate the beginning and the end of the sentence and the box in the middle is used to create the content. If the user clicks on it, then a menu is displayed that shows the different options for beginning a sentence. Since we want to talk about the protein *Bub1* we first insert 'Bub1' as a proper name. This looks as follows.

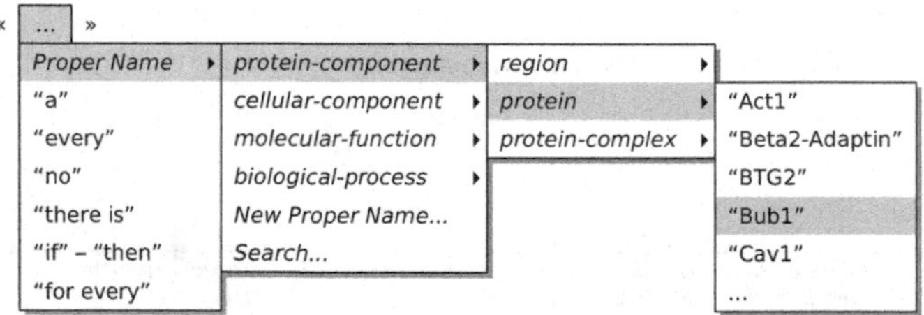

Proper names are hierarchically structured and the menu allows to navigate through this hierarchy. Alternatively, we can use the search option to find a certain term, or we can create a new proper name on-the-fly. In the next step we get

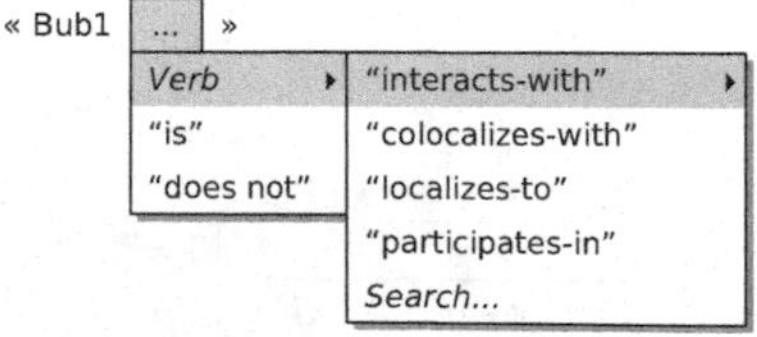

where the proper name 'Bub1' is now fixed as the beginning of the sentence, and we have a new menu with different entries. We want to express the interaction with another protein, and thus we choose the verb 'interacts-with'. Like proper names, verbs are hierarchically structured and we can navigate through this hierarchy. In the next step we get

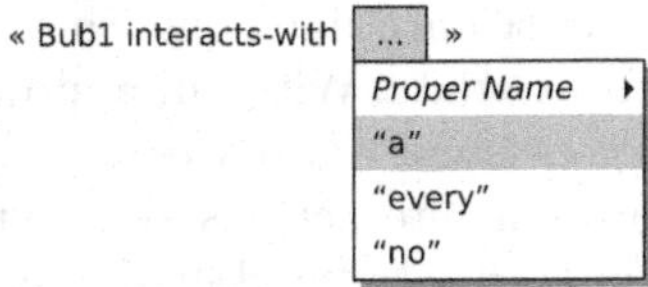

where we can define the second participant of the protein-interaction. Since we want to state that the interaction goes via a *trunk domain* of the protein *β2-Adaptin*, we first have to add the article 'a'. Then we get

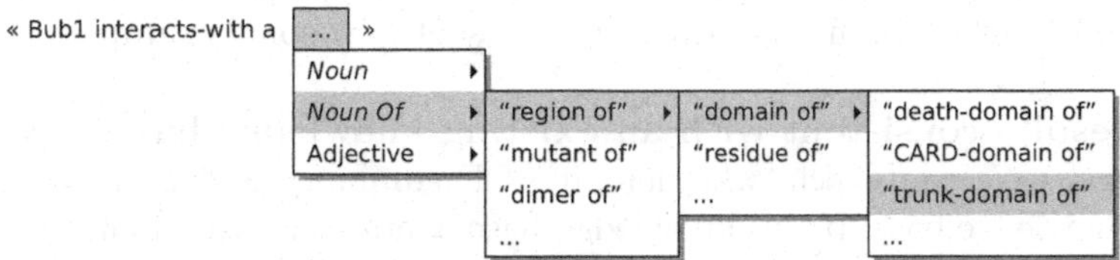

where we can choose the 'trunk-domain of'-relation. Like proper names, such of-relations are structured in hierarchies through which we can navigate (the same holds for nouns and adjectives). After that we get

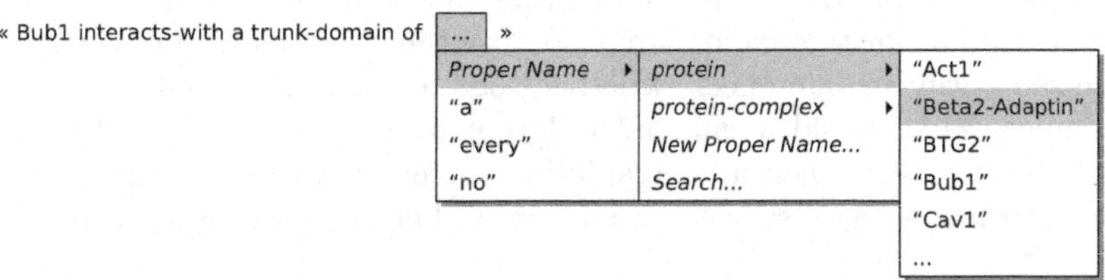

where we can specify the second protein 'Beta2-Adaptin'. Finally we get

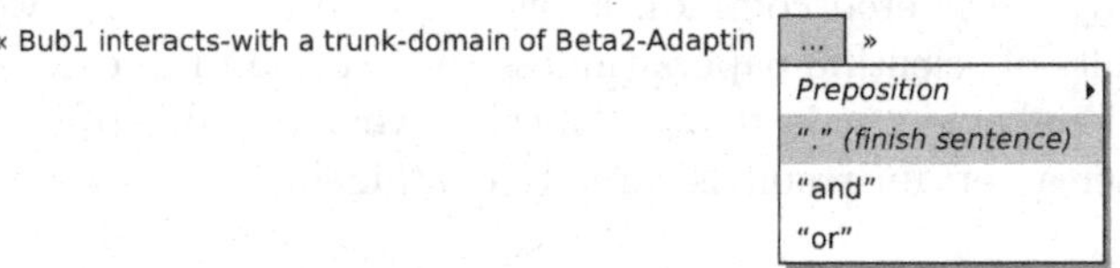

where we could use prepositions to add context information, e.g. to specify the organism in which the interaction takes place. In our example, we now finish the sentence with a full stop.

For the creation of this sentence we did not need any further knowledge about ACE. Every person that is familiar with English and knows how to handle a simple menu, is able to create ACE texts. However, to make such a tool really user-friendly we will need a lot of usability testing, as it is done – with promising results – for the *Ginseng*-system [1].

4 The Benefits of Our Approach

The preceding sections showed what needs to be done to express scientific results about protein interactions in ACE. Now it is time to take a look at the benefits.

Today there are many databases that contain life science data, but they are mostly unsynchronized, incomplete, and often not up-to-date. With our approach it would be much easier to provide complete and consistent databases.

Imagine that all the scientific papers about protein interactions summarize their results in ACE. We could use these formal summaries to build up a dynamically growing knowledge base about protein interactions. Of course, we would also have to collect all the knowledge that is contained in old papers. For these, we still need some form of classical text mining. But once we have such a knowledge base that is continuously updated with the results of new papers, then we would be able to answer many questions. We present now some examples.

Are some results consistent with an existing knowledge base or with other papers? We can check, whether an ACE summary is consistent with an existing knowledge base. If this knowledge base contains common knowledge, then the results should be consistent, or otherwise it can be seen as an appeal against the common knowledge.

Without formal declarations, it is impossible to check a paper for consistency. Probably there exist scientific papers that contain results which are inconsistent with the common knowledge. But since this can be very difficult to find out, neither the author nor the readers might realize the special status of the results.

In the same way we can check, whether there exist papers that contradict a certain paper. That would mean that different researchers claim contradictory results. Being aware of such a contradiction might lead to a dialogue between the corresponding scientists, which might entail better and consistent results.

Are some results (or parts of it) already known? With our formal approach we can check whether a certain result, or a part of it, is already known. Results that are already considered common knowledge are usually not worth to be described as results of scientific papers (unless they contain more detailed information or if additional evidence is given). Thus it is very valuable to be able to run a check, whether a certain result is already contained in the knowledge base or not.

Furthermore a researcher might want to check, whether there exists scientific literature that has arrived at the same or similar results. Altogether our approach would help the researchers to save a lot of time, since they would not need to search "manually" for the relevant literature.

Is there a known answer for a certain question? If someone – researcher or not – has a specific question about the domain (e.g. protein interactions), then we would be able to give automatically an answer.

Indeed, there exist already systems like *MEDIE*[13] that provide some sort of answer extraction using natural language processing. But such systems have serious shortcomings: There is always a trade-off between precision and recall, and only very simple queries are allowed. Furthermore, we cannot find answers that are spread over multiple articles.

[13] `http://www-tsujii.is.s.u-tokyo.ac.jp/medie/`

type	Bub1
supertypes	Kinase – Enzyme – Protein – Molecule
subtypes	BubR1
interacts directly with	Beta2-Adaptin, Cdc20, Mad3
interacts indirectly with	Mad2, APC
associates with	Cdc20, Mad3
phosphorylates	Bub1, Bub3
localizes to	Kinetochore, Chromosome
participates in	Cell-Communication, Signal-Transduction

Fig. 6. Overview over the object 'Bub1'

What is known about a certain object of interest? In some cases we do not want to ask a specific question, but we rather want to get an overview of a single object of interest (e.g. the protein *Bub1*). If we ask for information about such an object then we might get something as shown in Fig. 6. Such an overview could be used for a dynamic hypertext representation. This would allow us to navigate through the whole knowledge base, e.g. with an ordinary web browser. New papers that are submitted can be integrated *automatically* and thus such a web interface would be always up-to-date.

How are some objects of interest related? Instead of focusing on one single object, we might want to have an overview of the interrelations of a certain group of objects. We could extract, for example, the *interacts-with*-relations of all proteins and use this data for further examination, like the detection of clusters or hot-spots. Such examinations are already common in the research on proteins (e.g. [10], [16]), but only with restricted data. With our approach we could consider every interaction that has been published.

5 Outlook

We suggest an approach of using controlled natural language for making the results of scientific papers readable and – to some degree – understandable by computers. But in order to achieve this goal, there is still a lot of work to do. For example, we need an authoring tool as sketched in Sect. 3.3, that would support the authors of scientific papers in the creation of ACE summaries. A prototype of such a tool does already exist. Furthermore, we need tools for the definition of ontologies and for the collection and management of knowledge.

Besides all these technical requirements, there are also political ones. There must be a commitment among the scientists of the corresponding field of research – or at least among a large part of them – that scientific articles get summarized in ACE. If such a summary is optional then there is little hope that it gets established.

This is the point where the publishers and editors have to come into play. The publishers would have to make ACE summaries a mandatory part of the articles, and the editors would have to check whether these summaries are correct and complete. The creation of a formal summary should be an additional requirement to consider when writing a scientific article, besides all the requirements that already exist today (e.g. about the abstract, the keyword list, and the reference list). The formal summaries can also be seen as a robust indicator for the value of a scientific paper. Information that is already known and redundant information could be ignored automatically, and wrong statements are likely to be detected at some later point in time. Thus we could use the formal summaries to quantify and qualify the contribution of a certain author, institute, or journal.

Due to the immense benefits such a system would bring along, we believe in the great potential of our approach. It could be a first step towards better communication and persistence of biomedical knowledge.

References

1. Abraham Bernstein, Esther Kaufmann, Christian Kaiser. *Querying the Semantic Web with Ginseng: A Guided Input Natural Language Search Engine.* Department of Informatics, University of Zurich, 2005
2. Grady Booch, James Rumbaugh, Ivar Jacobson. *The Unified Modeling Language User Guide*, First Edition. Addison Wesley, 1998
3. Aaron M. Cohen, William R. Hersh. *A survey of current work in biomedical text mining.* In *Briefings in Bioinformatics*, 6(1):57-71, 2004
4. Nikolai Daraselia, Anton Yuryev, Sergei Egorov, Svetalana Novichkova, Alexander Nikitin, Ilya Mazo. *Extracting human protein interactions from MEDLINE using a full-sentence parser.* In *Bioinformatics*, 20(5):604-611, 2004
5. Andreas Doms, Michael Schroeder. *GoPubMed: exploring PubMed with the Gene Ontology.* In *Nucleic Acids Research*, 33:W783-W786, 2005
6. Norbert E. Fuchs, Stefan Hoefler, Kaarel Kaljurand, Tobias Kuhn, Gerold Schneider, Uta Schwertel. *Discourse Representation Structures of ACE 4 Sentences*, Technical Report ifi-2006.07. Department of Informatics, University of Zurich, 2006, `ftp://ftp.ifi.unizh.ch/pub/techreports/TR-2006/ifi-2006.07.pdf`
7. Norbert E. Fuchs, Kaarel Kaljurand, Gerold Schneider. *Attempto Controlled English Meets the Challenges of Knowledge Representation, Reasoning, Interoperability and User Interfaces.* The 19th International FLAIRS Conference (FLAIRS'2006), 2006
8. Norbert E. Fuchs, Uta Schwertel, Rolf Schwitter. *Attempto Controlled English – Not Just Another Logic Specification Language.* In *Logic-Based Program Synthesis and Transformation*, Eighth International Workshop LOPSTR'98, Lecture Notes in Computer Science 1559, Springer, 1999, `http://www.ifi.unizh.ch/attempto/publications/papers/LOPSTR98.pdf`
9. Melvin Fitting. *First-Order Logic and Automated Theorem Proving*, Second Edition. Springer, New York, 1996
10. L. Giot, J. S. Bader, C. Brouwer, A. Chaudhuri, et al. *A Protein Interaction Map of Drosophila melanogaster.* In *Science*, 302(5651):1727-1736, 2003
11. Thomas R. Gruber. *Toward Principles for the Design of Ontologies Used for Knowledge Sharing.* In *International Journal of Human-Computer Studies*, 43 (5-6):907-928, 1995

12. Lynette Hirschman, Jong C. Park, Junichi Tsujii, Limsoon Wong, Cathy H. Wu. *Accomplishments and challenges in literature data mining for biology.* In *Bioinformatics Review,* 18(12):1553-1561, 2002
13. Stefan Hoefler. *The Syntax of Attempto Controlled English: An Abstract Grammar for ACE 4.0*, Technical Report ifi-2004.03. Department of Informatics, University of Zurich, 2004, `ftp://ftp.ifi.unizh.ch/pub/techreports/TR-2004/ifi-2004.03.pdf`
14. Deborah L. McGuinness, Frank van Harmelen. *OWL Web Ontology Language Overview.* W3C Recommendation, 2004, `http://www.w3.org/TR/2004/REC-owl-features-20040210/`
15. Daniele Nardi, Ronald J. Brachman. *An Introduction to Description Logics.* In *The Description Logic Handbook: Theory, Implementation, and Applications*, Cambridge University Press, 2003
16. Benno Schwikowski, Peter Uetz, Stanley Fields. *A network of protein-protein interactions in yeast.* In *Nature Biotechnology*, 18:1257-1261, 2000
17. Rolf Schwitter, Anna Ljungberg, David Hood. *ECOLE: A Look-ahead Editor for a Controlled Language.* In *Proceedings of EAMT-CLAW03, Controlled Language Translation*, Dublin City University, 141-150, 2003
18. Rolf Schwitter, Marc Tilbrook. *Let's Talk in Description Logic via Controlled Natural Language.* To be presented at: Logic and Engineering of Natural Language Semantics 2006 (LENLS2006), Japan, 2006
19. Craig W. Thompson, Paul Pazandak, Harry R. Tennant. *Talk to Your Semantic Web.* In *IEEE Internet Computing*, 9(6):75-79, 2005
20. Mike Uschold, Michael Gruninger. *Ontologies: Principles, Methods and Applications.* In *Knowledge Engineering Review*, 11(2), 1996
21. Alexander Yeh, Alexander Morgan, Marc Colosimo, Lynette Hirschman. *BioCreAtIvE Task 1A: gene mention finding evaluation.* In *BMC Bioinformatics*, 6, 2005

SNP-Converter: An Ontology-Based Solution to Reconcile Heterogeneous SNP Descriptions for Pharmacogenomic Studies

Adrien Coulet[1,2], Malika Smaïl-Tabbone[2], Pascale Benlian[3], Amedeo Napoli[2], and Marie-Dominique Devignes[2]

[1] KIKA Medical, 35 rue de Rambouillet 75012 Paris, France
[2] LORIA (UMR 7503 CNRS-INPL-INRIA-Nancy2-UHP), Campus scientifique, BP 239, 54506 Vandoeuvre-lès-Nancy, France
{coulet, malika, napoli, devignes}@loria.fr
[3] Université Pierre et Marie Curie - Paris6, INSERM UMRS 538, Biochimie – Biologie Moléculaire, Paris, France
pascale.benlian@sat.ap-hop-paris.fr

Abstract. Pharmacogenomics explores the impact of individual genomic variations in health problems such as adverse drug reactions. Records of millions of genomic variations, mostly known as Single Nucleotide Polymorphisms (SNP), are available today in various overlapping and heterogeneous databases. Selecting and extracting from these databases or from private sources a proper set of polymorphisms are the first steps of a KDD (Knowledge Discovery in Databases) process in pharmacogenomics. It is however a tedious task hampered by the heterogeneity of SNP nomenclatures and annotations. Standards for representing genomic variants have been proposed by the Human Genome Variation Society (HGVS). The SNP-Converter application is aimed at converting any SNP description into an HGVS-compliant pivot description and vice versa. Used in the frame of a knowledge system, the SNP-Converter application contributes as a wrapper to semantic data integration and enrichment.

1 Introduction

One of the great challenges in the post-genomic area consists in exploring the involvement of individual genomic variations in biological processes. Technical advances in high-throughput genotyping enable rapid sampling of thousands of genotypes. Among the large amount of individual variations (more than 10 millions displaying a frequency higher than 1% in studied populations) dispersed all along the genome, very few are known to have an obvious pathological effect. These are named *mutations*. More general terms, such as *polymorphism* or *variant*, are preferred to characterize the general concept of variation [1]. Around 90% of the genome variations are limited to one-nucleotide substitutions (for example a guanine replaces a thymine at a given position in the genome) designated as single nucleotide polymorphism or SNP.

The challenge mentioned above, i.e. to explore the involvement of individual genomic variations in biological process, can be considered as a data mining problem.

U. Leser, F. Naumann, and B. Eckman (Eds.): DILS 2006, LNBI 4075, pp. 82 – 93, 2006.

Knowledge discovery in databases (KDD) is a process aimed at extracting from large databases information units that can be interpreted as knowledge units [2]. This process comprises three major steps: (i) the selection and preparation of data, (ii) the data mining operation, and finally (iii) the interpretation of the extracted units. Various integration problems may arise along the process. The first step often requires to integrate data from public and private databases in order to guide the selection step or to enrich the selected set of data. The last step also necessitates to assess the extracted information units with respect to existing knowledge [3]. In both cases, integration tasks will consist in establishing equivalence, consistency or discrepancy between data or concepts, as well as classifying new data or concepts among existing ones. This type of integration should therefore rely on a semantic conceptual frame in which reasoning mechanisms are available. Indeed, ontologies contribute to build such an environment [4].

An ontology is a formalization of a conceptualisation [5], that is to say the definition and the representation for a given domain of concepts and their relationships allowing human and machine agents to share knowledge about this domain, and to reason with respect to this knowledge. By providing a semantic conceptual frame to a data mining process, an ontology should play a valuable role to facilitate data integration as well as knowledge acquisition.

Pharmacogenomics is a multi-dimensional domain where genome variations, phenotypic data and drug properties can be mined together in order to find out possible associations of variations with individual good or adverse drug responses [6]. More and more pharmaceutical firms are willing to include the exploration of particular genomic variants in their drug clinical trials in order to detect relationships between the following three summits of the pharmacogenomics triangle (Figure 1): (1) drug (properties and administration), (2) phenotype (biological and clinical data), and (3) genotype (genome variations).

Integration of the genotype dimension in clinical trials is not straightforward partially because of the large number of variants present all along the genome. Indeed many genes contain more SNPs than can be conceivably genotyped in current studies. Thus the choice of a relevant subset of SNPs to be included in studies should be somehow guided. A knowledge base called PharmGKB participates in this effort by offering a repository for storing experimental data sets related to pharmacogenomic studies [7].

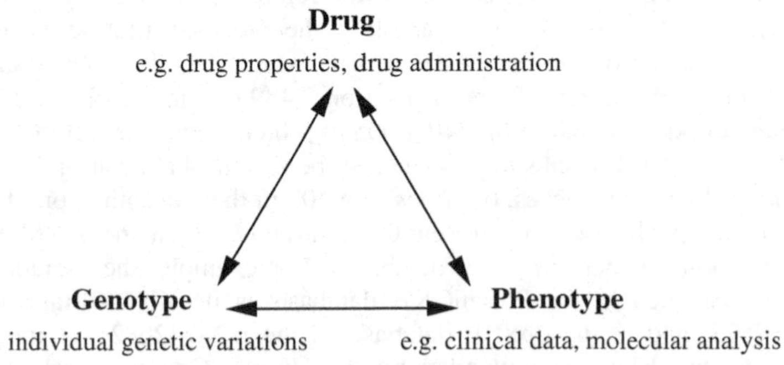

Fig. 1. Triangular schematization of the pharmacogenomics domain

The present research work focuses on the genotype summit of the pharmacogenomics triangle since its complexity is often underestimated, and since major difficulties arise when locally observed genotype data have to be confronted to existing data in public databases. Particularly the nomenclatures used to describe the SNPs are heterogeneous within the public databases themselves (dbSNP, UCSC genome browser, HapMap, PharmGKB), and when compared to private data sources, so that variant identification and correspondence between two heterogeneous sources is not easy to achieve [8].

In [9], we have introduced the SNP-Ontology represented in the OWL language as a contribution towards building a semantic frame for pharmacogenomic studies. Our purpose is to use this ontology to formally represent the knowledge on genomic variants (i.e., SNP-knowledge base) as the first step of a KDD process as in [10, 11]. We thus developed the SNP-Converter application which acts as a wrapper for entering variant individuals in the SNP-knowledge base, starting from data extracted from various SNP-related databases as in [12, 13].

The section 2 introduces the various SNP representations and the existing attempts for integration. Section 3 presents the SNP-Converter application: rationale and functionalities. Usage of this application in the frame of the SNP-knowledge base is described in section 4. Section 5 discusses the issues of the solution presented here, and the perspectives of this work in terms of contribution to future pharmacogenomic studies are proposed.

2 Heterogeneity and Integration of Genomic Variations Data

2.1 Heterogeneous Representations of Genomic Variations

By definition, a genomic variation is originally associated to a position in a genomic (chromosome) sequence. However, when it affects a transcribed region, it is propagated to transcript sequence and, if the position is in a coding region, to protein sequence. Variation databases indifferently represent variations in DNA, RNA or protein. Thus, they represent as well the original variation and its repercussions. For illustration, the substitution of a guanine by a thymine can be represented by G/T in the DNA sequence, GGC/GTC in the affected codon, g/u in the corresponding RNA, Gly/Val in the translated protein. In addition, the representation within the databases of the variant position differs depending on the reference sequence (and its version) used to locate it. Let us take an example : the G/T substitution is at position 11,087,877 in the chromosome 19 sequence, which has the accession number NC_000019 in the RefSeq database, at position 2,489,679 in the NT_011295 contig sequence, and at position 565 in the NP_000518 protein sequence (on the second nucleotide of the codon). The substitution can also be localized at position 26,747 in one of the associated gene sequences, or at position 108 in the eleventh exon of this gene. Various syntaxes can be used to represent these variants, which are also often referred by their accession numbers in given databases. For example, the variant described above would be cited in the PharmGKB database as the G/T variant at position chr19:11087877, and in the dbSNP database as the rs28942082 polymorphism. A generic syntax has been recommended by the Human Genome Variation Society

(HGVS). According to this proposed standard, our variant should be described by the following expression: NC_000019.8:g.11087877G>T, where NC_000019.8 is the unique accession number (in the NCBI RefSeq database) of the sequence used to position the variant, the letter g means that the sequence is genomic, by opposition to p for instance which is used for a protein sequence, 11087877 corresponds to the position in the referred sequence, and G>T describes the substitution itself (http://www.hgvs.org/mutnomen/recs.html) [14]. However this nomenclature has not been universally adopted yet. Previous nomenclatures sometimes subsist for historical reasons. For example our variant is still found in OMIM as the "FH NAPLES" or "Gly544Val", that is to say with denominations related to the historical context of its discovery. In addition, private and disease- or locus-specific databases continue using non-conventional representations that enlarge the set of possible nomenclatures. Figure 2 illustrates the numerous alternative manners of designating a unique genome variant in private and public databases. It is worth noting that some of the non-conventional notations (c) are ambiguous: the first one does not mention the reference nucleotide, the third and fourth ones refer to two different versions of the same protein.

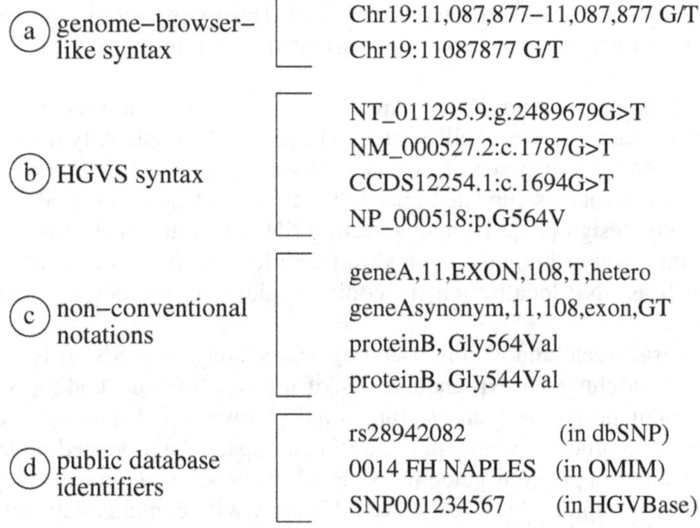

Fig. 2. Various notations or references for the same variant

Finding intersection between several genomic variation databases is a critical issue for genetic diagnosis and "variome" exploration [15, 16]. However, as shown above, this task is not easy because of the amount of alternative and equivalent representations. Thus a system capable of establishing equivalence i.e. aligning between the different representations of a given variant is needed for investigating genome variations, and for being a basis for further pharmacogenomic studies.

2.2 Integrated Solutions

A first solution for solving the problem of heterogeneous representation of genomic variations is to build integrated databases providing a single access to variants pertaining from various sources. The NCBI dbSNP database lists over 9 million human polymorphisms, and constitutes the largest source of variants over the web [17]. Indeed, together with directly submitted SNP data, dbSNP integrates data from other large public databases of variants such as the NCI CGAP-GAI database, the TSC (The SNP Consortium, Ltd) variation initiative, HGVBase, HapMap, PharmGKB, Perlgen. Furthermore, dbSNP is fully integrated with NCBI databases (GenBank, PubMed, LocusLink, Human Genome Project Data) leading to a rich set of data.

HGVbase (Human Genome Variation Database, formerly HGBase) is the product of a collaboration between the Karolinska Institute (Sweden) and the European Bioinformatics Institute (UK). It has been constructed as a means for gathering polymorphisms from all possible public sources [18]. Thanks to both collection and submission, this relational database is cataloguing more than 8 million polymorphisms and proposes interesting text-based search facilities. HGVbase has been interfaced with SRS (Sequence Retrieval System). An originality of this work is that the authors propose the first controlled vocabulary, the Mutation Event Controlled Vocabulary[1], to facilitate polymorphism data integration. Each HGVbase record contains all the information necessary to re-construct the variant description in the HGVS standard syntax.

TAMAL (Technology And Money Are Limiting) is based on a materialized data warehouse that integrates five SNP sources (HapMap, Perlgen, Affymetrix, dbSNP and the UCSC genome browser), and that offers querying facilities through current versions of these resources (updated quarterly) in view of facilitating SNP selection for genetic study design [19]. To help selecting SNPs that are likely involved in the genetic determination of human complex traits, various properties of SNP have been integrated such as SNP localisation (in coding regions, in promoters) or haplotype tagging.

LS-SNP (Large-Scale annotation of coding non-synonymous SNPs) is an original work aimed at enriching dbSNP annotations of non-synonymous coding SNPs with information about protein sequences, functional pathways and comparative protein structure models in order to predict polymorphism impacts on produced proteins [20]. This resource can be a precious guide for SNP selection before a clinical study.

The pharmacogenomics knowledge base (PharmGKB) contains data sets linking genotype and phenotype information [7]. This integrated resource presents two major interests. First, original polymorphisms are directly submitted to PharmGKB as results of clinical trials, enabling to link them to individual clinical data. Second, PharmGKB allows extended navigation through cross-referenced sources such as NCBI databanks, UCSC Genome Browser and Gene Ontology. This makes PharmGKB a valuable resource for interactively enriching annotations on given variants. PharmGKB data are structured according to an XML schema that defines the relationships between the different handled objects. However, as far as we know, PharmGKB is not exploitable for automatic data extraction and mining.

---

[1] http://www.ebi.ac.uk/mutations/recommendations/mutclass.txt

This brief panorama of integrated databases in the domain of genomic variations shows that each project has to solve in some way the problem of integrating heterogeneous variant representations. Methods used are rarely explicited since they must fit the data model associated to the database, and cannot be reused for other purposes. More general propositions have been made to promote integration of variant representations. A controlled vocabulary (the Mutation Event Controlled Vocabulary quoted above) has been proposed by the HGVbase. The Polymorphism Markup Language offers the possibility of exchanging data on sequence variations [21]. The associated DTD (Document Type Definition) describes polymorphism variation, frequency, population, assay, submitter and publication. DDBJ and JBIC recommend the use of PML for interoperability of data on SNPs and other genomic variations. Under the supervision of the Object Management Group, the SNP object has been precisely specified [22]. This work takes into account a large view of the data linked to SNPs in existing data sources. The HGVS participates in this effort of knowledge representation as one of the rare propositions looking at the genetic variation concept, and not simply at the representations of variants in databases [14]. It should be noted that the unequivocal identification of genomic variants does not mean here unique identifier, since the generic syntax proposed by HGVS allows multiple references to various types of sequences (chromosomes, contigs, transcripts, proteins). Finally, the XML PharmGKB schema presented as an ontology by the authors includes the representation of domain concepts and their relationships in a structured formalism [23].

2.3 Semantic Integration

Converting one SNP format into another one and establishing equivalence between variants displaying different representations calls for explicit domain knowledge about gene structure, transcript definition, and genetic code. This is one reason leading us to the design of the SNP-Ontology [9]. Indeed a specific ontology in the field of genomic variations is useful, because it embodies the abstract knowledge required for data integration and analysis. Existing initiatives mentioned above such as the PharmGKB ontology and the OMG SNP specification contributed to the early stages of this work. Several additional concepts were defined to provide the SNP-Ontology with the capacity of hosting any variant, whatever their description, as individuals instantiating the ontology concepts and properties. The SNP-Ontology has been coded with the OWL (Web ontology language) formalism and edited with the Protégé knowledge base framework [24][25]. OWL is the standard representation language for the semantic web. Its foundations are both description logics and web standard languages (XML and RDF-S). It allows building a knowledge base equipped with reasoning mechanisms such as subsumption, classification, consistency checking and instantiation. These mechanisms once plugged in Protégé lead to new inferred knowledge that can enrich the knowledge base. However integrating variant descriptions requires handling of concrete data (e.g. string, integer) that is not yet fully allowed by the description logic framework [26]. Thus, we have developed the SNP-Converter application that can be used either as a standalone application for format conversion purposes, or in the frame of an ontology-driven knowledge base for integrating datasets of genomic variants.

3 The SNP-Converter

3.1 Inputs and Outputs

A variant is considered as an observed variation located at a specific position along a sequence. The observed variation can be a nucleotide or an amino acid variation depending whether the sequence serving as reference for localisation is nucleic acid or protein. This definition, that follows the HGVS nomenclature standard, leads to represent a variant by four features:

(i) the identifier of a reference sequence (i.e. its accession number in a public sequence database) ;
(ii) the type of concerned sequence (genomic, coding : cDNA, mRNA or protein coded respectively by g., c., r. or p. according to HGVS standards) ;
(iii) the position of the variant in the reference sequence ;
(iv) the observed variation (G/T, G >-, ->T, GT>AG, g>u, Gly>Val, etc.).

Conjunction of these four features yields an unequivocal representation of the variant. As mentioned above, a given variant can be represented by several sets of features depending on the selected reference sequence identifier. The core of the SNP-Converter application takes a set of four features as input, and converts them into an alternative set of four features representing the same variant. Because most representations do not explicitly provide the input features, a data preparation step is embedded in the SNP-Converter application. Present implementation of the preparation step allows the extraction of input features from dbSNP, HapMap, HGVBase records (in XML format) and from flat files or spreadsheets of two private databases that follow non conventional notations such as the first and second ones in figure 2.c. Reciprocally the converted output features may be processed to comply with the output format adapted to their envisaged usage. The SNP-Converter is currently able to produce several output formats: a simple text file using HGVS nomenclature, dbSNP XML and submission file formats. A typical scenario of SNP-Converter usage is the conversion of interesting SNPs from a private database into the dbSNP submission file format.

3.2 The Conversion Process

The SNP-Converter process, shown in Figure 3a, can be decomposed into 4 steps: (1) data preparation, (2) conversion of the four input features into pivot features, (3) an optional additional conversion into specific output features, and finally (4) the edition of output data. A simple instantiation of this process is illustrated in Figure 3b.

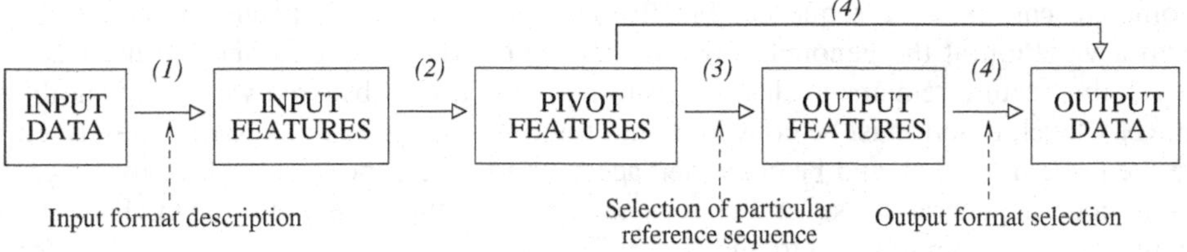

Fig. 3a. The SNP-Converter global process

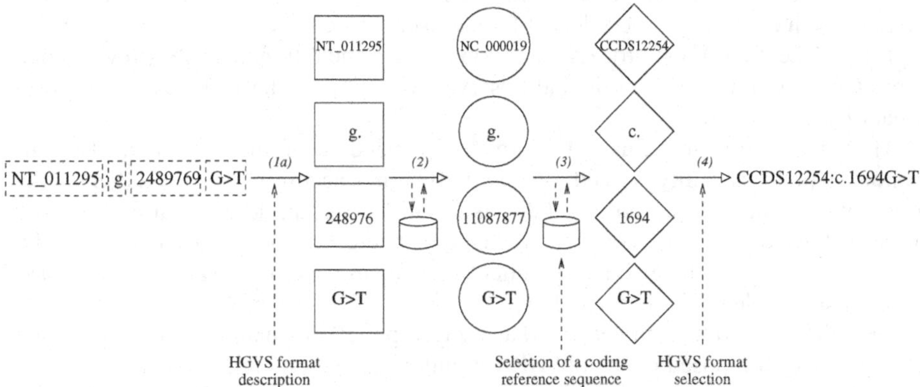

Fig. 3b. Illustration of the enactment of the SNP-Converter process on a given variant representation

(1) The data preparation step consists in extracting the four input features from input data and depends on each specific source format. This preparation step also depends on whether the variant description is explicit (e.g. Genome-browser-like syntax or HGVS syntax) or implicit (e.g. database identifier). (1a) When the description is explicit, the four input features can be directly extracted by parsing the description according to a format-specific scheme. (1b) When the description is implicit, input data should first be completed in view of extracting input features. For example if the input data is a dbSNP identifier, it can be used to query the database and extract from the variant record the explicit data composing the input features.

(2) Pivot features consist in the particular set of features obtained for a given variant when the reference sequence is the complete chromosome sequence (RefSeq accession number, e.g. NC_000019.8) that includes the input reference sequence. Since the pivot sequence type is genomic, the variant position and the nature of observed variation must be computed. The input reference sequence is first localized on the complete chromosome sequence using alternative data sources. For instance the relative position of a gene can be found thanks to the gene symbol in the RefSeq complete chromosome entry ("FEATURES/gene" section). Exon genomic positions can also be retrieved in the "FEATURES/mRNA" section. When the variant position is expressed relative to translation start (ATG), genomic coordinates of coding sequence can be retrieved from the NCBI CCDS database. The appropriate coordinate conversion can then be computed to finally produce the position of the variant relative to the

complete chromosome sequence. Finally the observed variation must be converted into a variation at the genomic level. If the input variation is described at the DNA level, this feature remains unchanged. Alternatively, if the observed variation is at the mRNA level, uracil must be converted into thymine. An observed variation described at the protein level should be converted according to the genetic code. Due to the genetic code degeneration, several codons can code for the same amino acid. Thus the conversion from amino acid to nucleic acid variation can lead to more than one set of features. The SNP-Converter outputs all these possibilities.

(3) The next step of reference features conversion is optional since it appears unnecessary when the desired output format fits to the pivot features. If not, the output reference sequence should be selected by the user, and can be DNA, cDNA, mRNA, or protein. The conversion process then follows the same rationale as the previous one to produce new relative position and observed variation in the new reference representation.

(4) Output data should finally be formatted depending on the purpose of the conversion. A first possibility is to edit the output features according to the HGVS syntax or any other syntax. A second possibility is to build a variant description in a specific format for database submission. Finally, another interesting possibility is to enter the output data in a knowledge-base-compliant formalism such as OWL to allow its assertion in a knowledge base (see below).

The SNP-Converter is implemented as a Java application, and has been tested on a set of variants composed of the dbSNP variants mapping on chromosome 19, and of variants extracted from a private database. For this purpose, dbSNP variants were extracted from downloaded dbSNP XML file and other variants from private text files. The goal was to find which variants from the private database were missing in the dbSNP database. The SNP-Converter application allowed us to compare the pivot features of the private variants with those of dbSNP variants. This experience allowed us to determine the overlapping coverage of both databases, and to identify several variants that were not yet submitted to dbSNP.

4 The SNP-Converter as a Wrapper for Semantic Integration

The SNP-Ontology (see section 2.3) plays the role of a coherent domain-specific global schema for a knowledge-base. We have made a mapping between the four features handled by the SNP-Converter and the SNP-Ontology concepts allowing the SNP-Converter to assert variants as individuals in the knowledge base. Since these four features are extracted from input data, the whole process leads to an indirect mapping of source schemas on the ontology. In practice we found relevant for any new variant, to insert in the SNP-knowledge base, not only its original set of features (for sake of traceability), but also the pivot features computed by the SNP-Converter. Thanks to these pivot features, the SNP-Converter is capable of qualifying as equivalent variants initially represented by distinct descriptions (see Fig. 4). The equivalence checking performed by the SNP-Converter is used here as a procedural extension of description-logics-based reasoners, aimed at enriching the knowledge base.

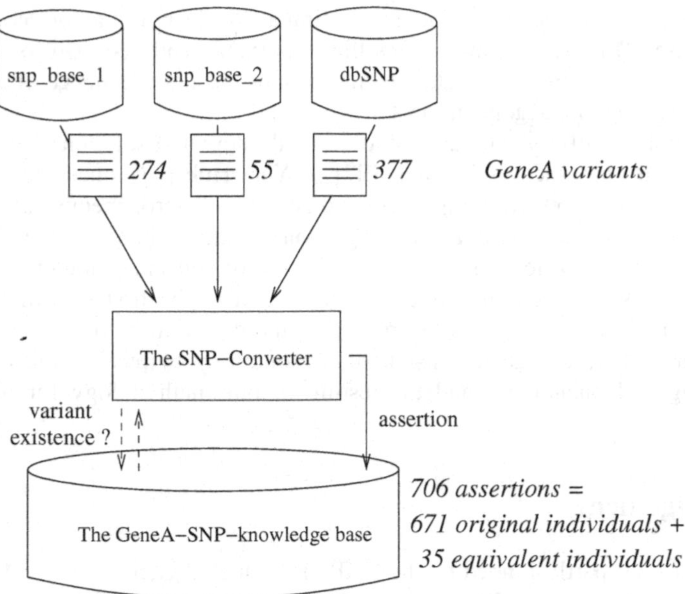

Fig. 4. Schematization of the use of the SNP-Converter application as a wrapper coupled to a knowledge base

Figure 4 also shows the result of an experience carried on variants of a specific gene (named here geneA). Three sets of data were processed by the SNP-Converter application : 274 and 55 variants from private databases snp_base_1 and snp_base_2 respectively, and 377 variants from dbSNP. Among the 706 assertions created by the wrapper, 671 could be qualified as original individuals, and 35 were found equivalent to existing individuals.

5 Conclusion and Discussion

The SNP-Converter application has been developed to face the heterogeneity problem in genomic variation representation. The SNP-Converter can be used standalone to pass from one variation description to another. As such it constitutes a valuable tool for several use-cases: confronting private and public variant data, preparing submission of new variants to public databases, facilitating variant annotation retrieval from heterogeneous databases, guiding the choice of relevant variants to include in clinical trials, etc. The core of the SNP-Converter was designed to be generic thanks to the mapping with the SNP-Ontology. However, the handling of new sources requires some ad hoc adaptations for driving the extraction of the input features. This task will be facilitated by an administration interface. It should be noted that the SNP-Converter works with constant RefSeq versions and therefore is faced to the common problem of managing sequences pertaining to different assemblies.

With respect to the KDD process, our objective is to settle a semantic frame facilitating semantic data integration, data mining and incremental knowledge acquisition.

In particular we consider semantic data integration as the design of an ontology-based knowledge base. This work demonstrates the importance and necessity of adequate wrappers preceding the semantic data integration stage as a consequence of the limits of existing knowledge management tools.

Our methodology differs from already described integrated solutions (see Sect. 2) and more general ones such as BioMart [27] or YeastHub [28] since most of these approaches are limited to facilitating integrated access to heterogeneous data whereas our goal is to facilitate data mining and integration of data-mining results in a knowledge base. The work reported here constitutes a proof of concept limited to one of the pharmacogenomics triangle summits (see Fig. 1), and to the first step of the KDD process. Nevertheless it allows us to proceed in the data mining process. Complete demonstration will necessitate extension of the ontology to include the two other summits (drug and phenotype) and the testing of our methodology for these two domains.

Acknowledgement

This work has been partly funded by the EUREKA-labeled GenNet research and development contract between KIKA medical, PhenoSystems and Loria-CNRS. AC benefits from a CIFRE fellowship. Special thanks to Romain Demoustier from KIKA medical and to David Atlan from Phenosystems for stimulating discussions.

References

1. Kruglyak,L., Nickerson,D. Variation is the spice of life. Nat Genet. 27, 3 (2001) 234-6.
2. Frawley,W., Piatetsky-Shapiro,G., Matheus,C. Knowledge Discovery in databases: An Overview. Knowledge Discovery in Databases, AAAI/MIT Press.(1991) 1-30.
3. Janetzko,D., Cherfi,H., Kennke,R., Napoli,A., Toussaint,Y. Knowledge-based Selection of Association Rules for Text Mining. 16h European Conference on Artificial Intelligence, ECAI'04, Valencia (2004).
4. Vetere,G., Lenzerini,M. Models for Semantic. Interoperability in Service Oriented Architectures, IBM Systems. Journal, 44 (2005).
5. Gruber,T.R. A Translation Approach to Portable Ontology Specifications. Knowledge Acquisition. 5 (1993) 199-220.
6. Evans,W., Relling,M. Pharmacogenomics: moving toward individualized medicine, Nature. 29 (2004) 464-468.
7. Klein,T., Chang,J., Cho,M., Easton,K., Fergerson,R., Hewett,M., Lin,Z., Liu,Y., Liu,S., Oliver,D. et al. Integrating genotype and phenotype information: an overview of the PharmGKB project. Pharmacogenom. J. 1 (2001) 167–170.
8. Marsh,S., Kwok,P., McLeod,H. SNP databases and pharmacogenetics: great start, but a long way to go. Hum Mutat. 20, 3 (2002) 174-9.
9. Coulet,A., Smaïl-Tabbone,M., Napoli,A., Benlian,P., Devignes M.D. SNPOntology for semantic integration of genomic variation data. ISMB 2006, Fortaleza. [OnLine]. https://hal.inria.fr/inria-00067863
10. Anand,S., Bell,D., Hughes, J. The role of domain knowledge in data mining, Conference on Information and Knowledge Management CIKM'95, Baltimore, USA (1995).

11. Euler,T., Scholz,M. Using Ontologies in a KDD workbench, ECML/PKDD'04 Workshop on Knowledge Discovery and Ontologies (KDO'04), Pisa (2004).
12. Catarci,T., Lenzerini,M. Representing and using inter-schema knowledge in cooperative information systems. Journal of Intelligent and Cooperative Information Systems. 2 (1993) 375-398.
13. Levy,A. Logic-Based Techniques in Data Integration Logic Based Artificial Intelligence. Jack Minker. Kluwer Publishers (2000).
14. den Dunnen,J., Antonarakis,S. Mutation nomenclature extensions and suggestions to describe complex mutations: a discussion. Hum Mutat. 15 (2000) 7–12.
15. den Dunnen,J., Paalman,M. Standardizing mutation nomenclature: why bother? Hum Mutat. 22 (2003) 181–182.
16. Cotton,R.G.H., Kazazian,H.H. Toward a human variome project. Hum Mutat. 26,6 (2005) 499.
17. Sherry,S., Ward,M., Sirotkin,K. dbSNP—Database for Single Nucleotide Polymorphisms and Other Classes of Minor Genetic Variation. Genome Res. 9 (1999) 677–679.
18. Fredman,D., Munns,G., Rios,D., Sjoholm,F., Siegfried,M., Lenhard,B., Lehvaslaiho,H., Brookes,A. HGVbase : a curated resource describing human DNA variation and phenotype relationships. Nucleic Acids Res. 32 (2004) D516-9.
19. Hemminger,B., Saelim,B., Sullivan,P. TAMAL: an integrated approach to chosing SNPs for genetics studies of human compex traits. Bioinformatics. 22 (2006) 626-627.
20. Karchin,R., Diekhaux,M., Kelly L., Thomas D., Pieper,U., Eswar,N., Haussler,D., Sali,A. LS-SNP: large-scale annotation of coding non-synonymous SNPs based on multiple information sources. Bioinformatics. 21 (2005) 2814-2820.
21. Sugawara,H., Mizushima,H., Kano,T., Shigemoto,Y., Hashimoto,Y., Tomabechi,I., Sakagami,N. et al Polymorphism Markup Language (PML) for the interoperability of data on SNPs and other sequence variations, 19th International CODATA Conference (2004).
22. OMG Single Nucleotide Polymorphisms specification (2005) [Online] http://www.omg.org/cgi-bin/doc/dtc/05-02-06.pdf.
23. Oliver,D., Rubin,D., Stuart,J, Hewett,M., Klein,T., Altman,R. Ontology development for a pharmacogenetics knowledge base. Pac Symp Biocomput. (2002) 65-76.
24. Horrocks,P., Patel-Schneider,F., van Harmelen,F. From SHIQ and RDF to OWL: The making of a web ontology language, Journal of Web Semantics, 1, 1 (2003) 7-26.
25. Noy,N., Sintek,M., Decker,S., et al. Creating Semantic Web contents with Protege-2000. IEEE Intelligent Systems 16. (2001) 60-71.
26. W3C Web Ontology Working Group (WOWG), (2004) Owl web ontology language semantics and abstract syntax. W3C recommendation [Online]. http://www.w3.org/TR/owl-ref/.
27. Kasprzyk,A., Keefe,D., Smedley,D., London,D., Spooner,W., Melsopp,C., Hammond,M., Rocca-Serra,P., Cox,T. Birney,E. EnsMart: A Generic System for Fast and Flexible Access to Biological Data. Genome Res. 14 (2004) 160-169.
28. Cheung,K.H., Kevin Y. Yip,K.Y., Smith,A., deKnikker,R., Masiar,A., Gerstein,M. YeastHub: a semantic web use case for integrating data in the life sciences domain, Bioinformatics.21 (2005) i85-i96. 28.

SABIO-RK: Integration and Curation of Reaction Kinetics Data

Ulrike Wittig, Martin Golebiewski, Renate Kania, Olga Krebs, Saqib Mir, Andreas Weidemann, Stefanie Anstein, Jasmin Saric, and Isabel Rojas

Scientific Databases and Visualization Group, EML Research gGmbH,
Schloss-Wolfsbrunnenweg 33, 69118 Heidelberg, Germany
Ulrike.Wittig@eml-r.villa-bosch.de
http://sabiork.villa-bosch.de/

Abstract. Simulating networks of biochemical reactions require reliable kinetic data. In order to facilitate the access to such kinetic data we have developed SABIO-RK, a curated database with information about biochemical reactions and their kinetic properties. The data are manually extracted from literature and verified by curators, concerning standards, formats and controlled vocabularies. This process is supported by tools in a semi-automatic manner. SABIO-RK contains and merges information about reactions such as reactants and modifiers, organism, tissue and cellular location, as well as the kinetic properties of the reactions. The type of the kinetic mechanism, modes of inhibition or activation, and corresponding rate equations are presented together with their parameters and measured values, specifying the experimental conditions under which these were determined. Links to other databases enable the user to gather further information and to refer to the original publication. Information about reactions and their kinetic data can be exported to an SBML file, allowing users to employ the information as the basis for their simulation models.

1 Introduction

The biosciences have undergone some dramatic changes in the last few years. Novel lab approaches like high-throughput methods enable scientists to rapidly produce an enormous amount of data. For researchers this poses problems connected with retaining an overview of these data and accessing them. Thus one of the biggest challenges in biological science at present is to achieve data comparability and ease of access for the scientific community. To attain this goal, experimental data from different sources need to be standardized and integrated into databases.

At the moment only a small number of databases exist which contain information about biochemical reaction kinetics. The BRENDA enzyme database [1] offers a comprehensive list of kinetic parameters based on literature information. UniProt [2] started to include kinetic parameters as comments related to biophysicochemical properties, also manually extracted from publications. The BioModels database [3] stores published mathematical models of biological interest annotated and linked to relevant data resources (e.g. publications or databases). The models include kinetic

U. Leser, F. Naumann, and B. Eckman (Eds.): DILS 2006, LNBI 4075, pp. 94 – 103, 2006.

laws and their parameters represented in SBML (Systems Biology Mark-up Language) format [4] and can be used for simulations of biochemical reactions or networks.

In order to compare kinetic data and develop biochemical network models, kinetic parameters need to be consistently described and related to kinetic mechanisms, equations representing the kinetic laws and environmental conditions. The known mechanisms of biochemical reactions should be reflected in mathematical formulas, which have to be linked to the corresponding parameters, such as kinetic constants and concentrations of each reaction participant. As kinetic constants highly depend on environmental conditions, they only can be specified completely by describing these conditions used for determination. Data sets based on an experiment assayed under similar experimental conditions should be associated to each other to facilitate the comparison. Therefore, users interested in information about reaction kinetics require databases that merge and structure all these data.

The SABIO-RK (**S**ystem for the **A**nalysis of **Bio**chemical Pathways - **R**eaction **K**inetics) database is designed to meet these requirements and to support researchers interested in information about biochemical reactions and their kinetics. This report will mainly focus on the database content, integration and curation processes. Modelling of the database and the retrieval of data by database searching will be briefly discussed.

2 Data Integration

SABIO-RK represents an extension of the SABIO biochemical pathway database also developed at EML Research [5]. Figure 1 represents a simplified schema of the main database objects and their relations in SABIO and SABIO-RK. SABIO contains information about biochemical pathways, reactions and their participants (enzymes, reactants etc.). These data are connected with specific protein information, organisms or cellular locations. SABIO-RK combines the general data about biochemical reactions stored in SABIO with information about their kinetic properties. The type of kinetic law and its representation in a formula is given if provided in the literature. This also includes effectors (e.g. cofactors, activators or inhibitors) of the reactions and their type of interaction (e.g. competitive or non-competitive inhibition). The kinetic laws are represented with their parameters, including their measured values. Since many of the publications only contain kinetic constants (e.g. Km, kcat or Vmax) but have no description of the kinetic law type, these parameter values are also inserted independent from a kinetic law type.

Additionally the database contains descriptions of the experimental conditions (e.g. pH, temperature, and buffer) for the measured parameter values. In the buffer description all components of the assay are represented including coupled enzyme assays.

In order to establish a broad information basis, data from different sources are integrated into SABIO-RK (Figure 1). Most of the reactions, their associations with biochemical pathways and their enzymatic classifications (EC classifications [6]) are downloaded from the KEGG (Kyoto Encyclopedia of Genes and Genomes) database [7] and stored in SABIO. In contrast, the kinetic data contained in SABIO-RK are manually extracted from scientific articles and verified by curators. At the moment it is very difficult to extract this information automatically, such as by the use of text mining technologies, given that most of the data are highly scattered through various publications and are frequently found in tables, formulas or graphs. However, we are working on the development of support tools, one of them for the identification of synonyms of chemical compound names, as we will describe in section 3.

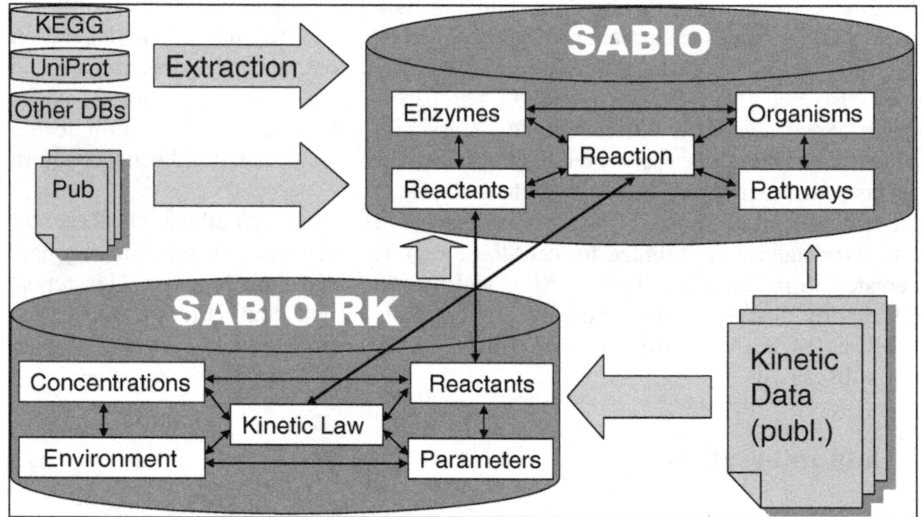

Fig. 1. Population, content and schematic relation of SABIO and SABIO-RK. SABIO contains general information about biochemical pathways and reactions in different organisms, including details about corresponding enzymes and reactants. Most of these data are collected from other databases like KEGG and UniProt. SABIO-RK extends SABIO by storing information about the reactions' kinetic properties, such as the kinetic laws with their corresponding parameters and environmental conditions under which they were determined.

As standards for publishing data of biochemical reactions and reaction kinetics are lacking, the curators are faced with problems like synonymic or aberrant notations of compounds and enzymes, multiplicity of parameter units and missing information about assay procedures and experimental conditions. During the curation process, we unify and structure the data consistently in order to facilitate the comparison of the kinetic data obtained under different experimental conditions or from different organisms, tissues etc. Furthermore structured data enable the user to understand the behaviour of a biochemical reaction under environmental changes like for example increase of temperature or pH variations.

The information source of each database entry is clearly shown and linked to the PubMed [9] database in order to allow the user to refer to the original paper to obtain additional information about the experiment described.

Systematic names of organisms named in the publication are identified according to the NCBI taxonomy [8] and additional information about specific strains of organisms is stored in the database. If the enzyme of the original organism was expressed in another organism, the host organism is represented in the general comment line of an entry.

The extraction work is done by students using a web-based interface to enter the extracted information into a temporary SQL database. A list of publications, expected to contain kinetic data was obtained by keyword searches in the PubMed database and is used as the basis for data extraction. Before transferring the data to the final database SABIO-RK, they are checked, complemented and verified by a team of biological experts to eliminate possible errors and inconsistencies.

As of May 2006, data from more than 1400 publications were evaluated from which 820 were found to contain useful kinetic data that were extracted and inserted into the intermediate database. About 65% are already curated and inserted into the SABIO-RK database. From one publication more than one database entry can arise if different reactions, enzymes, environmental conditions etc. are connected to measured parameter values. Currently SABIO-RK contains about 5100 curated single database entries referring to about 190 organisms, 1100 different reactions, and 320 enzymes catalysing these reactions. Each database entry comprises at least one kinetic parameter. Since there are many publications processed where no information about the kinetic law is addressed, currently in SABIO-RK only 30% of all entries are related to a kinetic law formula. Of the about 20.000 chemical compound names included in the database, 13.470 refer to different compounds, i.e. the average number of alternative names per compound is 1.5.

SABIO-RK only refers to the original source of kinetic data compared parameter values extracted from a referenced paper are not linked to this publication. To avoid redundancies, copying of errors and linking to disparate experimental conditions, the original source of the referenced values is included in the database as separate entries. However, a comparison of the parameter values is possible since entries from different sources are linked by the same reaction or enzyme, assuming that the experimental conditions are the same.

Links to the UniProt database enables the user to gather further information about proteins corresponding to the enzymes.

3 Curation Process

The curation of extracted data is used to achieve correctness and consistency within the database. Already existing standards for data formats are applied as well as new standards are defined if necessary. For example, the unification of parameter units or chemical compound names involves existing standards as the SI system for unit notation or the nomenclature recommendations for chemical compounds of the International Union of Pure and Applied Chemistry (IUPAC). In contrast to, for enzyme specifications (mutants, isoforms, etc.) database-internal norms are assigned additionally to the enzyme classification (EC) system of the International Union of Biochemistry and Molecular Biology (IUBMB). Already existing controlled vocabularies are used for the representation of organisms [8], tissues [1], cell locations [1] etc.

During the curation process, most of the data are unified and structured, with the exception of some information which is stored as comment lines or descriptions, such as in the case of the buffers' compositions. The description of a buffer can be very complex containing for example information about coupled enzyme reactions and synthetic or labelled derivatives of physiological compounds. Therefore, currently information about the buffer composition is stored as a free text. Additional comment lines also contain information about host organisms in which proteins are expressed (e.g. recombinant enzymes expressed in Escherichia coli), or information about the enzyme proteins, especially the protein name if no EC classification is known.

The fact that chemical compounds often have multiple alternative names complicates the work of the database curators. They need to find out whether a compound described in the publication is already contained in the database, possibly with synonymic names, or it is necessary to include this new compound in the database. To address this problem we have developed a tool for the linguistic analysis of chemical terminology, more precisely the names of organic compounds, named CHEMorph [10]. CHEMorph analyses systematic and semi-systematic names, class terms, and also otherwise underspecified names, by using a morpho-syntactic grammar developed in accordance with IUPAC nomenclature [11]. It yields an intermediate semantic representation of a compound which describes the information encoded in a name. The tool provides SMILES strings [12] for the mapping of names to their molecular structure and also classifies the terms analysed. The general process together with an example analysis is shown in Figure 2. The systematic compound name *7-hydroxyheptan-2-one* is transformed into a semantic representation which describes the following: The compound *compd* with its three *parameters* in parentheses, which are the formal descriptions of (i) the skeleton structure, (ii) the name's prefix, and (iii) the name's suffix. From this semantic expression, the corresponding SMILES string [CC(=O)CCCCCO] and the class list is calculated. Currently by matching the yielded SMILES strings, CHEMorph can be used to identify synonymous compound names in order to check if a compound is already contained in SABIO-RK. Future developments will include the matching of chemical structures generated from SMILES strings. With the help of additional Natural Language Processing (NLP) methods, the existing compounds in SABIO-RK can be analyzed for wrong synonyms and multiple entries. By this, the completeness and consistency of the compound data can still be improved.

Also organisms often have synonymic names. They can be described in the literature by their common or systematic name. The SABIO-RK database refers to the NCBI taxonomy and uses systematic names to be able to compare data. Since some authors only give the common name of an organism the curator has to deduce the systematic name. For example the organism described as "rat" can be transferred to Rattus sp. or Rattus norvegicus where the latter is mainly used in laboratories. The curators have to decide if the general organism name is used or not.

Units of kinetic parameters and concentrations can be written in different ways and often have multiple scales. Different systems of standardisation exist in parallel, for example enzyme activities can be noted in *katal* (*mol/s*), *international units* (*µmol/min*), *mg/min* or similar units. This makes the comparison of the data quiet difficult. Therefore a list of scaled and standardized units was established within SABIO-RK based on the recommendations of the International System of Units (SI) [13]. All parameter units and concentrations stored refer to a list that relates synonymic notations with the correct SI standards.

For consistency and to avoid duplicate entries, lists of compounds, organisms, tissues, compartments, kinetic law types and parameter units already existing in the SABIO-RK database are provided for selection at the input interface. These lists also contain synonyms referring to the same content to enable the search for alternative names of compounds, tissues etc. These may mean that the information presented to the database user is not exactly that included in the paper because the entries are presented with recommended names, however the user can always obtain the synonyms of the entries with multiple names. Already existing reactions can be searched in the database by defining one or two reaction participants.

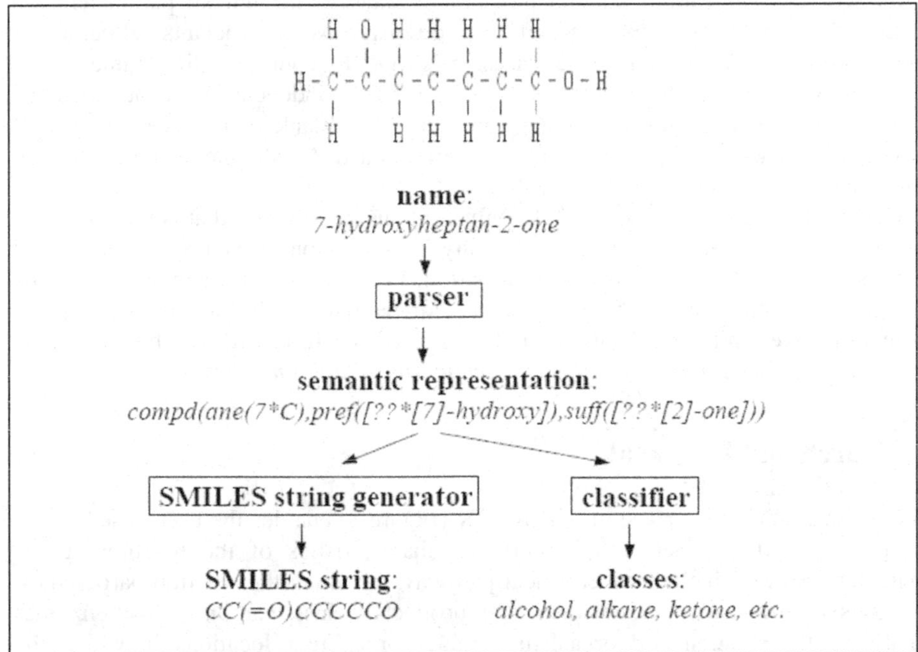

Fig. 2. Overview of the CHEMorph system to support manual database curation. A chemical compound name is parsed and gets a semantic representation assigned, which is taken as a basis to calculate a SMILES string and the classes the compound belongs to.

Enzymes variants catalysing the reactions are distinguished by a description of their subform, like wildtype or mutant protein species. They are named as *wildtype* or *mutant* followed by their name. Different isoforms of an enzyme are also named as *wildtype* followed by the name or abbreviation of the isoform. Furthermore, by including the specification of one or more accession number(s) of the UniProt database (if available), a direct link is provided to the properties of the enzyme.

Additionally, the curators are confronted with missing or only partial information in the literature. For example a reaction definition can be incomplete supposing that only substrates of reactions are named without a definition of the reaction products. Knowing the chemical mechanism of the enzymatic reaction an equation could be completed manually by biochemical experts, but this work could be very time-consuming and furthermore the result could be imprecise. Therefore general compound classes representing specific chemical properties are used as reaction participants. A tool developed for the SABIO biochemical pathway database allows for the classification of chemical compounds based on their functional groups using SMILES strings [14]. Different levels of compound classes based on this compound classification system can deduce more information about the unknown products of a reaction. For compounds for which a SMILES string cannot be assigned, e.g. underspecified or class names such as *deoxysugar* CHEMorph can be used for a classification based on the name.

Sometimes, publications contain incomplete datasets, i.e. not all parameters are measured or initial concentrations of reaction participants (reactants, effectors or enzymes) are missing. For these cases, SABIO-RK contains all parameters or concentrations required for a complete kinetic formula, independent of the existence of the corresponding values. Missing values are left blank or represented by null values. In this way, when exporting the information in SBML the user maintains a reference to all parameters, with or without values.

One major point is that the database only contains information that is mentioned in the corresponding paper. There is neither any interpretation of data by the biological experts, nor the addition of further information. For example, if the authors describe the kinetic mechanism of the reaction as competitive inhibition and no explicit formula is given in the publication, the SABIO-RK database will not show a kinetic formula but the kinetic law type named *Competitive inhibition*.

4 Search and Retrieval

The web-based user interface of SABIO-RK (Figure 3) enables the user to search for reactions and their kinetics by specifying characteristics of the reaction. These characteristics may include biochemical pathways in which the reaction participates, reactants of the reaction (substrates and products), classification of the enzymes catalysing the reaction and organisms, tissues or cellular locations in which the reaction takes place. The search for kinetic data can be further specified by the experimental conditions used for their determination (currently only pH value and temperature), which are considered solely for the retrieval of kinetic data. The system retrieves all entries satisfying the given criteria and indicates whether kinetic information for the associated reaction under the search criteria is specified (organism, tissue, cellular location and experimental conditions). Apart from this, the system also indicates whether there are kinetic data available for the enzymes catalysing each reaction. This approach has been selected to support the variations in the definition of the reactions composing a pathway, e.g. where a reaction can be substituted for by a very similar reaction with a slight change in the reactants. The next version of the interface will also enable the user to search for networks or paths of reactions between two compounds or enzymes. The kinetic data can then be viewed and selected for export in SBML format. Reactions with no kinetic data can also be included in the SBML file.

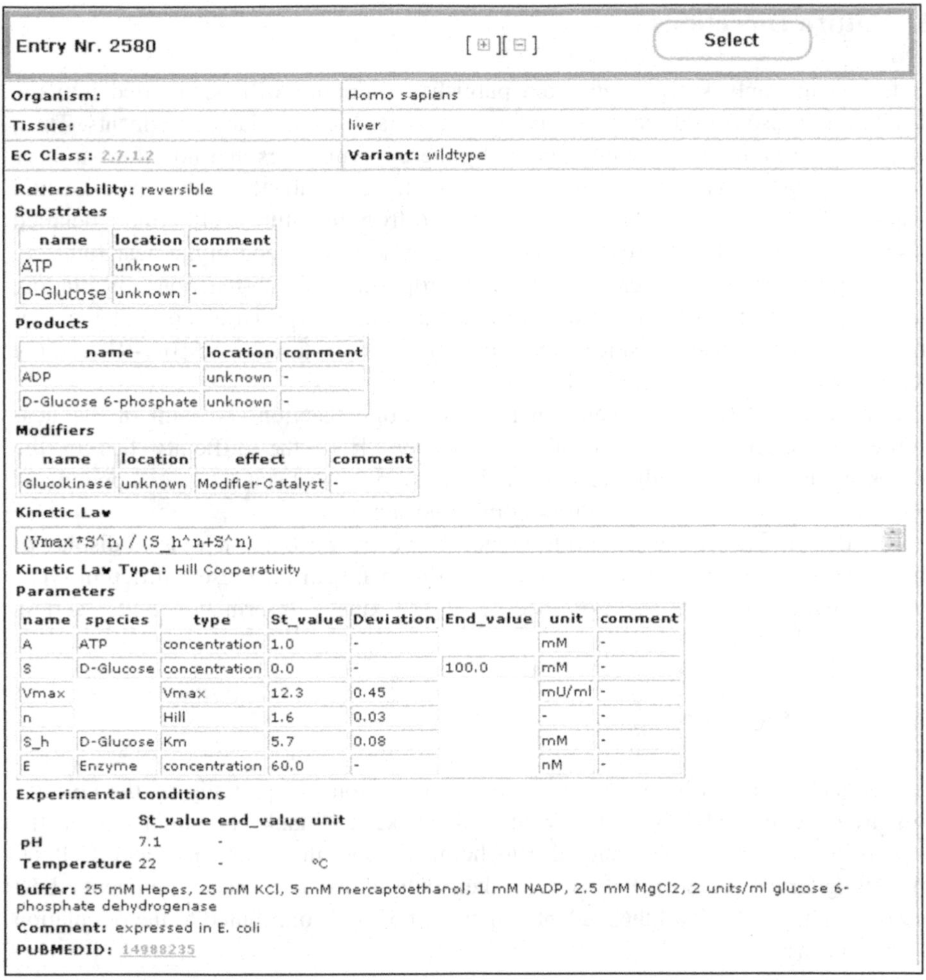

Entry Nr. 2580 [⊞][⊟] Select

Organism:	Homo sapiens
Tissue:	liver
EC Class: 2.7.1.2	**Variant:** wildtype

Reversability: reversible

Substrates

name	location	comment
ATP	unknown	-
D-Glucose	unknown	-

Products

name	location	comment
ADP	unknown	-
D-Glucose 6-phosphate	unknown	-

Modifiers

name	location	effect	comment
Glucokinase	unknown	Modifier-Catalyst	-

Kinetic Law

(Vmax*S^n)/(S_h^n+S^n)

Kinetic Law Type: Hill Cooperativity

Parameters

name	species	type	St_value	Deviation	End_value	unit	comment
A	ATP	concentration	1.0	-		mM	-
S	D-Glucose	concentration	0.0	-	100.0	mM	-
Vmax		Vmax	12.3	0.45		mU/ml	-
n		Hill	1.6	0.03		-	-
S_h	D-Glucose	Km	5.7	0.08		mM	-
E	Enzyme	concentration	60.0	-		nM	-

Experimental conditions

	St_value	end_value	unit
pH	7.1	-	
Temperature	22	-	°C

Buffer: 25 mM Hepes, 25 mM KCl, 5 mM mercaptoethanol, 1 mM NADP, 2.5 mM MgCl2, 2 units/ml glucose 6-phosphate dehydrogenase

Comment: expressed in E. coli

PUBMEDID: 14988235

Fig. 3. SABIO-RK database entry. An example data set represents a specified reaction including kinetic data, experimental conditions and additional information extracted from a publication.

5 Summary

The SABIO-RK database has been designed to meet the Systems Biology community requirements. It aims to support modelers with high quality data in setting-up in-silico models describing biochemical reaction networks. The database enables complex searches for reactions, parameters, etc. and uses existing or defines new standards for data formats. Selected kinetic data can be exported in SBML format to build models for the simulation of complex biochemical processes. SABIO-RK also bundles information for researchers interested in comparing reaction kinetic data originating from different sources.

6 Future Directions

In future, not only kinetic data from published literature will be inserted into the database but also data directly entered by scientists doing the lab experiments. Thus, all the needed information can be given by the experimenters and no information is lost. In doing so, users would be able to directly compare their own experimental results in SABIO-RK with kinetic data extracted from literature. Furthermore detailed descriptions of the kinetic reaction mechanism will be included in the near future to give the opportunity to represent kinetic properties of sub-reactions or binding mechanisms of enzymes in the database. Finally, data export functions of the user interface will be expanded, since a lot of the information stored in SABIO-RK can not yet be formally described in SBML.

Currently SABIO-RK contains mainly metabolic reactions we aim in the near future to incorporate more signalling reactions. Here the difficulty lies in the representation of the signalling reactions, i.e. multiplicity of states of a compound and general descriptions of compounds or compound families.

In order to improve the support to modeller, we are working on a visual display of the networks been set-up by the users, using this platform for present different types of information, such as the existence or not of kinetic information under certain experimental conditions.

Acknowledgement

The project is funded by the Klaus Tschira Foundation and partially by the German Research Council (BMBF). We would also like to thank the members of the Bioinformatics and Computational Biochemistry and the Molecular and Cellular Modelling Groups of EML Research for their helpful discussions and comments. Last but not least, we thank all the student helpers, who have contributed to the population of the database.

References

1. Schomburg I, Chang A, Ebeling C, Gremse M, Heldt C, Huhn G, Schomburg D (2004) BRENDA, the enzyme database: updates and major new developments. *Nucleic Acids Res*, 32, D431-3
2. Bairoch A, Apweiler R, Wu CH, Barker WC, Boeckmann B, Ferro S, Gasteiger E, Huang H, Lopez R, Magrane M, Martin MJ, Natale DA, O'Donovan C, Redaschi N, Yeh LS (2005) The Universal Protein Resource (UniProt). *Nucleic Acids Res*, 33, D154-D159
3. Le Novere N, Bornstein B, Broicher A, Courtot M, Donizelli M, Dharuri H, Li L, Sauro H, Schilstra M, Shapiro B, Snoep JL, Hucka M (2006) BioModels Database: a free, centralized database of curated, published, quantitative kinetic models of biochemical and cellular systems. *Nucleic Acids Res*, 34, D689-91

4. Hucka M, Finney A, Sauro HM, Bolouri H, Doyle JC, Kitano H, Arkin AP, Bornstein BJ, Bray D, Cornish-Bowden A, Cuellar AA, Dronov S, Gilles ED, Ginkel M, Gor V, Goryanin II, Hedley WJ, Hodgman TC, Hofmeyr JH, Hunter PJ, Juty NS, Kasberger JL, Kremling A, Kummer U, Le Novere N, Loew LM, Lucio D, Mendes P, Minch E, Mjolsness ED, Nakayama Y, Nelson MR, Nielsen PF, Sakurada T, Schaff JC, Shapiro BE, Shimizu TS, Spence HD, Stelling J, Takahashi K, Tomita M, Wagner J, Wang J (2003) The systems biology markup language (SBML): a medium for representation and exchange of biochemical network models. *Bioinformatics*, 19, 524-31
5. Rojas I, Bernardi L, Ratsch E, Kania R, Wittig U, Saric J (2002) A database system for the analysis of biochemical pathways. *In Silico Biol* 2,0007
6. IUBMB: http://www.chem.qmul.ac.uk/iubmb/enzyme/
7. Kanehisa M, Goto S, Hattori M, Aoki-Kinoshita KF, Itoh M, Kawashima S, Katayama T, Araki M, Hirakawa M (2006) From genomics to chemical genomics: new developments in KEGG. *Nucleic Acids Res*, 34, D354-7
8. NCBI Taxonomy: http://www.ncbi.nlm.nih.gov/entrez/query.fcgi?db=Taxonomy
9. PubMed: http://www.pubmed.gov
10. Anstein S, Kremer G, Reyle U (2006) Identifying and Classifying Terms in the Life Sciences: The Case of Chemical Terminology. *Proceedings of the Fifth International Conference on Language Resources and Evaluation (LREC)*. To appear
11. IUPAC: http://www.chem.qmul.ac.uk/iupac/
12. Weininger D (1988) SMILES, a chemical language and information system. 1. Introduction to methodology and encoding rules. *J Chem Inf Comput Sci*, 28, 31-36
13. International System of Units (SI): http://www.bipm.fr/en/si/
14. Wittig U, Weidemann A, Kania R, Peiss C, Rojas I (2004) Classification of chemical compounds to support complex queries in a pathway database. *Comp Funct Genom,* 5, 156-62

SIBIOS Ontology: A Robust Package for the Integration and Pipelining of Bioinformatics Services

Malika Mahoui[1], Zina Ben Miled[2], Sriram Srinivasan[3], Mindi Dippold[1], Bing Yang[2], and Li Nianhua[2]

[1] School of Informatics, IUPUI
{mmahoui, mimeie}@iupui.edu
[2] Department of Electrical and Computer Engineering, IUPUI
{zmiled, niali, yanbing}@iupui.edu
[3] Department of Computer and Information Sciences, IUPUI
srsriniv@iupui.edu

Abstract. The recent technological advancements in biological research have allowed researchers to advance their knowledge of the domain far beyond expectations. The advent of easily accessible biological web databases such as NCBI databases and associated tools such as BLAST are key components to this development. However, with the growing number of these web based biological research tools and data sources, the time necessary to invest in becoming a domain expert is immense. Therefore, it is important to allow for easy user deployment of the wealth of available data sources and tools necessary to conduct biological research. In this paper we discuss an approach to create and maintain a robust ontology knowledge base that serves as the core for SIBIOS, a workflow based integration system for bioinformatics tools and data sources. Further, deployment of the ontology in various components of SIBIOS is discussed.

Keywords: Data integration, scientific workflows, ontologies, fault tolerance.

1 Introduction

Data integration and service discovery in the Life Sciences are key challenges that impede discoveries in biology and bioinformatics. The necessity in retrieval of available data that are generated by the technologically evolving field of biology and bioinformatics has resulted in introduction of more supporting tools. For example, a user may be interested in a particular gene such as *BRAC1 Human Gene* [1]. The user may use *GENBANK* [2], a public nucleotide sequence repository to retrieve the gene sequence. The results of this search can be given to BLAST [3] to find additional genes with similar conserved regions. The next step may be the translation of the gene sequence found into 6 reading frames by *TRANSEQ* [4] to find proteins of interest. Finally, the structure and functional motifs of the protein may need to be studied via services such as *PRINTS* [5] and *FINGERPRINTSCAN* [6] in order to find additional information related to the effects of mutations in the *BRAC1 gene* [1]. The process is known as in-silico experiment and involves a mixture of database and tools deployed in a workflow fashion. In-silico experiments take time and require sophisticated expertise from biologists due to two main reasons. In one hand, data sources and

U. Leser, F. Naumann, and B. Eckman (Eds.): DILS 2006, LNBI 4075, pp. 104 – 113, 2006.

bioinformatics tools hereafter referred as *bioinformatics services* lack a registry mechanism by which researchers are able to retrieve the services needed for their experiments, just by relying on a set of metadata for service description provided as part of service registration. As a result only a handful list of databases such as GenBank [2] and SwissProt [7], and a limited number of bioinformatics tools such as Blast [3] are used by the research community; while hundreds of other services [8] offering valuable quality data and analysis capabilities remain under-utilized. On the other hand, the distributed nature and heterogeneity of bioinformatics services at the syntactic as well as at the semantic levels render their manipulation and deployment cumbersome and time consuming. Hence researchers need to continuously work on the interoperability between services by data copying and pasting, and when necessary performing data formatting and data filtering operations. Therefore there is a great value in automating the process of service selection, service composition and service invocation when working with in-silico experiments.

Several integration systems are being proposed to assist researchers in conducting their analyses [9, 10, 11, 12, 13, 14]. These initiatives can be broadly classified under either a warehouse approach or a wrapper based approach. A global schema is used to reconcile service heterogeneity in warehouse solutions by having copies of bioinformatics services at the server hosting the integrating system. Solutions based on the wrapper approach often use ontologies as the basis for their integration solutions [10, 11, 15, 16, 17]. SIBIOS, the system for the integration of bioinformatics services falls into the latter approach [13, 15, 18].

This paper describes the main challenges faced during the design and the maintenance of SIBIOS ontology. It also describes the ontology features to support easy deployment of SIBIOS system by researchers.

Section 2 briefly describes the main features of SIBIOS. The ontology design will be detailed in Section 3. Section 4 describes how the ontology is deployed within SIBIOS system. Discussion on related and future work is presented in Section 5.

2 Overview of SIBIOS Architecture

SIBIOS operates in a distributed client-server environment in order to facilitate service selection and dynamic execution of workflows [18]. The main components of SIBIOS architecture are highlighted in figure 1. Workflow building is aided by the service composition module. The tasks of service composition are twofold: assist the user in selecting the appropriate services which compose the workflow and ensuring correct pipelining of services. Correct pipelining of services ensures that a service *s2* can be composed after service *s1* only if service s2 is able to use the output of service *s1* as input parameters. Service selection offers two options for the user to select services for the workflow. The semi-automated mode of service selection offers a step by step process where the user selects the services that are needed and assembles the workflow. In the second alternative the user will submit a high level description of the workflow that will be passed on to the automated service composition. This latter module will generate a list of potential workflows from which the user will select the most appropriate one for execution. The fault tolerance module enhances system reliability during workflow execution by the workflow enactment.

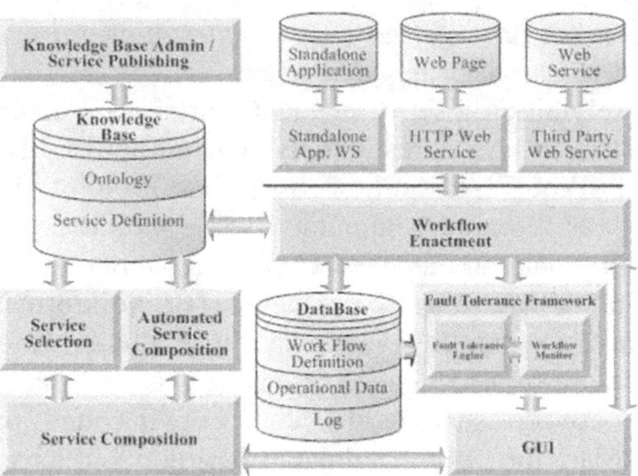

Fig. 1. SIBIOS Architecture

The ontology design and supporting applications described in the following sections are advancements to the SIBIOS system described in [13, 15, 18]. These enhancements aim into facilitating the ontology maintenance as new services are added to the system. In addition the ontology is deployed to support new components added to SIBIOS system.

3 Ontology Design

3.1 Service Oriented Approach

As in SIBIOS bioinformatics services constitute the building blocks of workflows, a service centric approach is adopted for the design of SIBIOS ontology. The purpose of the ontology is to define services at different levels so that the description can support both service composition and service invocation.

The role of the ontology at the service invocation level is to provide a common terminology to describe all input and output parameters of services. For example, by mapping one of the output parameters of Genbank [2], namely *accession,* and one of the input parameters of SwissProt [7], namely *EC*, to the ontology term *EC_number*, it is possible to compose a workflow where SwissProt service can be scheduled to execute on the input of GenBank service.

In the context of service composition, the role of the ontology is (1) to support a service description that allows researchers with different bioinformatics expertise to find the services needed for their experiments; (2) and to serve as a mapping model to ensure correct composition between services. To allow researchers to search for services beyond just browsing through a list of names, the description of services has to be enriched with metadata that can be used as the basis for formulating their service needs. For example consider a user who wants to find a service that performs multiple sequence alignment. The ontology should be modeled in a way so that it allows the user to perform an "advanced" search feature where he/she can search for services based on the task they perform.

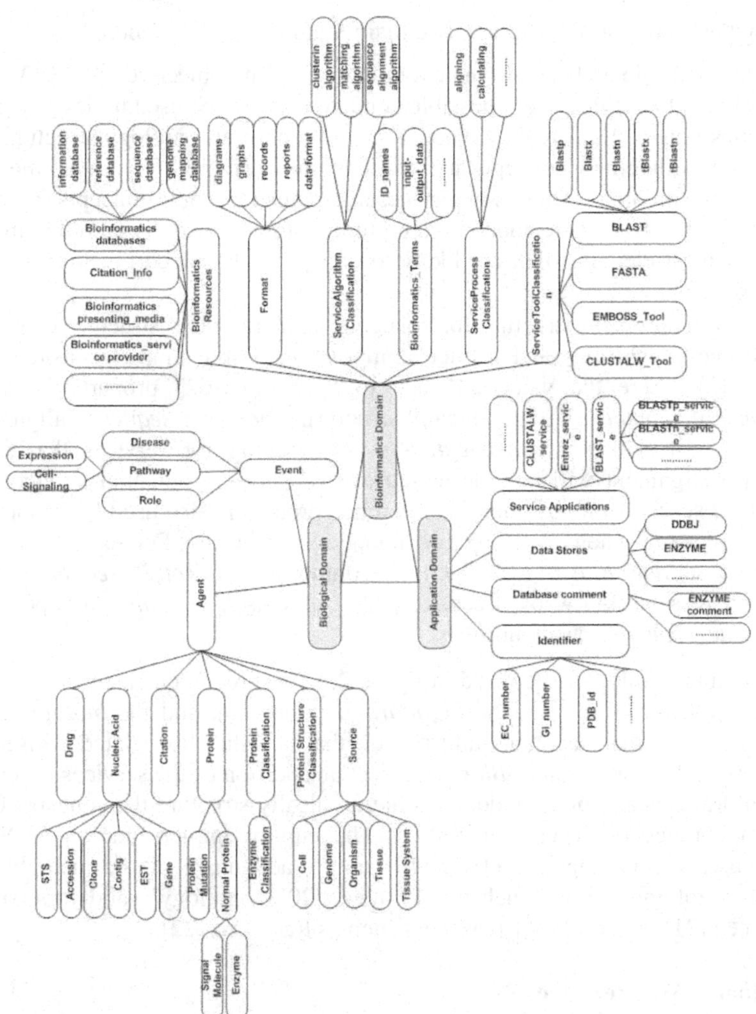

Fig. 2. Domain Ontology Hierarchy

In the context of ontologies, properties are used to describe the set of features by which a service can be searched for. Based on the study undertaken in [16, 17] classifying bioinformatics tasks; as well as our survey of both existing integration systems and a large number of services in areas such as sequence analysis and database searches, five properties emerged for service description. These properties relate to the task performed by the service, the input it accepts, the output it returns, the resources (mainly databases) it uses, and the algorithm (or function) it implements. Note that not all of these five properties will be utilized for any given service. For example SwissProt database does not use "*Function_of*" property as this property is more specific to bioinformatics tools.

During the design of the ontology two main features were considered:

- Provide a scalable and easy-to maintain ontology: The objective in SIBIOS design is clearly not to include every possible concept that will be used to describe service properties ranges. Rather, the objective is to propose an ontology structure which can accommodate new concepts and services. Furthermore, to facilitate the task of updating the ontology when new services and potentially new concepts (needed for their properties range) are added, it is preferable to organize the ontology in such a way that most elements that are likely to change be localized together within the ontology structure.
- Provide a hierarchical structure for concepts: A hierarchical structure can be used to enhance capabilities and service composition. For example, consider *Blastn* service [3] where the value assigned to "*perform_task*" property is *pairwise sequence alignment*. By hierarchically structuring *pairwise sequence* alignment as a subclass of *sequence alignment*, a user searching for services that perform sequence alignment will be able to retrieve *Blastn* as a potential service. In the context of service composition, this inference process is also useful to improve the search precision when matching two connecting services. For example, one can find that service *Blastn* for which the output is *nucleotide_sequence* can be followed in a workflow by a service *s* that accepts a *sequence*, a superclass of *nucleotide_sequence*, as input.

The resulting ontology is depicted in figure 2. with three main categories at the top level: the *application* domain, the *bioinformatics* domain, and the *biology* domain. The application domain contains information that is tightly associated with specific services or applications such as *Blastn* service; an addition of the services description. The bioinformatics and biology domains hierarchically structure the concepts used to describe the range of service properties. The current implementation of SIBIOS ontology uses OWL language OWL [19] taking advantage of the available tools including ontology editors such as Protégé [20], ontology validators such as WonderWeb [21], and ontology reasoners such as RacerPro [22].

3.2 Ontology Maintenance

SIBIOS ontology is designed as an open ontology to be able to cope with users' requirements to add new services and corresponding bioinformatics and biological concepts as they are needed for building new workflows. Service deletion is also a desirable property to have as services become obsolete over time and are replaced by more powerful ones.

Despite the availability of powerful editor tools such as protégé [20] to help users manipulate ontologies, the process of ontology update remains a complex process for scientists as several issues need to be considered. Consider for example the procedure of adding a new service using Protégé editing tool. First scientists have to define the new service as a subclass of service application domain. Comments for the new service are optional but highly recommended. In next step, service restrictions (i.e. implementation of service properties) need to be created. This is considered as the most complex part in the entire procedure. To find all the required biological and bioinformatics concepts describing the range of the properties or relationships, scientists have to check through all the unstructured/flatted concepts located at different imported ontology files. Furthermore other restrictions need to be defined in order to enable the reasoning capabilities of the ontology. A procedure inverse to the insert procedure is required for removing an existing service.

The Ontology Writer component, part of the knowledge base administration module, allows scientists to easily update/upgrade the ontology without having to be exposed to low level basic operations associated to each service update or delete as described above.

4 Ontology Deployment in SIBIOS

4.1 Service Selection

With the wide array of knowledge levels among biologists and bioinformaticians, it is necessary to allow a flexible system for service selection. Service browsing, the most common exploratory method supports classification of services by properties such as input, output, and task [17]. This approach provides a process that is adaptable to users of all levels. However, in order to provide a more robust querying interface, additional service descriptions and querying capabilities are necessary such as those supported by [12].

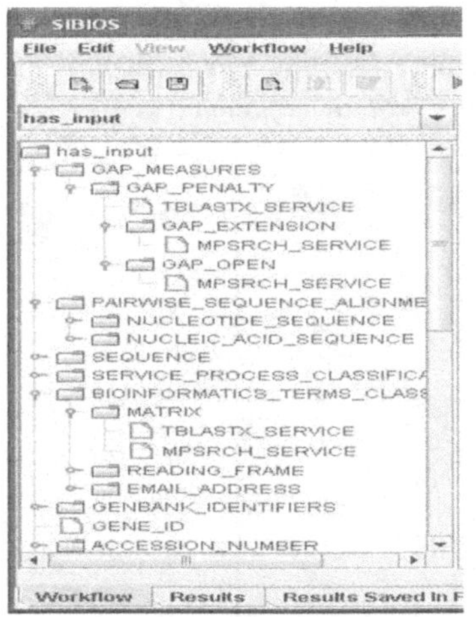

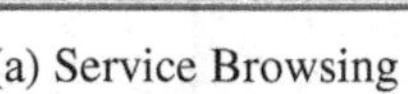

(a) Service Browsing

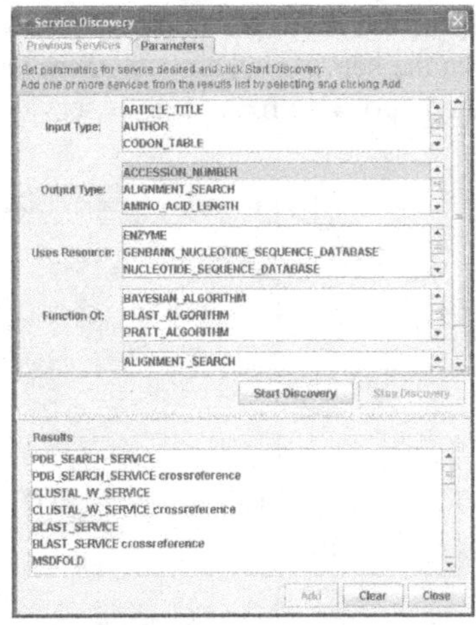

(b) Service Discovery

Fig. 3. Service Selection Interface

Service selection in SIBIOS utilizes both service browsing and advanced querying. RACERPro [22] with its reasoning capabilities is used to retrieve information from the ontology. Using browsing capabilities users can browse through the existing services categorized using either one of the available service properties. Figure 3.a shows part of the services classified using "*has_input*" property. For each property a hierarchical structure is built based on the list of biological and bioinformatics terms involved in describing the property. The names of services are attached as leaf nodes to the constructed hierarchy. Note that a service can be located at different levels of the hierarchy, if it accepts more that one parameter for the given property.

The property based classifications are built dynamically each time a user connects to SIBIOS server. SIBIOS also supports the requirements of more sophisticated users who often want to discover services by combining more than one property (e.g. service input and task performed). The example in figure 3b illustrates how SIBIOS supports the advanced query (also called service discovery). Similarly to service browsing, SIBIOS leverages on the reasoning capabilities of RacerPro to attach to each listed biological or bioinformatics concept the list of services for which the concept can be inferred as its domain.

4.2 Service Composition

Service composition is the process of connecting services into a meaningful workflow. SIBIOS ontology is deployed each time a service *s* is proposed to be scheduled after pervious services. Such composition is allowed only if a subset of the input parameters of *s* matches a subset of the output parameters of each targeted previous service. To leverage on the hierarchical structure that characterizes the description of bioinformatics and biological terms, we utilize reasoning capabilities of Racerpro [22] to support an inclusive matching rather than an exact matching between input/output parameters. For example if a previous service to s has *sequence* as output and one of the input parameters of service s is *protein sequence* then the composition between the services can occur. Similarly if the input of service *s* is *sequence*, then it should compose with a previous service that outputs *protein sequence*.

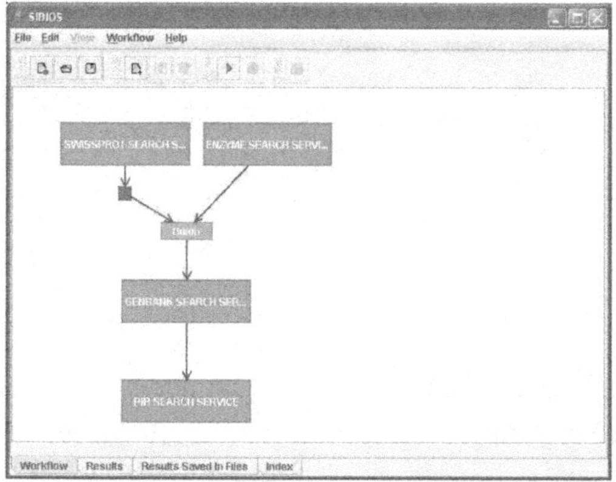

Fig. 4. An example of workflow composition in SIBIOS

To perform the service composition five connectors are provided that specify how results from previous services can be combined. These connectors are: primitive connector, UNION connector, INTERSECT connector, MINUS connector and CROSS connector. The primitive connector is used when a service *s* has only one antecedent service. The semantic of the last four connectors is compatible with the relational algebra set operators. To generate the links between services, the researcher may select the services using the browsing option; and then make use of the connectors to establish the link between services. The other option consists of using the service discovery option of service selection to perform the connection at the same time the service is selected (Figure 3.b). As a final note it is to be seen that the features offered by service browsing, service connector and service discovery can be used in tandem to build a workflow.

An example of workflow generated by SIBIOS is shown in figure 4, where Genbank [2] search service is scheduled to execute using the output generated by Enzyme [23] and Swissprot [7] services. The output of Genbank search service is fed as input to PIR search service [24]. The services were selected using service discovery and service browsing. The orchestration of the services to form the workflow was achieved using service composition of SIBIOS.

4.3 Fault Tolerance Framework

The workflow enactment module of SIBIOS is a complex engine with distributed systems collaborating with each other to achieve the goal of the scientific workflow. The role of the fault tolerance module is to increase the reliability of the workflow enactment module and therefore minimizing failures of workflow executions. A detailed description of the fault tolerance framework is beyond the scope of the paper. This section details the fault tolerance strategy that can be used when a service in the workflow fails and how the ontology is deployed for this purpose. There are currently three ways by which a service failure can be resolved in SIBIOS-the mirror service, equivalent replacement service and the nested/sub workflow, invoked in this default order to deal with service failures. The nested/sub workflow is still in research phase and will not be considered in this section.

Mirror service is the first option considered to recover from a service failure. This option provides minimal change to the initial workflow described by the user, so as for the internal workflow specification (only the service URL changes). In case the mirror service is not an option or is itself subject to failure, a replacement service strategy is considered. Service replacement utilizes the properties used to describe services namely input, output, resources, algorithm and task performed to decide the replacement of a failing service. In a perfect case, a service *s2* is elected as a placement of service *s1* if and only if the range values of each of *s2* properties "match" those of *s1* properties. For example if *s1* has *EC_Number* and *sequence_length* as input parameters, then service *s2* needs to include *EC_number* and *sequence_length* as input parameters in order to be considered as a replacement service. However, in practice different bioinformatics services are unlikely to be identical in terms of all describing properties. Hence in order to broaden the search domain to find equivalent replacement services we identified four parameters, that when combined, provide different levels for replacing a failed service. First, for each property of the failed service, we consider only the parameters used for the service invocation (e.g. *EC_number*); instead of considering all parameters used to describe the property's range. Second, we distinguish between an exact match for a property range where all parameters (e.g. *EC_Number* and *sequence_length*) used for invoking the failed service are also found in the replacement service; and a partial match where some of the parameters (e.g. *sequence_length*) may not be present in the replacement service. In terms of the service properties, we assign different weights to the four properties defining a service. For example, while the replacement service has to "match" the failed service in terms of the *input* and *output* parameters, this criterion is relaxed for properties such as the *resources* property. Finally, we use the subsumption relationship that characterizes the properties range to broaden the search for service replacement. For example, a service that has *Identifier* as input parameter will be considered as replacement for a failed service with *EC_number* as an input parameter, since the former is a superclass of the latter.

In summary, equivalent replacement service procedure relies on the ontology and the reasoning capabilities of RacerPro to search for a replacement of a failed service.

5 Discussion

Most integration systems that do not adopt a warehouse approach use ontologies as the basis for their integration solution. Ontologies are becoming popular in the domain of bioinformatics used in TAMBIS [25], BACIIS [26], BIOMOBY [9] with sub-projects MOBY-S and Semantic-MOBY, myGrid [10], and PROTEUS [11]. MOBY-S and Mygrid are most relevant to SIBIOS as they were designed during the same period when SIBIOS ontology was designed. The description of the services in terms of a set of properties is the common feature shared by the ontologies. This is used by the service selection and service composition modules. The difference that distinguishes SIBIOS from myGrid and BIOMOBY projects is that SIBIOS does not reply on the service providers or third parties to register their services. Fault tolerance strategies in scientific workflow systems have just caught the attention of researches. Fault tolerance is addressed as future work in [12] as 'Some computational environments are less reliable than others'. SIBIOS fault tolerance component uses the ontology and its reasoning capabilities to improve the reliability of the system especially when dealing with http based bioinformatics services.

Future extensions include leveraging the features of the recent semantic Web language, OWL-S [27], to unify service description currently defined at the ontology level; and service wrappers currently implemented as an XML specification. This can be achieved using OWL-S as this latter allows a multi-layer description of services that can be deployed in various components of SIBIOS.

Acknowledgements. This project was supported in part by NSF CAREER DBI-DBI-0133946 and NSF DBI-0110854.

References

1. Abel KJ, Xu J, Yin GY, Lyons RH, Meisler MH, Weber BL. Mouse Brca1: localization sequence analysis and identification of evolutionarily conserved domains. Human Molecular Genetics. 1995 Dec 4(12): 2265-73.
2. (Website) Genbank-http://www.ncbi.nlm.nih.gov/Genbank
3. (Website) Blast-http://www.ncbi.nih.gov/BLAST/

4. (Web site) Transeq, EMBOSS tool for translating DNA/RNA into protein. http://www. ebi. ac.uk/emboss/transeq/
5. (Web site) PRINTS. http://umber.sbs.man.ac.uk/dbbrowser/PRINTS/
6. Scordis P, Flower DR, Attwood TK. FingerPRINTScan, "intelligent searching of the PRINTS motif database". Bioinformatics. 1999 Oct, 15(10):799-806.
7. (Web site) SwissProt. http://us.expasy.org/sprot/
8. Galperin M., "The Molecular Biology Database Collection". 2006 update. Nucleic Acids Research, vol. 34, pp. D3-D5, 2006.
9. Wilkinson M, Schoof H, Ernst R, and Hasse D, "BioMOBY Successfully Integrates Distributed Heterogeneous Bioinformatics Web Services". The PlaNet Exemplar Case, Plant Physiol. 2005 May; 138(1):5-17. http://www.biomoby.org/
10. Wroe C, Stevens R, Goble C, Boberts A, Greenwood M, "A Suite of DAML+OIL ontologies to Describe Bioinformatics Web Services and Data". *International Journal of Cooperative Information Systems*, Vol 12, No 2, 2003.
11. Cannatoro M, Comito C, Schiavo FL and Veltri P, "Proteus, a Grid based Problem Solving Experiment for Bioinformatics: Architecture and Experiments". IEEE Computational Intelligence Bulletin. Feb 2004. Vol. 3. No. 1.
12. Ludascher B, et Al., "Scientific Workflow Management and the KEPLER System", Concurrency and Computation: Practice & Experience, Special Issue on Scientific Workflows, to appear, 2005.
13. Ben Miled Z, Gao N, Bukhres O, Lu L, Li N, He Y, Mahoui M and Chen J: SIBIOS, "A System for the Integration of Bioinformatics Services", Proc. of the Second International Workshop on Challenges of Large Applications in Distributed Environments. IEEE. 2004.
14. (Website) Pegasys-http://bioinformatics.ubc.ca/pegasys/
15. Ben Miled Z, Mahoui M, Gao N, Lu L, Chen J and He Y, "A Service Discovery Approach in Support of Web Service Integration", BIBE'04, Proc. of the 5th IEEE Symposium on Bioinformatics and Bioengineering 2004.
16. Stevens R, Goble CA and Bechhofer S, "Ontology-based Knowledge Representation for Bioinformatics". Briefings in Bioinformatics, 1(4):398-416, November 2000.
17. Stevens R, Goble CA, Baker PG, Brass A, "A classification of tasks in bioinformatics". Bioinformatics 17:1:180-188, 2001.
18. Mahoui M, Lu L, Gao N, Li N, Chen J, Bukhres O and Ben Miled Z, "A Dynamic Workflow Approach for the Integration of Bioinformatics Services", Cluster Computing Journal, pp. 279-291, 2005. http://sibios.engr.iupui.edu/
19. Jim H, Eric M: Web Ontology Language. 2004. http://www.w3.org/2004/OWL/
20. (Web Site) The Protégé Ontology Editor and Knowledge Acquisition System. http://protege. stanford.edu/
21. (Web Site) WonderWeb OWL Ontology: http://phoebus.cs.man.ac.uk:9999/OWL/Validator
22. Michael W and Ralf M, "A High Performance Semantic Web Query Answering Engine". http://www.franz.com/products/racer/
23. Website: Enzyme Search Service- http://www.expasy.org/enzyme/
24. Website: PIR-"The Protein Information Resource (PIR)". *Nucl. Acids. Res.* 2000 28: 41-44.
25. Stevens R, Baker PG, Bechhofer S, Ng G, Jacoby A, Paton NW, Goble CA and Brass A, "TAMBIS: Transparent Access to Multiple Bioinformatics Information Sources", Bioinformatics, 16:2 PP.184-186, 2000.
26. Ben Miled, Z., Webster, Y., Li, N., Liu, Y., "An Ontology for the Semantic Integration of Life Science Web Databases," International Journal of Cooperative Information Systems, Vol. 12, No. 2, June 2003.
27. (Web Site) DAML Services – Tools, http://www.daml.org/services/owl-s/tools.html

Data Structures for Genome Annotation, Alternative Splicing, and Validation

Sven Mielordt[1,*], Ivo Grosse[1], and Jürgen Kleffe[2]

[1] Leibniz Institute of Plant Genetics and Crop Plant Research (IPK), 06466 Gatersleben, Germany

[2] Charité-Universitätsmedizin Berlin - Campus Benjamin Franklin Institut für Molekularbiologie UND Bioinformatik, Arnimallee 22, 14195 Berlin, Germany

mielordt@ipk-gatersleben.de, grosse@ipk-gatersleben.de, juergen.kleffe@charite.de

Abstract. To establish a clean basis for studying alternative splicing and gene regulation in life science projects, a powerful data modeling and also a strict validation procedure for assigning levels of reliability to given gene models is essential. One common problem of public genome databases are insufficiently organized and linked description data, which make it difficult to study relations of the alternative isoforms of a gene that are relevant for medicine and plant genome research. This is a severe obstacle for the integration of biological data and motivated us to establish a new modeling instance and that we call splice template or sTMP. Every sTMP has a unique splicing pattern, but the length of the first and the last exon remains undefined. This allows to model different gene isoforms with the same splicing pattern. By utilizing this more fine-grained data structure, many cases of plurivalent mRNA-CDS relations are uncovered. There are more than 3,000 extra CDSs in the human genome compatible with the categories sTMP, mRNA and CDS, which exceed the classical one-to-one relations of mRNAs and CDSs. In one case, 11 extra CDSs are compatible with one mRNA. Crosslinks between mRNAs derived from different sTMPs leading to the same CDS are now accessible as well as disease-related ruptures in UTR regions. This allows discovering and validating disease and tissue specific differences in alternative splicing, gene expression and regulation. Another problem in public databases is a too much relaxed standard for labeling genes "confirmed by ESTs and full-length-cDNAs." We provide a pipeline that handles gene annotations from different sources, integrates them into complex gene models and assigns strict validation tags, constrained by a local low-error model for the alignments of genome annotation and transcripts. The data structures are being implemented and made publicly available at the Plant Data Warehouse of the Bioinformatics Center Gatersleben-Halle (http://portal.bic-gh.de/sTMP).

Keywords: Gene and genome annotation, alternative splicing, data integration, splice template, validation and confirmation, quality control, Fasta-XML format.

[*] Corresponding author.

U. Leser, F. Naumann, and B. Eckman (Eds.): DILS 2006, LNBI 4075, pp. 114 – 123, 2006.

1 Introduction

Gene annotation and prediction is still a challenging task. On average less than 40% of the *ab-initio* gene predictions for Arabidopsis are error-free and hence genes with complete full-length cDNA support are important for testing and training gene prediction software [1]. Studying alternative splicing and gene regulation in animal and plant genome projects not only demands a strict validation procedure, but also a powerful meta-data modeling. Insufficiently organized and linked description data makes it difficult to express and uncover relations between the alternative isoforms of a gene. This is a severe obstacle for the integration of genome data.

The EnsEMBL and TIGR-XML Annotation. EnsEMBL [2] and TIGR-XML [3] model protein-coding genes on the genomic DNA as fixed hierarchical tree structures as shown in Fig. 1. A gene locus may have one or more splice isoforms (mRNAs). Each mRNA splits into the protein coding CDS region and the two untranslated regions 5'UTR (upstream) and 3'UTR (downstream). This topology cannot deal with alternative start codons for the same mRNA or other alternative mRNAs with the same splicing pattern, such as mRNAs with alternative transcription start sites or alternative polyadenylation sites. Moreover, there are no crosslinks given between the different CDSs of a gene, although these would be instructive, since often alternatively spliced mRNAs differ only in the UTR regions but lead to the same CDS and therefore code for the same protein.

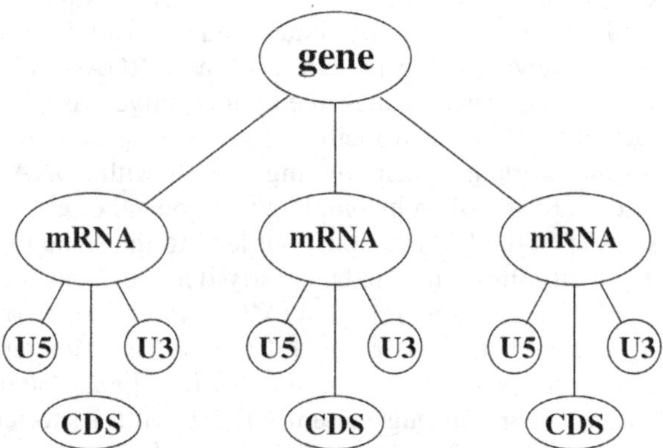

Fig. 1. The EnsEMBL and TIGR-XML entities model protein-coding genes on the genomic DNA as fixed hierarchical tree structures

The GenBank (RefSeq) Annotation. In contrast, NCBI [4] GenBank (RefSeq) annotations do not care for the relationship between mRNAs and CDSs and are therefore more general. However, they provide only unrelated lists of mRNAs and CDSs for each gene, as can be seen in **Fig. 2**. The users are left alone to build and run their own programs for matching mRNAs and CDSs in order to find the relations between mRNAs and CDSs.

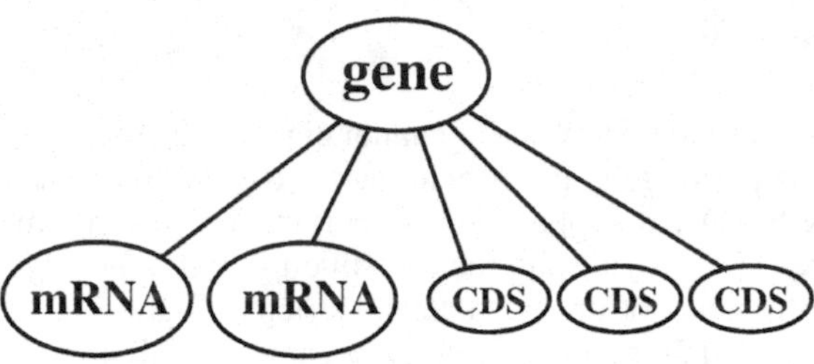

Fig. 2. GenBank (RefSeq) annotations do not take into account the relationship between mRNAs and CDSs and therefore provide only unrelated mRNA and CDS lists

The present situation motivated us to develop a better model or data structure. It provides a network graph instead of a tree and offers direct access to problems like "which mRNAs lead to the same CDS" or "which mRNAs are likely to be disease-related due to ruptures or extensions in UTR regions" or "which Arabidopsis splice template belongs to a specific barley EST."

2 Results and Discussion

Splice Templates: A Data Structure for Genes and mRNAs. We propose a new data structure for a gene locus and its mRNAs that we call splice template or sTMP, shown as a toy example in Fig. 3. Every sTMP has a unique splicing pattern, but the length of the first and the last exon remains undefined. This allows modeling different gene isoforms with the same splicing pattern, such as mRNAs with alternative start codons , alternative transcription start sites due to alternative promoters, or alternative polyadenylation sites. The sTMP also facilitates uncovering new or prolonged UTR regions for known splice variants when aligning sTMPs with cDNA transcripts. This is important, because UTRs are often incompletely annotated or even missing.

If mRNAs only differ in the UTR regions, but lead to the same CDS and therefore code for the same protein, this can now be clearly derived from the developed data model. In the example presented in **Fig. 3**, sTMP 1 splits into two mRNAs, which differ only by the length of the first exon. mRNA 1 has a longer first exon than mRNA 2 and therefore mRNA 1 can be translated into two proteins by using two alternative start codons, giving a long variant CDS 1 with a shorter 5' UTR and a short CDS 2 with a longer 5' UTR. mRNA 2 only encodes the short CDS 2, because the first exon contains only the downstream ATG. mRNA 3 is like mRNA 2, but differs in the 3'UTR region. Thus, it encodes the same protein (with the same CDS 2) as mRNA 2. Differences in the 3'UTR region can influence the in-vivo stability of the mRNA, and very long 3'UTRs make the mRNA subject to "nonsense-mediated-mRNA decay" (NMD) degradation [5]. Frequent alternative usage of 3'UTR introns and gene models with up to ten such introns might be due to a number of disease related transcripts showing cases of disrupted NMD.

Noncoding and regulatory RNAs are important and move more and more into the focus of research efforts. These are small interfering RNAs (siRNAs), "misc_RNAs," and many other types like snoRNA genes, some of which are polycistronic. In the developed data model, we subsume them all as "regulatory RNA" or regR. **Fig. 3** shows two of them under mRNAs 3 and 4 in the example and they easily fit into the data structure.

Additionally, we connect to orthologous and paralogous genes via similarity or identity links, depending on the level of similarity. Combined with an ontology database, this settles our gene network data structure for complex queries. The Plant Data Warehouse ontology database fosters the investigation of gene families and subnets of interest to the users.

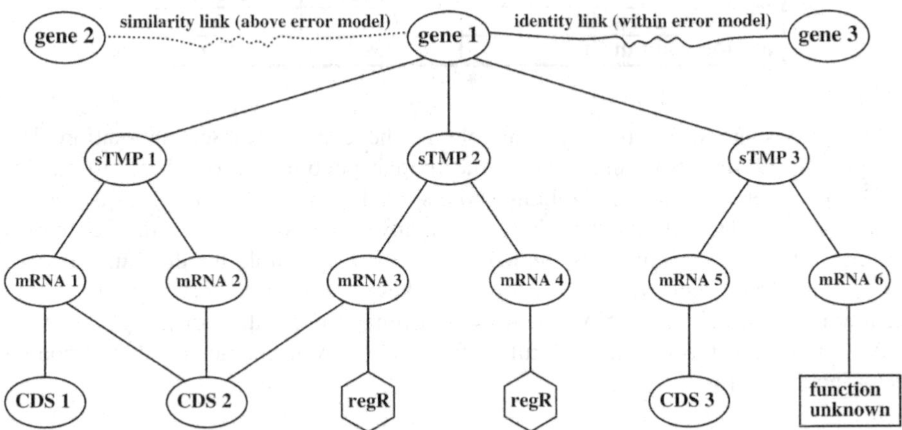

Fig. 3. Every sTMP has a unique splicing pattern, but the length of the first and the last exon remains undefined. The data structucture makes it possible to model different gene isoforms with the same splicing pattern. Crosslinks between genes refer to homologous genes. Noncoding and regulatory RNAs easily fit into the data structure.

Application of Splice Templates. To estimate how many genes have plurivalent or missing mRNA-CDS relations, we analysed the human genome, NCBI RefSeq 35.1 build. As shown in **Table 1**, there are 26796 genes, 27542 mRNAs, and 28063 CDSs annotated. There is no compatible mRNA for 674 CDSs; in these cases, we generate an extra mRNA that is identical to the CDS, leading to 28216 mRNAs overall. There are also 199 mRNAs without a compatible CDS. In such cases, we assign the tag "function unknown." In total, 28181 sTMPs are required as upper nodes for all mRNAs and CDSs.

Many cases of plurivalent mRNA-CDS relations are uncovered by the sTMP data structure as shown in **Table 2**. There are more than 3,000 extra CDSs compatible with the categories sTMP, mRNA, and CDS, which exceed the classical one-to-one relations between mRNAs and CDSs. In the most outstanding case, 11 extra CDSs are compatible with one mRNA. Specifically, 12 CDSs, which differ only in the position of the start codon, are compatible with one mRNA. Hence, these transcripts have the same splicing pattern, but differ in the functional assignment of UTR and CDS.

Table 1. RefSeq genes and isoforms in the human genome

Human Genome	# of items	
	RefSeq	revised
genes	26796	26796
sTMPs	0	28181
mRNAs	27542	28216
CDSs	28063	28063
CDSs without mRNA	674	0
mRNAs without CDS	199	199

RNA genes are only sparsely available in the current RefSeq annotation. For example, all 79 snoRNAs annotated in the human genome are on chromosome 15. Since from other species (e.g. Arabidopsis) a spread over all chromosomes is known, we speculate (oder expect) that there are many more such genes in the human genome, but that up to now only the team that was responsible for the annotation of chromosome 15 considered them worthwhile. These RNAs are involved in processing and modification of other RNAs, such as ribosomal and small nuclear spliceosomal RNAs. SnoRNAs form a large family of relatively well-characterised non-coding RNAs (ncRNAs) [6].

Table 2. Plurivalent relations in the human genome

Human Genome	number of items		
		for all	
	sTMPs	mRNAs	CDSs
extra CDSs	3115	3079	3077
extra mRNAs	117	80	3044
		max. for one	
	sTMP	mRNA	CDS
extra CDSs	8	8	11
extra mRNAs	3	4	11

2.1 Splice Templates and the Quality of Annotations

While collecting and validating large sets of genes, we learned that gene annotations in the public databases are often unreliable. Gene annotation databases usually provide the accession numbers of supporting ESTs and full-length-cDNAs, but no detailed alignments. However, these would be important for the annotators of the public databases and help to increase the relaxed standards currently in use. For example, the NCBI handbook [4] suggests allowing a global error rate of up to 5% and requires only 50% or a coverage of 1000 bp of the gene model in order to assign the label "confirmed."

Moreover, the currently used error models do not take into consideration local clusters of errors. We think that sufficiently spaced single nucleotide mismatches or gaps can be tolerated in larger numbers than dense blocks of errors, which might indicate alternative splice sites or a wrong gene annotation. Here we propose a stricter standard by accepting a global error rate of not more than 1% and at most two errors within a sliding window of length 20 bp. Such almost perfect matching is required to distinguish NAGNAG acceptor isoforms, which differ only by a single codon. Their subtle effect on mRNA and protein structure is often overlooked but may have a great impact on biology and disease [7].

Confirmed untranslated regions (UTRs) are important for improving promoter recognition and our understanding of gene regulation. These elements were not incorporated in previous training-sample oriented studies, but we include them now.

2.2 Differences of RefSeq and EnsEMBL Annotations

The human NCBI RefSeq GenBank annotation version 35.1 and the EnsEMBL version hg17 both rely on the same genomic assembly from May 2004. Surprisingly, the annotation of genes and mRNAs differs in many cases. Although using the same genomic template, many mRNAs are annotated with different spliced alignment patterns in both databases, leading to incompatible solutions. Another important annotation source is the AltSplice database from EBI [8]. It is based on the EnsEMBL version tiling path and can be seen as a revised and improved version for many genes. Therefore, we compare the RefSeq and the AltSplice annotation.

One of the standard techniques for gene annotation is a spliced alignment of ESTs and cDNAs [9], [10] against the genomic template as shown in **Fig. 4**. It is slow and often unreliable in finding short exons, long introns, the true start codon, or the correct strand of single exon genes. In contrast, the sTMP data structure allows a simple query that checks existing gene annotations by in-silico splicing and aligning the resulting sTMPs with full-length cDNA. It allows to quickly find all low-error matching mRNA-cDNA pairs for whole genomes by calculating seed triggered [11] branch-and-bound k-band alignments. This was done with the separately downloaded RefSeq transcripts as cDNA database against the sTMPs derived from the RefSeq genome or the AltSplice annotation.

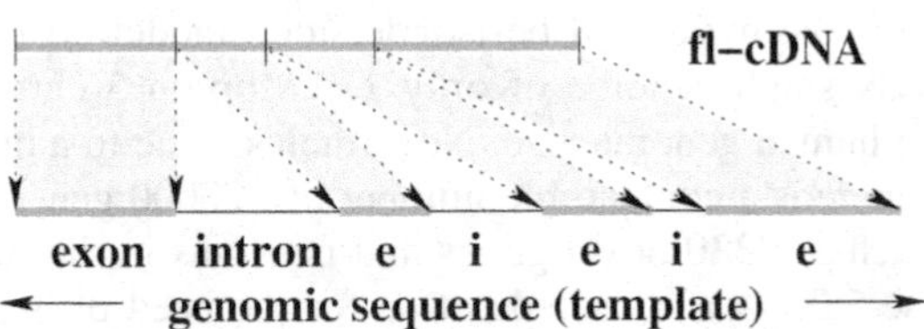

Fig. 4. Spliced alignment of an EST or full-length cDNA ("fl-CDNA") shows exons and introns on the genomic DNA sequence

As shown in **Table 3**, 6,614 conflicts exist between the RefSeq and the AltSplice annotation. Most of them consider only a few nucleotides at the boundaries of the first or the last exons, but 1,660 are severe exon count conflicts, meaning that the spliced alignment solution differs in the number of exons. Generally, the average quality of the AltSplice database is higher. 1,056 AltSplice models and 3559 RefSeq models exceed a global error threshold of 1%. 3307 AltSplice models and 5829 RefSeq models exceed the local error threshold of 2 errors in a sliding window of length 20 bp. These annotations should be furthermore scrutinized in order to correct a possibly wrong spliced alignment.

Table 3. Conflicts in the RefSeq and AltSplice human genome annotation

conflicts between refSeq and AltSplice	6614
exon count conflicts between refSeq and AltSplice	1660
numberOfBadGlobalErrorRate_AltSplice	1056
numberOfBadGlobalErrorRate_refSeq	3559
numberOfBadLocalErrorRate_AltSplice	3307
numberOfBadLocalErrorRate_refSeq	5829

It shall be emphasized that these conflicts are artefacts of different annotation tools, lazy alignment rules and insufficient data structures. By theory, such should never happen.

2.3 Finding Novel Genes and Isoforms

The splice templates might also be useful for uncovering novel splice isoforms. Often transcripts are mapped with a high local but low global error rate to known splice templates, which indicates minor but important deviations. Because the alignment and refinement procedure between a splice template and a transcript is very fast, we can iterate the algorithm as long as needed to find the true isoform and map it onto the genomic DNA. The results are new splice templates and mRNAs for known gene loci.

2.4 Novel Genes and Isoforms in the Arabidopsis Genome

Arabidopsis thaliana is the major and best-understood model organismn in the plant world with a relatively small genome of only 119 Mbp on 5 chromosomes. This is only about 3% of the human genome size. Nevertheless, due to a higher gene density and smaller introns, there is a comparable number of ~28000 genes. Surprisingly, our pipeline found as much as 2330 novel genes and isoforms in the *Arabidopsis* TIGR-XML genome release 5.0 — a clear indication for the need of improved annotation pipelines. The data structures are being implemented at the Plant Data Warehouse [12] of the Bioinformatics Center Gatersleben-Halle (http://portal.bic-gh.de/sTMP).

2.5 The Fasta-XML Format – Combining the Advantages of XML and Fasta

The advantage of the XML format is its highly structured descriptive power. This makes it perfectly machine readable, and the validity of the syntax is checkable by means of a DTD definition file. Unfortunately, other aspects make the format unfavorable for many purposes: files become inflated in size and are difficult to read for humans. Moreover, DOM parsers easily run out of memory when constructing the complete data tree for huge input files (try parsing the 100 Mbyte TIGR XML version 5.0 file with the Arabidopsis chromosome 1 on a 32-bit PC with 2 GB of memory).

The advantage of the Fasta format is its simple structure as a flat file for one or multiple sequences. A single line of description after the '>' symbol preceding each sequence gives a very compact format, which is widely used as input and output for many bioinformatics tools. The disadvantage is its lack of structure in the description line.

Our solution is splitting the XML data into four Fasta format files for genes, sTMPs, mRNAs, and CDSs. Only the locally relevant part of the well-formed XML-like description data is provided in a single line followed by the sequence itself. The user can easily parse the relevant part. A toy example is presented in appendix The Fasta-XML Descriptor Format, and a detailed description can be found at http://portal.bic-gh.de/fasta-xml/.

We provide strictly validated ready-to-use spliced gene isoforms (i) as a data source for alternative splicing, (ii) as large training samples for training and benchmarking gene prediction software, and (iii) for an efficient detection of novel gene isoforms.

We also provide a converting tool, which transforms traditional XML and the Fasta-XML format. The data structure can also serve for updating ER models or for adding information to classical GenBank format files.

3 Conclusions

To establish a clean basis for studying alternative splicing and gene regulation in life science projects, a powerful data modeling and also a strict validation procedure for assigning levels of reliability to given gene models is essential. We introduce a new data structure that we call splice template or sTMP. Every sTMP has a unique splicing pattern, but the length of the first and the last exon remains undefined. This allows to model different gene isoforms with the same splicing pattern. Crosslinks between mRNAs derived from different sTMPs but leading to the same CDS become accessible as well as disease-related ruptures and extensions in UTR regions. We present a pipeline that handles gene annotations from different sources, integrates them into complex gene models, and assigns strict validation tags. This allows uncovering and validating disease and tissue specific differences in alternative splicing, gene expression, and regulation. The XML-Fasta format combines the advantages of XML and Fasta. All data are publicly available at http://portal.bic-gh.de/sTMP/.

Acknowledgements. We thank Friedrich Möller for valuable discussions, three unknown reviewers for valuable comments, and the German Ministry of Education and Research (BMBF Grant No. 0312706A) for financial support.

References

1. Haas,B.J., Volfovsky,N., Town,C.D., Troukhan,M., Alexandrov,N., Feldmann,K.A., Flavell,R.B., White,O. and Salzberg,S.L. Full-length messenger RNA sequences greatly improve genome annotation. Genome Biology 2002, 3(6):research0029.1–0029.12
2. EnsEMBL/UCSC Golden Path gene annotation. URL: http://genome.ucsc.edu/goldenPath/
3. TIGR (2004). The Arabidopsis thaliana genome TIGR/NCBI revision 5.0 from Frebruary 19, 2004. URL: http://www.ncbi.nlm.nih.gov/mapview/map_search.cgi?taxid=3702
4. NCBI (2004-2006). URL: http://www.ncbi.nlm.nih.gov/
5. Schell,T., Kulozik,A., Hentze1,M.W. Integration of splicing, transport and translation to achieve mRNA quality control by the nonsense-mediated decay pathway. Genome Biology (2002), doi:10.1186/gb-2002-3-3-reviews1006.
6. Scottish Crop Research Institute (2004). Computational Biology http://bioinf.scri.sari.ac.uk/cgi-bin/plant_snorna/introduction (snoRNAs)
7. Hiller,M., Huse,K., Szafranski,K., Jahn,N., Hampe,J., Schreiber,S., Backofen,R., Platzer,M. Widespread occurrence of alternative splicing at NAGNAG acceptors contributes to proteome plasticity. Nature Genetics 36, 1255 - 1257 (2004)
8. Thanaraj T. A., Stamm, S., Clark, F., Riethoven J.-J., Le Texier, V., Muilu, J. ASD: the Alternative Splicing Database Nucleic Acids Research, 2004, Vol. 32, Database issue D64-D69 (2004-2005)
9. Usuka,J., Zhu,W., Brendel,V. (2000) Optimal spliced alignment of homologous cDNA to a genomic DNA template. Bioinformatics 16, 203-211.
10. Kent, W. James. BLAT—The BLAST-Like Alignment Tool. Gen. Res. 12:656–664 (2002)
11. Kleffe,J., Möller,F., Wessel,R., Wittig,B. (2006). Identification of perfect matches in large sets of sequences, submitted. (ClustDB)
12. Grosse,I., Funke,T., Kuenne,C., Neumann,S., Stephanik,A., Thiel,T., Weise,S. Integrative Datenanalyse mit dem Plant Data Warehouse Vorträge für Pflanzenzüchtung, 70:50-53, 2006.

Appendix: The Fasta-XML Descriptor Format

Since some plurivalent relations imply circles and therefore convert the tree into a graph, an extension to standard XML is needed. This is the reason for using extra enumerating tags for the mRNA and CDS layer. Due to space restrictions, only a very simplified sketch can be presented here. For details, please visit http://portal…/fasta-xml/. The four model layers usually give four Fasta-XML files but can also serve for supplementing GenBank files or for establishing ER models.

1) gene layer: the core element specifies the unspliced genomic DNA, extended by some 5,000 bp extra sequence (specified by the o5 and o3 elements for upstream and downstream sequence. Multiple contig sources can be given, and differences of the sources may be denoted. The offset 'off' element is only needed here and for the sTMP (next node layer) to give the start on the genomic contig. A negative offset value applies for genes on the reverse strand. The following toy example stands for GeneID 123456 on the Homo sapiens chromosome 21:
<gene>Hs::21::123456<o5>5000</o5> <core>1194156</core> <o3>5000</o3>

<src1>NCBI::35.1<con>NT_011512.10<off>15646575</off></con><id>123456</id><name>TPTE89</name></src1><src2>GoldenPath::17<con>chr3<off>15681575</off></con><id>ENSG000000514241</id><diff1>126674a-,785144ac</diff1> </src2></gene>

2) splice template layer: the core element specifies the spliced genomic DNA, extended by some 5,000 bp extra sequence (specified by the o5 and o3 elements for upstream and downstream sequence). Exon and intron lengths are also given: <sTMP>Hs::21::123456::1<off>0</off><o5>5000</o5><core>924<ex>347,91,178,116,202</ex><in>245,34778,572,99</in></core><o3>5000</o3> </sTMP>

3) mRNA layer: the syntax is quite similar to the splice templates, but needs additional tags to handle plurivalence. There is no offset needed, because the sequence is the same as the splice template: <mRNA>Hs::21::123456::1::1<o5>5000</o5><core>924<ex>347,91,178,116,202</ex><in>245,34778,572,99</in></core><o3>5000</o3></mRNA>

4) CDS layer: exons lengths with dots mean one and the same exon for both adjacent parts in neighboring regions. Example: the first exon of length 91 bp in the mRNA has a part of 88 bp in the u5 and a part of 3 bp in the core (only the ATG start codon): <CDS>Hs::21::123456::1::1::2,2,2,2<o5>5000</o5><u5>435<ex>347,88.</ex><in> 245</in></u5><core>399<ex>.3,178,116,102.</ex><in>34778,572,99</in></core> <u3>100<ex>.100</ex></u3><o3>5000</o3></CDS>

An intron with a bracketed length of 34,778 bp exactly separates u5 (5'UTR) and core (CDS). Here the 5'UTR is 3 bp longer and the CDS is 3 bp shorter than in CDS#2. An alternative start codon at the beginning of the next exon with a length of 178 bp is used:<CDS>Hs::21::123456::1::1::3,3,3,3<o5>5000</o5><u5>438<ex>347,91</ex> <in>245,(34778)</in></u5><core>396<ex>178,116,102.</ex><in>(34778),572,99</in><core><u3>100<ex>.100</ex></u3><o3>5000</o3></CDS>

BioFuice: Mapping-Based Data Integration in Bioinformatics

Toralf Kirsten and Erhard Rahm

University of Leipzig, Germany
tkirsten@izbi.uni-leipzig.de, rahm@informatik.uni-leipzig.de

Abstract. We introduce the BioFuice approach for integrating data from different private and public data sources and ontologies. BioFuice follows a peer-to-peer-like data integration based on bidirectional mappings. Sources and mappings are associated with a domain model to support a semantically meaningful interoperability. BioFuice extends the generic iFuice integration platform which utilizes specific operators for data fusion and workflow-like script programs. BioFuice supports explorative data analysis and query and search capabilities. We outline the integration approach by an illustrating scenario, the architecture of BioFuice and its query interface.

1 Introduction

Many biological and medical applications require access to a variety of molecular-biological objects, such as genes, proteins, their interrelationships and functions, and their correlations with phenotypical effects. These objects are maintained in a high number of diverse web-accessible data sources [Ga05] as well as in local (private) data sources, e.g. specific analysis results such as a particular list of genes or medical data on patients participating in clinical trials. Typically, such data is highly diverse so that their integration is laborious and error-prone and difficult to perform by domain experts.

Traditional data integration approaches like data warehousing and mediators are often applicable but also time-consuming to deploy and may lack sufficient support for features such as explorative data analysis. These integration approaches typically require a unified global schema to obtain a consistent view over data from different sources. However, creating such a schema for more than a few data sources is almost impossible due to the high diversity, complexity and fast evolution of sources. Each new source to consider may require adapting the global schema as well as applications built upon this schema.

A promising alternative to the traditional data warehousing and mediator solutions using a global schema are so-called peer-to-peer approaches for data integration [Ha03]. They are based on bilateral mappings between autonomous data sources, called data peers, instead of mappings between data sources and a global schema. Adding a new data source can thus be achieved by mapping it to only one existing peer instead of adapting the global schema and mapping the source to it. In bioinformatics, a peer-to-peer approach seems especially appropriate since bilateral mappings can often be derived from existing cross-references between objects of different

U. Leser, F. Naumann, and B. Eckman (Eds.): DILS 2006, LNBI 4075, pp. 124 – 135, 2006.

sources. Such cross-references refer to so-called accessions, i.e. unique object identifiers, and are omnipresent in public data sources. The cross-references are typically maintained by domain experts and thus of high quality. However, they are currently used mostly for manual web navigation which is unsuitable for evaluating large sets of objects, e.g. for gene expression analysis. Moreover, the semantics of the cross-references is typically not made explicit making it difficult for the user to find and correctly use all relevant sources and mappings for a given application task.

iFuice (information Fusion utilizing instance correspondences and peer mappings) [Ra05] is a recently proposed approach for peer-to-peer data integration. It utilizes mappings, e.g. sets of cross-references, to combine or fuse information from different sources. Sources and mappings are related to a domain model to support semantically meaningful information fusion. The iFuice architecture incorporates a mapping mediator offering both interactive and script-driven, workflow-like access to the sources and their mappings. The script programmer can use powerful generic operators to execute and manipulate mappings and their results. iFuice is a generic data integration approach which is not targeted for a specific application domain. An initial use case of iFuice was to combine bibliographic data for a citation analysis of database publications [Ra05, RT05].

In this paper we describe how iFuice and its extension BioFuice can be used for data integration in bioinformatics applications. Key characteristics of BioFuice include:

- *Peer-to-peer integration:* By following the iFuice paradigm BioFuice aims at utilizing instance-level cross-references which already exist, e.g. as web links, or can be generated by bioinformatics tools, such as BLAST. New sources can be dynamically integrated as needed by mapping the new source to (at least) one already integrated source.
- *Semantic integration:* To address semantic integration, BioFuice utilizes a high-level domain model containing domain-specific object types and mapping types. The domain model is used to categorize specific sources and mappings so that they can be selected and accessed according to current application requirements.
- *Comprehensive query capabilities:* BioFuice utilizes the high-level operators and scripting facility of iFuice to perform data access, mapping execution and data fusion. This infrastructure makes it possible to react to new application needs and to support complex data integration and analysis workflows. BioFuice substantially extends the generic iFuice facilities by providing a graphical query interface for explorative analysis and automatically generating script programs from interactively specified queries. Both predefined queries as well as keyword searches are supported.
- *Local data sources:* BioFuice integrates both public and local (private) data sources. In particular, query and script results or copies of entire sources may be stored within a local database for later reuse. BioFuice can also be operated in an offline mode (e.g. on a notebook) by only evaluating local data sources.

The rest of the paper is organized as follows. In the next section we introduce the basic idea of the BioFuice approach by using an illustrating scenario. We also outline selected high-level operators and their usage. In Section 3, we introduce the BioFuice system architecture. Section 4 describes the interactive query and search capabilities for explorative analysis. We discuss related work in Section 5 before we conclude in Section 6.

2 Illustrating Scenario

To illustrate our data integration approach we consider an analysis task on human expressed sequence tag (EST) sequences. Typically, such ESTs are short DNA sequences of a specific organism that are generated by sequence machines. We assume that the analysis application needs to classify EST sequences into three classes which are represented by the following queries:

Query 1 (Q1): Return all unaligned EST sequences of a given set, i.e. EST sequences for which no corresponding DNA sequence can be found.

Query 2 (Q2): Return all EST sequences of a given set that are associated with protein-coding DNA sequences.

Query 3 (Q3): Return all EST sequences of a given set that are associated with non-coding DNA sequences.

Such a classification is a typical data integration problem since the given set of EST sequences has to be combined with further molecular-biological data on genes and proteins to decide which sequences fall into which class.

Figure 1a shows four (physical) sources and associated mappings we want to use for this scenario within a so-called *source mapping model (SMM)*. There are three public data sources (Ensembl [Bi04], NetAffx [Li03], SwissProt [Bo03]) and one local data source, MyEstSet, specifying the EST sequences of interest. A *physical data source (PDS)*, e.g. Ensembl, may offer objects of different types. We call the object types of one PDS the *logical data sources (LDS)*. The notation <ObjectType>@<PhysicalDataSource> is used to denote a specific LDS, e.g. Gene@Ensembl or Protein@SwissProt. Each LDS has an identifying attribute (accession) plus additional attributes.

Object types of any source are represented in the abstract domain model (Figure 1b). Each mapping between source instances has a mapping type which is also represented in the domain model. For example, mappings of type GeneCodedProteins relate gene instances to their associated proteins. The domain model is used to semantically categorize data sources and mappings and at a much higher conceptual level than a global schema. We do not include attributes for object types to accommodate a large variety of data sources and to make it easy to construct the domain model. In many cases, we expect a small set of object and mapping types to be sufficient.

New sources can be flexibly included by adopting relevant metadata, i.e. the physical source name and its associated object type (building the new LDS) as well as definitions for querying and searching within the source. In addition, at least one mapping should be defined to connect the new LDS to an existing LDS so that the new LDS can be used together with others. The domain model is automatically adapted to the changes made within the source mapping model.

Mappings can often be represented by sets of cross-references between objects/instances of different LDS. For instance, the mapping between Gene@Ensembl and Protein@SwissProt can be derived from the existing SwissProt references in the Ensembl gene instances. Alternatively, mappings can be derived on demand by executing queries or a program (script). In our example, the private source MyEstSet contains ESTs that are only described by a sequence. Hence, neither the PDS MyEstSet nor the public PDS Ensembl provide correspondences between the instances they offer. In this case, a BLAST[1]-like tool can be used to determine for a set of EST sequences the most similar DNA sequences within Ensembl. This creates a mapping between EstSequence@MyEstSet and SequenceRegion@Ensembl thereby integrating the local source into the peer-to-peer network represented by the SMM. Special mapping types are so-called *same-mappings* interrelating semantically equivalent instances of the same object type. In Figure 1a there is one same-mapping on genes between Ensembl and NetAffx.

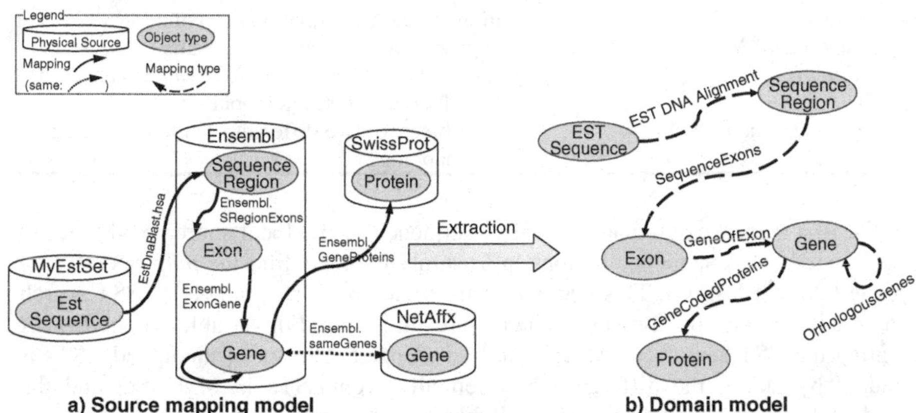

Fig. 1. Data integration scenario

To process data and mappings, iFuice and thus BioFuice offer a set of high-level operators which can be combined within script programs. Table 1 shows a selection of these operators that are relevant for the examples in this paper; the full definitions are given in [Ra05]. The operators typically operate on a set of input objects, e.g. an entire LDS, and generate a set of output objects which can be used as the input of further operators. In Table 1, OI denotes a set of object instances from one object type; objects are identified by their ids which are assumed to also identify the LDS the objects belong to.

To solve the EST classification problem posed in the beginning of this section we can use the following simple script determining three sets of EST sequences:

```
$alignedEstMR:=map(MyEstSet,{EstDnaBlast.hsa});
$unalignedEstOI:=diff(MyEstSet,domain($alinedEstMR));
$codingEstMR:=compose($alignedEstMR,
                map(range($alignedEstMR),{Ensembl.SRegionExons}));
$proteinCodingEstOI:=(domain($codingEstMR));
$nonCodingEstOI:= diff (domain($alignedEstMR), $proteinCodingEstOI);
```

[1] BLAST stands for Basic Local Alignment Search Tool [Al90].

Table 1. Selected iFuice script operators (OI = set of object instances)

Operator	Description
OI:=queryInstances(LDS, query condition)	executes a query on the specified LDS and returns object instances (OI) meeting the query condition
OI:=searchInstances(LDS, {keywords})	executes a keyword search on the specified LDS and returns object instances containing at least one of the specified keywords
OI:=getInstances(OI)	returns complete instances for objects identified by their id values
OI:=traverse(OI,{mapping names})	traverse specified mappings on input instances; multiple mappings are automatically composed
OI:=traverseSame(OI,PDS)	traverse same mappings to target PDS
MR:=map(OI,{mapping names})	returns a mapping result MR (mapping table) that associates each input object to the corresponding output objects by executing specified mappings
OI:=domain(MR)	returns the domain (input objects) of a MR
OI:=range(MR)	returns the range (output objects) of a MR
MR:=compose(MR,MR)	composes two given mapping results
OI:=diff(OI,OI)	returns the difference set of object instances between the first and second input set
AO:=aggregateSame(OI,PDS)	fuses objects of two different PDS interrelated by a same mapping

In the first step, we associate each EST sequence of the local source MyEstSet to the associated DNA sequence regions in Ensembl by executing the *map* operator on mapping EstDnaBlast.hsa. This mapping was created by performing a BLAST search during the integration of MyEstSet. The map result is stored in variable *alignedEstMR* indicating the EST sequences which could be mapped. The set of unaligned ESTs is computed by taking the difference between all given ESTs in MyEstSet and the aligned ESTs in (the domain of) *alignedEstMR* as shown in Step 2 (answer to query Q1). To further distinguish between protein-coding and non-coding aligned EST sequences, we consider that genes typically consist of multiple intron and exon sequences. Usually, intron sequences are spliced out before the protein coding (translation) process starts. Therefore, sequence regions within introns are typically non-coding sequences. Conversely, exon sequences are highly involved in the protein coding process. In Step 3 we apply the mapping Ensembl.SRegionExons to determine the exons associated with the aligned DNA sequence regions. The domain of this composed mapping thus corresponds to the protein-coding aligned EST sequences (Step 4; answer to Q2). The set of non-coding and aligned EST sequences can be derived as the difference set between all aligned ESTs and all protein-coding and aligned EST sequences (Step 5; answer to Q3).

The example illustrates the power of the set-oriented operators for interconnecting data from different sources. The operators make it also easy to react to new analysis needs. For instance, we can associate the found protein-coding EST sequences not only to SwissProt proteins but also to genes of Ensembl and NetAffx, e.g. for microarray-based gene expression analysis. This is achieved by the following script extension.

```
$codingEstProteinMR:=compose($codingEstMR, map(range($codingEstMR),
                        {Ensembl.ExonGene,Ensembl.GeneProteins}));
$codingEstGeneOI:=traverse(range($codingEstMR),{Ensembl.ExonGene});
$fusedGeneAO:=aggregateSame($codingEstGeneOI,NetAffx);
```

The first statement returns a mapping result associating each protein-coding EST to the corresponding protein in SwissProt, the second statement determines all associated genes available in the Ensembl data source. This set of genes can further be fused with information on genes of the NetAffx source by traversing the corresponding same mapping. The fused gene information contains the attributes of both LDS, Gene@Ensembl and Gene@NetAffx.

3 BioFuice Architecture

Figure 2 gives an overview of the BioFuice system architecture. It consists of two main components, the iFuice core and the BioFuice query component. BioFuice utilizes the generic iFuice core to execute script programs and fuse data from several sources. The BioFuice query component provides interactive query functionality, supports local storage of analysis results and meets specific bioinformatics requirements, e.g. to export genomic sequences in specific formats for later use in other analysis tools. BioFuice query and iFuice core may run on the same machine or different machines. The BioFuice query component can also be run in stand-alone mode independent of iFuice, e.g. offline on a laptop. In this case analysis and query processing are restricted to local data sources. BioFuice is already in use for different applications, in particular for gene expression analysis, protein interaction analysis and analysis of non-coding RNAs.

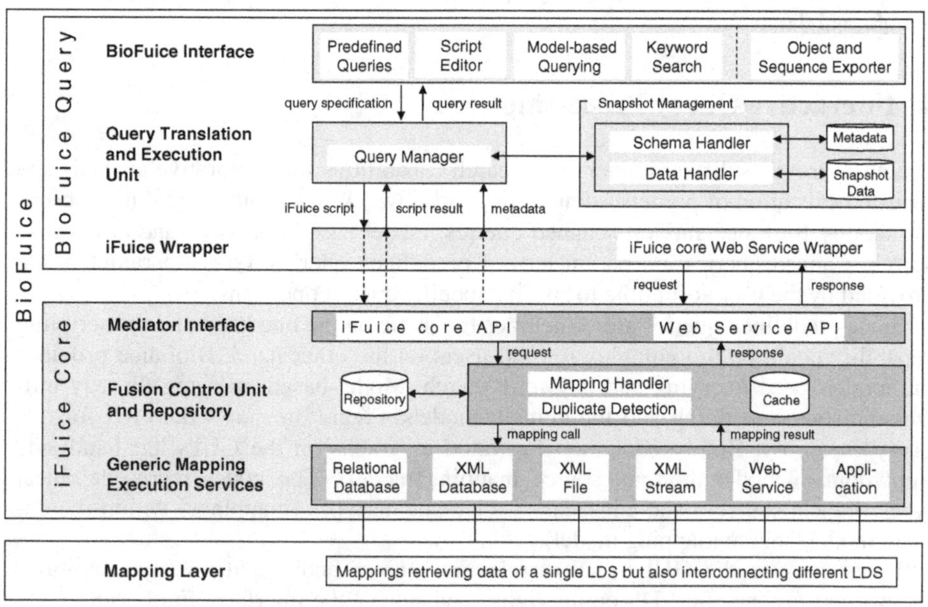

Fig. 2. BioFuice system architecture

The iFuice core consists of mapping execution services, a fusion control unit, a repository and the mediator interface. The mediator interface supports access to the iFuice functionality by a basic application interface but also by specific web service methods (so that iFuice core may run on a separate server machine). Typically, the interface is used to start a script combining multiple data and mapping operations. The mapping handler executes the script, temporarily caches the results and provides the results to the application via the mediator interface. The iFuice repository contains all metadata of the source mapping model and the domain model. In particular, it stores all LDS descriptions and mapping definitions. Each mapping definition is associated with mapping execution services that implement the specified mapping, e.g. a web service, Java application or SQL query. The spectrum of available mapping execution services supports mappings for sources of different formats, such as relational databases, XML-based sources but also application tools. The implementation of mappings by bioinformatics tools allows complex analysis and integration workflows, e.g. to perform search (blast) and data cleaning tasks.

Like the iFuice core the BioFuice query component is modularly structured. Subcomponents include a user interface, a query translation and execution unit and a local snapshot management unit. The user interface not only supports query specification but also visualization and export of query results. Query capabilities include predefined structured queries and unstructured keyword search, and are further described in the next section. The query manager translates interactively specified queries into an internal query format and automatically generates an iFuice script whenever the user decides to utilize the original (non-local) sources. Alternatively, queries may be restricted to local data, in particular materialized analysis results and copies (snapshots) of public sources. In this case, the query manager maps user queries to the corresponding statements on local sources, e.g. SQL statements for snapshots stored within a relational database.

4 Interactive Query Processing

BioFuice provides different query and search capabilities for explorative analysis and repeated execution of predefined analysis workflows. In addition to the iFuice scripting facility BioFuice supports canned queries, model-based querying, and a keyword search. Canned queries are parameterized predefined queries. Query parameters are provided by the user at runtime to specify specific query conditions.

Since predefined queries are sometimes too static on the one hand and the scripting capability could be too complex for end users on the other hand, BioFuice provides the model-based querying and keyword search. Model-based querying directly utilizes the source mapping and the domain models. Figure 3a shows the GUI for this query capability. Both models are illustrated as graphs on the GUI's left hand side (top: domain model, bottom: source mapping model). The nodes represent object types (logical sources) and edges stand for mapping types (mappings) within the domain model (source mapping model).

Users can use the GUI to select relevant sources and specify query conditions (keywords) for specific LDS on the right hand side of the interface. Furthermore, the query targets are specified, i.e. LDS for which object instances are finally retrieved.

Objects of target LDS sharing the same object type can be aggregated, i.e. attribute sets of corresponding instances from these LDS are merged. BioFuice automatically determines available mapping paths from the source mapping model connecting the selected LDS and query targets; the paths are visualized on the right hand side for user selection. Before the query is executed the user can also specify whether the original sources or the local snapshots should be used to answer the specified query.

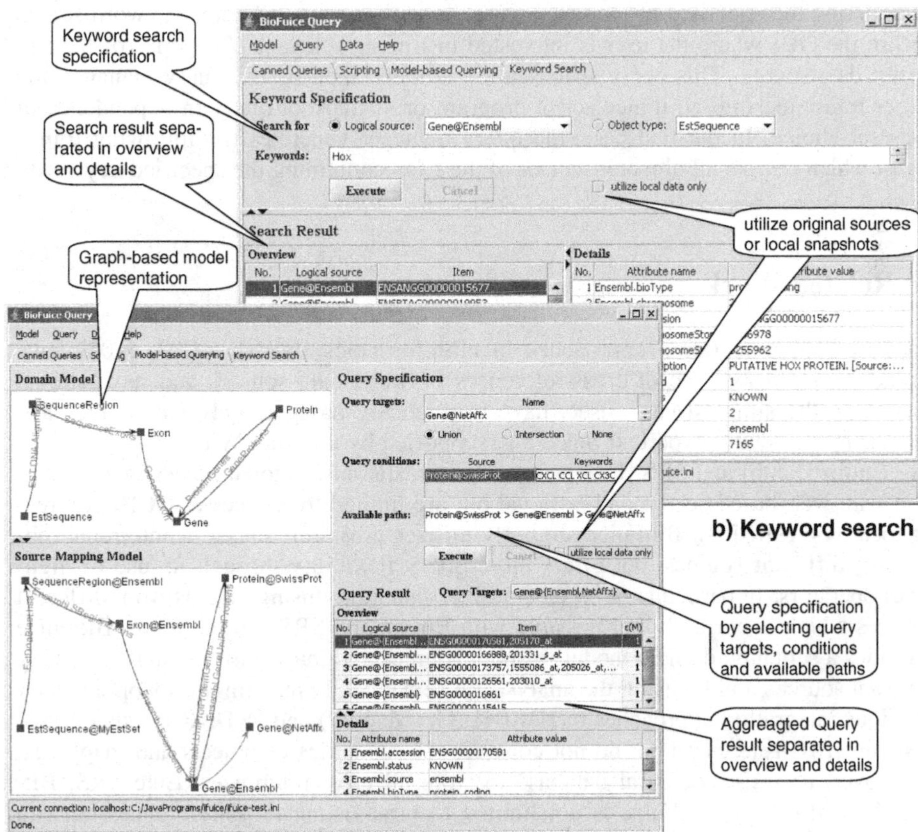

Fig. 3. Selected BioFuice query capabilities

Figure 3a illustrates such a query specification for a simple explorative analysis task. The goal is to find all genes of the source NetAffx corresponding to Chemokine proteins, i.e. special proteins that are responsible for cell-cell interactions. Based on the SMM of Figure 1, the plan is to use SwissProt to determine the relevant proteins and to traverse to the associated genes in Ensembl and then to the corresponding genes in NetAffx. BioFuice translates the interactively specified query into the following iFuice script:

```
$Proteins:=searchInstances(Protein@SwissProt,"CXCL CCL XCL CX3C");
$Genes:=traverse($Proteins,{Ensembl.ProtGenes});
$FusedGenes:=aggregateSame($Genes,NetAffx);
```

In Step 1, we utilize the LDS Protein@SwissProt to search for *Chemokine* proteins. [Ta05] provides a list of such proteins that are systematically classified in four groups *CXC*, *CC*, *XC*, and *CX3C*[2]. These group names can be used to search for relevant proteins since they are part of the protein name. In Step 2, the found proteins are associated with Ensembl genes which are fused with NetAffx genes in Step 3 as result.

In contrast to the model-based query capability, the keyword search looks either for objects of one selected LDS or of all LDS with the same chosen object type by considering the specified keywords. Figure 3b shows an example for keyword search within the GUI where the user is interested in finding relevant *Hox* genes of the Ensembl data source. This query specification is then taken by the query manager and either translated into an iFuice script program or a snapshot query in dependence of the user choice. In the first case, the query manager generates the following iFuice script which returns all object instances of the LDS containing the specified keyword.

```
$HoxGenes:=searchInstances(Gene@Ensembl,"Hox");
```

5 Related Work

Previous data integration approaches in bioinformatics [HK04, LC03, Iv05] have already made heavy use of cross-references between data sources and navigational access. In the simplest case, users have to manually navigate between sources and objects by following hypertext links, e.g. supported by a portal. Widely used systems like Entrez [Sc96] and SRS [Et03] support automatic navigational access in combination with web-based search and retrieval but are limited to sources at NCBI (Entrez) or local copies (SRS). BioFuice similarly utilizes cross-references to interrelate objects of different sources, but offers more query flexibility through its use of high-level operators. In particular, BioFuice utilizes operators fusing objects from different sources which is currently not possible with Entrez and SRS. Furthermore, BioFuice provides a semantic domain model for both sources and mappings helping to identify relevant sources and focusing the analysis on semantically meaningful mappings. Our previous integration approaches of [DR04] (GenMapper) and [KDKR05] also utilize existing cross-references but do not consider the semantics of objects and mappings. Moreover, they use a central database to store all cross-references. Like SRS, BIS [La03], and others, BioFuice is not limited to pre-existing cross-references but can also compute mappings, e.g. by queries or bioinformatics tools such as BLAST.

Many data warehouse and mediator systems utilize a global schema aiming at a consistent view over different data sources. As discussed in the introduction, such a schema is hard to create and maintain due to the large number of relevant sources and their high degree of heterogeneity. Some approaches such as ALADIN [LN05], HumMer [Bi05], AutoMed Toolkit [Ma05], and ANNODA [PC05] try to address this problem by using automatic schema matching. Other approaches simply take the union of the local schemas as the global schema. This exposes the heterogeneity and complexity to the user but still suffers from the need to change the global schema (and thus dependent mappings and applications) whenever one of the sources changes or a new source is added. In contrast to these approaches, BioFuice avoids the

---

[2] The additional character 'L' within the script identifies ligand proteins.

construction of a global schema but uses bidirectional peer mappings between sources. The domain model used is at a much higher abstraction level than a global schema and does not include details like the exact set of source attributes.

The BioFuice domain model can be seen as a domain-specific ontology. Ontologies provide a common understanding of a domain and thus are of interest for data integration. An overview of ontology-based approaches for data integration is presented in [Wa01]. While BioFuice currently utilizes user-defined object types as concepts and mapping types as their semantic relationships, other approaches reuse pre-defined ontologies [Me00] and focus on efficient query processing and rewriting by applying description logic [St03], different kinds of rules [NF05] and a quality model [He05]. Moreover, associating each relevant source attribute to an ontology concept need much more fine-grained ontologies than we use in BioFuice. Creating and maintaining fine-grained ontologies has similar problems than a global schema.

Peer-to-peer data management systems typically avoid the construction of a global schema. Database-oriented systems like Piazza [Ha03], PeerDB [Ng03] and Orchestra [Iv05] allow queries to be formulated on one peer and to be propagated through the system. Conversely, BioFuice can execute queries as well as mappings containing instance correspondences between sources, and can aggregate data from different sources. In contrast to Orchestra, BioFuice does not need to copy remote sources to local copies but uses the source schemas and their instances as provided.

6 Conclusions

We presented the BioFuice approach to integrate data from decentralized private and public data sources and ontologies. BioFuice follows a peer-to-peer-like data integration based on bidirectional mappings. Sources and mappings are associated with a domain model to support a semantically meaningful interoperability. BioFuice extends the generic iFuice integration platform which utilizes specific operators for data fusion and workflow-like script programs. BioFuice supports explorative data analysis and interactive query and search capabilities. BioFuice is operational and being used in different applications, such as for gene expression analysis, protein interactions analysis, and detection and analysis of non-coding RNAs.

Acknowledgements

The authors thank Andreas Thor, Nick Golovin and David Aumüller for useful discussions and their collaboration in developing the iFuice core component. We also thank the unknown reviewers for their constructive hints to improve the paper. The work is supported by the German Research Foundation, grant BIZ 1/3-1.

References

[Al90] Altschul, S. F. et al.: Basic Local Alignment Search Tool. Journal of Molecular Biology 215(3):403-10, 1990.

[Bi04] Birney, E. et al.: An Overview of Ensembl. Genome Research 14: 925-928, 2004.

[Bi05] Bilke, A. et al: Automatic Data Fusion with HumMer. Proc. 31st VLDB Conf., Demo description, 2005.

[Bo03] Boeckmann, B. et al.: The SWISS-PROT protein knowledgebase and its supplement TrEMBL in 2003. Nucleic Acids Research 31: 365-370, 2003.

[Et03] Etzold, T. et al.: SRS: An Integration Platform for Databanks and Analysis Tools in Bioinformatics. In [LC03]: 109-145.

[DR04] Do, H.-H.; Rahm, E.: Flexible Integration of Molecular-biological Annotation Data: The GenMapper Approach. Proc. EDBT Conf., 2004.

[Ga05] Galperin, M. Y.: The Molecular Biology Database Collection: 2005 Update", Nucleic Acids Research, 33, D5-D24, 2005.

[Ha03] Halevy, A. et al.: Piazza: data management infrastructure for semantic web applications. Proc. WWW, 2003.

[He05] Heese, R. et al: Self-extending Peer Data Management. Proc. Database Systems in Business, Technology and Web (BTW), 2005.

[HK04] Hernandez, T.; Kambhampati, S.: Integration of Biological Sources: Current Systems and Challenges Ahead. SIGMOD Record 33(3), 2004.

[Iv05] Ives, Z. et al.: Orchestra: Rapid, Collaborative Sharing of Dynamic Data. Proc. of Conf. on Innovative Data Systems Research (CIDR), 2005.

[KDKR05] Kirsten, T.; Do, H.-H.; Körner, C.; Rahm, E.: Hybrid Integration of molecular-biological Annotation Data. Proc. 2nd Int. Workshop on Data Integration in the Life Sciences (DILS), 2005.

[La03] Lacroix, Z. et al.: The Biological Integration System. Proc. 5th ACM Int. Workshop on Web Information and Data Management, 2003.

[LC03] Lacroix, Z.; Critchlow T. (Eds.): Bioinformatics: Managing Scientific Data. Morgan Kaufmann, 2003.

[Li03] Liu, G. et al.: NetAffx: Affymetrix probesets and annotations. Nucleic Acids Research, 31(1): 82-86, 2003.

[LN05] Leser, U.; Naumann, F.: (Almost) Hands-Off Information Integration for the Life Sciences. Proc. 2nd Conf. on Innovative Data Systems Research (CIDR), 2005.

[Ma05] Maibaum, M. et al.: Cluster based Integration of heterogeneous biological Databases using the AutoMed Toolkit. Proc. 2nd Int. Workshop on Data Integration in the Life Sciences (DILS), 2005.

[Me00] Mena, E. et al.: Observer: An Approach fro Query processing in Global Information Systems based on Interoperation across pre-existing Ontologies. Distributed and Parallel Databases 8(2): 223-271, 2000.

[NF05] Necib, C. B.; Freytag, J.-C.: Query Processing Using Ontologies. Proc. 17th Conf. on Advanced Information Systems Engineering (CAISE), 2005.

[Ng03] Ng, W. S. et al.: PeerDB A P2P-based System for Distributed Data Sharing. Proc. 19th Int. Conf. on Data Engineering, 2003.

[PC05] Prompramote, S.; Chen, Y.P.: Annonda: Tool for integrating molecular-biological Annotation Data. Proc. 21st Int. Conf. on Data Engineering (ICDE), 2005.

[Ra05] Rahm, E. et al.: iFuice - Information Fusion utilizing Instance Correspondences and Peer Mappings. Proc. 8th Int. Workshop on the Web & Databases (WebDB), 2005.

[RT05] Rahm, E.; Thor, A.: Citation analysis of database publications. SIGMOD Record 34(4), 2005.

[Sc96] Schuler, G. D. et al.: Entrez: Molecular biology database and retrieval system. Journal of Methods in Enzymology 266:141-62, 1996.

[St03] Stevens, R. et al.: Complex Query Formulation over diverse Information Sources in TAMBIS. In [LC03], 190-224, 2003.

[Ta05] Tanaka, Toshiyuki et al.: Chemokines in tumor progression and metastasis. Cancer Science 96(6): 317-322, 2005.

[Wa01] Wache, H. et al.: Ontology-based Integration of Information - A Survey of existing Approaches. Proc. Workshop on Ontologies and Information Sharing (IJCAI), 2001.

A Method for Similarity-Based Grouping of Biological Data

Vaida Jakonienė, David Rundqvist, and Patrick Lambrix

Department of Computer and Information Science
Linköpings universitet, SE-581 83 Linköping, Sweden

Abstract. Similarity-based grouping of data entries in one or more data sources is a task underlying many different data management tasks, such as, structuring search results, removal of redundancy in databases and data integration. Similarity-based grouping of data entries is not a trivial task in the context of life science data sources as the stored data is complex, highly correlated and represented at different levels of granularity. The contribution of this paper is two-fold. 1) We propose a method for similarity-based grouping and 2) we show results from test cases. As the main steps the method contains specification of grouping rules, pairwise grouping between entries, actual grouping of similar entries, and evaluation and analysis of the results. Often, different strategies can be used in the different steps. The method enables exploration of the influence of the choices and supports evaluation of the results with respect to given classifications. The grouping method is illustrated by test cases based on different strategies and classifications. The results show the complexity of the similarity-based grouping tasks and give deeper insights in the selected grouping tasks, the analyzed data source, and the influence of different strategies on the results.

1 Introduction

During the last decade an enormous amount of biological data has been generated and techniques and tools to analyze this data have been developed. Many of these tools use data clustering and classification techniques. For instance, these techniques are used to find similar sequences for predicting the functionality of new sequences [GH04], to find correlated genes based on microarray data [SS02], or to classify publications according to an ontology to locate relevant documents faster [DS05]. A basic task underlying these approaches is the computation of a similarity value between objects. Different techniques are developed to compute a similarity value between objects based on the object types. For instance, edit distance [Lev66] and n-gram [PPF95] are well-established techniques to define similarity between strings, while BLAST [AGMML90] can be used to define a similarity measure between DNA or protein sequences. Recently, a number of projects discussed methods to compute semantic similarity over terms in a Gene Ontology (GO) ontology (e.g. [CSC05] and [SFSZ05]). The similarity between GO terms can be used to compute a similarity between data entries that are annotated with these GO terms [LSBG03].

U. Leser, F. Naumann, and B. Eckman (Eds.): DILS 2006, LNBI 4075, pp. 136–151, 2006.

Data entries in biological data sources are often complex and store different types of information. Although most of the research has focused on organizing the data based on aspects, such as sequence similarity and function, we need to analyze data using different aspects and from different points of view to obtain deeper insights in the characteristics of the data and to discover new knowledge. This means that we need to be able to organize the data based on different attributes or different combinations of attributes. [KLKTB04] illustrates how a combination of attributes could be used to find data entries describing the same protein. In this case, search on sequence similarity is complemented with the analysis of sequence length, organism and the data source where the sequence was originally submitted. In this paper we use the term *grouping* to refer to the task of organizing the data according to a certain aspect or a combination of aspects. Further, we concentrate on the task of *similarity-based grouping*. During similarity-based grouping the analyzed data entries are compared with respect to a selected subset of attributes, and similarity functions that are relevant to the attributes are used to compute the similarity of the stored values.

Grouping of data entries in one or more data sources is an operation underlying many different data management tasks. Grouping can be used to structure and visualize search results in a convenient way for the user. This is especially important when large data sources are studied. The possibility to get an overview over the data may lead to the discovery of new knowledge or may allow biologists to locate the information of interest faster. The identification of similar data entries and their grouping are core operations when performing data cleaning activities [HGPWW04]. The identified groups of similar data entries can be further analyzed and merged into a single data entry. In the context of data integration, techniques underlying grouping are important to correlate data entries at different data sources. The grouping task can be narrowed to the duplicate detection task, where it is required that matched data entries represent the same real-world object. Duplicate detection can be both used for data cleaning [KLKTB04] and for data integration [BBBDN05].

A number of aspects influence the quality of the grouping results: the quality of the data sources, the selection of the grouping attributes and the algorithms implementing the grouping procedure. In some cases, given a grouping task, it can be difficult to decide on which attributes to perform grouping. Also, different sets of attributes may seem relevant to the grouping task, but lead to varying quality of the results [KLKTB04]. Further, suitable algorithms need to be selected to compute the similarity between data entries and to organize similar data entries into groups. Many methods exist, but it is often not clear which methods perform best for which grouping tasks. The study of the properties, and the evaluation and the comparison of the different aspects that influence the quality of the grouping results, would give us valuable insight into the best way to use the grouping procedures. It would also lead to recommendations on how to improve the current procedures and develop new procedures. To be able to perform such studies and evaluations we need environments that allow us to compare and evaluate different grouping procedures.

In this paper, as a first step towards the development of an environment to support grouping tasks, we propose a method that covers the main steps and components that should be included in such environments (section 2). The grouping method is illustrated by test cases based on different strategies and classifications (section 3). In subsections 3.1-3.5 we describe the grouping task, the test cases, the implementation approaches and the evaluation approach according to the method. In subsection 3.6 we analyze the evaluation and grouping results and show how we obtain deeper knowledge about the grouping tasks, the analyzed data source, and the influence of different strategies on the results. The paper concludes in section 4.

2 Method for Grouping Biological Data

In this section we describe a method that supports similarity-based grouping of biological data and that enables the development of grouping procedures. The components and the main steps of the method are illustrated in figure 1.

The method uses as input the data source on which the grouping is performed. Further, it uses similarity functions that can compute similarity values between data values, and grouping attributes on which we base the computation of the similarity of data entries in the data source. There may also be external sources to support the grouping task. Based on this input the method can generate groupings of data. In addition to this, the method also allows the evaluation and analysis of the grouping results. For this purpose we use a library of known

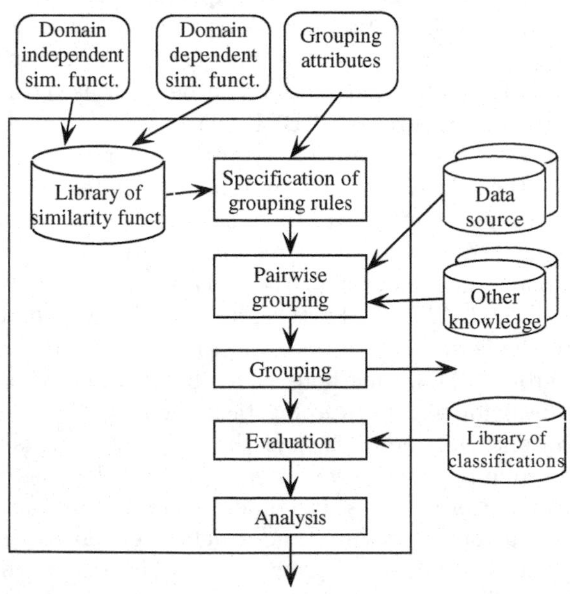

Fig. 1. Method for similarity-based grouping

classifications. The library stores selected sets of data entries organized into classes[1]. The method then returns the generated groups of data entries as well as reports from the evaluation and analysis.

The main steps in the method are: 1) specification of grouping rules that define how to identify similar data entries; 2) pairwise grouping by computing the similarity between pairs of data entries; 3) grouping similar data entries into groups; 4) evaluation of the quality of the generated groups with respect to given classes and 5) analysis of the grouping and evaluation results. We note that the library of similarity functions and the specification of grouping rules are often domain dependent, while the other steps are based on general techniques and approaches. Further, before the grouping procedure can be applied to a data source, the data source usually needs to go through a number of data transformation steps, such as merging of data and data translation from one format to another. In the remainder of this section we briefly describe the different components and steps of our method.

Library of Similarity Functions. This component represents a collection of functions that compute similarity scores between data values:

$SimFunc(v_1, v_2) \rightarrow [0, 1]$. We distinguish between domain independent and domain dependent similarity functions. The former group of similarity functions can be applied to any kind of data values, for instance, string-based functions. In the latter case, similarity functions can only be used to compare values of specific types of data values, for instance, protein sequences. The designer of similarity functions should be aware of and develop approaches for dealing with collection type values and missing values, which are often encountered in biological data.

Specification of Grouping Rules. In our method grouping rules are used to express conditions on which two data entries are compared for similarity. During this step the user defines a grouping rule or selects an already available grouping rule that is deemed to be relevant to the current grouping task. A grouping rule may combine different similarity functions applied to one or more grouping attributes. For instance, a grouping rule may specify that two data entries are similar if the sum of weighted similarity functions applied to certain grouping attributes is higher than a given threshold. In general, the specification of grouping rules is not a trivial task. We may need to do much experimentation and fine-tuning to specify rules that are adequate for a certain grouping task. Our method aims to support this.

Pairwise Grouping. Given a data source and grouping rules, the pairwise grouping step performs pairwise comparisons of data entries. Auxiliary domain knowledge may be used. The result of this step is the identification of the pairs of data entries that are similar. The pairwise grouping includes the following steps: a) selection of pairs of data entries in the data source for the comparison; b) comparison of data values of the selected grouping attributes by applying the

[1] In the rest of the paper we use *classes* to refer to given classifications and *groups* to refer to the results of grouping techniques.

defined similarity functions; and c) comparison of the selected data entries on the basis of given grouping rules. While for small data sources all pairs of data entries can be analyzed, for large data sources pruning techniques may be used to decrease the number of performed comparisons.

Grouping. The step takes as input pairs of similar data entries and organizes the data entries into a set of groups composed of similar data entries. Different techniques can be used to perform grouping and they may vary on a number of aspects. For instance, the groups can be allowed to overlap or may be required to be disjoint. Some approaches may require the transitivity property between similar data entries or they may allow to ignore some similarity relationships, e.g. in order to split a group into smaller groups. Also, there may be a restriction on the total number of groups or on the number of allowed data entries in a group. The generated groups can be seen as the final result of the grouping method or they can be used an input to the evaluation step.

Evaluation. During the evaluation step different measures are computed to evaluate the quality of the grouping results. Two groups of quality measures are distinguished [SKK00]: internal and external quality measures. Internal quality measures compare different groupings only based on information obtained during the grouping (e.g. pairwise similarity between data entries). External quality measures evaluate the grouping results with respect to known classes. As emphasized in [SKK00], to select the best grouping approach for the analyzed task, the approaches have to be compared with respect to a number of measures. In our method, the library of classifications is used to compute external quality measures.

Analysis. During this step the grouping and evaluation results are analyzed. Different forms and reports are generated providing support for exploring the results from different points of view. For instance, valuable insight could be gained by analyzing and studying the entries belonging to a single group, the correlation between groups and classes, and the influence of external knowledge on the results.

3 Test Cases

In this section we illustrate our method for similarity-based grouping of biological data. Further, we show how the method can be used to gain deeper knowledge about a particular setting, i.e. about the data source, the different steps in the method, and the influence of the various choices on the results. In our test cases we worked on two grouping tasks: grouping of proteins with respect to their biological function and with respect to what classes of isozymes they belong to. Proteins are isozymes (or isoenzymes) if they are enzymes that catalyze the same chemical reaction, but they may differ in their amino acid sequences [BTS02]. Isozymes differ in their kinetic properties, the way they are regulated by other proteins and quantities in which they are expressed in different tissues. For example, the enzyme *Lactate dehydrogenase* is built by two isozymic polypeptide chains: H and M. The H isozyme functions optimally in aerobic environments and is expressed highly in the heart, while the M isozyme works under anaerobic conditions and is expressed highly in skeletal muscle. For each grouping task we explore the impact of different grouping attributes and grouping rules, which use different types of similarity functions. Also, we study the influence of different grouping algorithms. We describe the setting of the experiments and discuss the grouping and evaluation results.

3.1 Data and Knowledge

Data Source. The data source contains 190 human proteins involved in glycolysis that we retrieved on the 6th of October 2005 from the Entrez retrieval system. The data entries have different origin. They either come from the data sources RefSeq, SWISS-PROT, PRF, PIR and PDB, or are translated from nucleotide sequences at GenBank, EMBL and DDBJ.

Grouping Attributes. The following types of data were selected as relevant to the given grouping tasks:

- DEFINITION is an attribute that describes a protein. It combines information on protein name, synonymous names, isozyme indicator and organism name, for instance, "`ATPase, H+ transporting, lysosomal 31kD, V1 subunit E isoform 1 [Homo sapiens].`".
- PRODUCT is an attribute that holds the name of the gene product and, in some cases, stores the isozyme indicator in a less complex way compared to DEFINITION, for instance, "`liver phosphofructokinase isoform a`".
- SEQUENCE[2] is the attribute where the amino acid sequence of a protein is stored, for instance,

```
  1 malsdadvqk qikhmmafie qeanekaeei dakaeeefni ekgrlvqtqr lkimeyyekk
 61 ekqieqqkki qmsnlmnqar lkvlrarddl itdllneakq rlskvvkdtt ryqvlldglv
121 ...
```

- GO[3] ANNOTATIONS. Some of the data entries are annotated by one or more GO terms, which are denoted by their GO id:s, for instance, "`go:0005524 | go:0004396`". Several attributes in a document may contain GO terms. For our experiments we found GO terms in the attribute DBSOURCE for SWISS-PROT data entries and in the "note" property in the "CDS" field under the FEATURES attribute for the other data entries.

Other Knowledge. Of the 190 data entries, only 71 data entries were originally annotated by GO terms, which we refer to by GO_{ann}. To increase the number of

[2] In the original file this attribute is called ORIGIN, but for the sake of readability we use term SEQUENCE.

[3] The GO ontologies are de facto standard ontologies that describe the roles of genes and proteins in different organisms [GO00]. The three independent publicly available ontologies are: biological process, molecular function and cellular component. Today, many different bio-data sources are annotated with GO terms. The terms in GO are arranged as nodes in a directed acyclic graph, where multiple inheritance is allowed.

data entries with GO terms we used mappings between data values and ontological terms found on the web pages of the GO Consortium. In the experiments, we used the mapping *spkw2go* to translate values of the KEYWORDS attribute into GO terms, which we refer to by GO_{sw}. Also, we used the mapping *ec2go* to translate values of the EC-NUMBER attribute into GO terms, which we refer to by GO_{ec}. During the grouping process, knowledge in the GO ontology was used to compute similarity scores. To explore the quality of the available mappings, we decided to analyze GO_{ann}, GO_{sw} and GO_{ec} in different combinations. All these combinations resulted in variants of our original data source that included only data entries annotated by these terms. As a result, the number of analyzed data entries differed among the test cases. For instance, GO_{ann} and GO_{sw} annotations were available for 75 data entries, while GO_{ann} and GO_{ec} occurred in 92 data entries.

Classifications. The data entries were, for the whole data source, manually classified into 28 disjoint classes according to biological function. For instance, all data entries in one of the classes relate to *Phosphofructokinase* (which is the enzyme responsible for turning Fructose-6-phosphate into Fructose-1, 6-biphosphate). Data entries belonging to the same class may describe the same real-world protein, proteins having similar function, fragments of proteins having the same or similar function, and hypothetical proteins that are strongly believed to have the same or similar function. In the classification the two largest classes consist of 56 and 53 data entries, while 13 classes consist of a single data entry. The largest classes represent the enzymes *Pyruvate kinase* and *Phosphofructokinase*, which are the most prominent regulatory enzymes in glycolysis.

The isozyme classification was constructed by further dividing the classes in the function-based classification. For example, the data entries in the *Phosphofructokinase (PFK)* class were distributed into three classes: *Liver-type PFK*, *Platelet-type PFK* and *Muscle-type PFK*. The classification resulted in 52 disjoint classes, where the two largest classes contained 29 and 27 data entries, while 31 classes contained a single data entry.

3.2 Library of Similarity Functions

We used domain independent and domain dependent similarity functions.

$EditDist(v_1, v_2)$ is a function that computes similarity based on the edit distance between strings. The distance between strings v_1 and v_2 is defined by the least number of operations needed to turn v_1 into v_2. The allowed operations are insertions, deletions and replacements. The distance is transformed into a similarity score by the function: $score = 1 - \frac{distance}{MaxLength(v_1,v_2)}$.

$SeqSim(v_1, v_2)$ is a function that performs pairwise sequence alignment and returns a similarity score between sequences. We use a sequence alignment tool implemented in Java, JAligner [JAligner], to compute an alignment between the sequences. The tool implements an improved version of the Smith-Waterman algorithm for producing gapped alignments between sequences. The similarity score is defined as the number of matches in the alignment divided by the length of the alignment.

$SemSim(v_1, v_2)$ is a function that computes the similarity between two sets of GO terms. To evaluate the distance between two GO terms we use an edge-based algorithm that counts the number of edges needed to traverse the GO hierarchy from one term to another. The algorithm counts the number of is_a relationships needed to go up in the hierarchy u, the number of is_a relationships needed to go down in the hierarchy d and the number of other relationships o. The similarity between two GO terms is then defined as $score = e^{-0.5 \cdot ((\frac{u}{p_u})^2 + (\frac{d}{p_d})^2 + (\frac{o}{p_o})^2)}$, where p_u, p_d and p_o are weights for the different types of edges. In the test cases we used p_u=2, p_d=1 and p_o=1. Two sets of GO terms are defined to be similar if each term of one set is similar to a term in the other set.

Table 1. Test cases. Grouping on protein function. n^e - number of analyzed entries, n^g - number of groups, n^c - number of classes, p - purity, E - entropy, F - F-measure, MI - mutual information.

Test case	Grouping rule	n^e	n^g	n^c	p	1-E	F	MI
1	$SemSim(GO_{ann})> 0.95$ GO_{ann} for component, process, function domains	71	23	24	0.90	0.93	0.88	0.86
2	$SemSim(GO_{ann})> 0.95$	67	26	23	1.00	1.00	0.97	0.91
3	$SemSim(GO_{ann} + GO_{sw})> 0.95$	75	23	24	0.80	0.87	0.79	0.79
4	$SemSim(GO_{ann} + GO_{ec})> 0.95$	92	26	25	1.00	1.00	0.99	0.88
5	$SemSim(GO_{ann} + GO_{sw} + GO_{ec})> 0.95$	93	26	25	0.86	0.93	0.88	0.81
6	$SemSim(GO_{ann} + GO_{sw} + GO_{ec})> 0.95$; parent GO terms removed	93	26	25	0.86	0.93	0.88	0.81
7	$SemSim(GO_{ann})> 0.95$ or $SemSim(GO_{sw})> 0.95$ or $SemSim(GO_{ec})> 0.95$	93	14	25	0.48	0.65	0.51	0.59
8	$SemSim(GO_{ann})> 0.95$ or $SemSim(GO_{ec})> 0.95$	92	26	25	1.00	1.00	0.99	0.88
9	$SemSim(GO_{ann} + GO_{ec})= 1$	92	26	25	1.00	1.00	0.99	0.88
10	$SemSim(GO_{ann} + GO_{ec})> 0.85$	92	21	25	0.70	0.78	0.71	0.68
11	$SemSim(GO_{ann} + GO_{ec})> 0.95$ grouping algorithm: cliques	92	29	25	1.00	1.00	0.84	0.88
12	$EditDist(definition)> 0.9$, for $GO_{ann} + GO_{ec}$	92	67	25	1.00	1.00	0.59	0.77
13	$EditDist(definition)> 0.7$, for $GO_{ann} + GO_{ec}$	92	55	25	0.96	0.97	0.66	0.76
14	$SeqSim(sequence)> 0.85$, for $GO_{ann} + GO_{ec}$	92	44	25	1.00	1.00	0.74	0.81
15	$EditDist(definition)> 0.85$	190	94	28	0.97	0.98	0.54	0.57
16	$EditDist(product)> 0.85$	190	105	28	0.99	0.99	0.49	0.57
17	$EditDist(definition)> 0.7$	190	68	28	0.81	0.87	0.56	0.50
18	$EditDist(product)> 0.7$	190	78	28	0.95	0.98	0.64	0.58
19	$EditDist(definition)> 0.9$ or $EditDist(product)> 0.9$ or $(EditDist(definition)> 0.6$ and $EditDist(product)> 0.6)$	190	64	28	0.94	0.96	0.70	0.58
20	$SeqSim(sequence)> 0.85$	190	59	28	0.99	0.99	0.66	0.62

Table 2. Test cases. Grouping on isozymes. n^e - number of analyzed entries, n^g - number of groups, n^c - number of classes, p - purity, E - entropy, F - F-measure, MI - mutual information.

Test case	Grouping rule	n^e	n^g	n^c	p	1-E	F	MI
21	$EditDist(definition) > 0.85$	92	67	47	0.89	0.95	0.73	0.85
22	$SemSim(GO_{ann} + GO_{ec}) > 0.95$	92	26	47	0.59	0.79	0.65	0.79
23	$EditDist(product) > 0.85$	92	56	47	0.83	0.92	0.73	0.84
24	$SeqSim(sequence) > 0.85$	92	44	47	0.91	0.96	0.90	0.91
25	$EditDist(definition) > 0.85$	190	94	52	0.87	0.93	0.63	0.67
26	$EditDist(product) > 0.85$	190	105	52	0.88	0.94	0.58	0.68
27	$SeqSim(sequence) > 0.85$	190	59	52	0.95	0.97	0.91	0.75

3.3 Specification of Grouping Rules

The studied grouping rules are collected in the second column of table 1 and table 2 for the grouping tasks on function and isozymes, respectively. The similarity functions in the tables are shown with one argument representing the type of the compared values. When exploring grouping on function, we developed a number of test cases based on various combinations of GO_{ann}, GO_{sw} and GO_{ec} (test cases 1-11). All these cases, except test case 1, use only function-related terms in the GO ontology. In the test cases 12-20 we analyzed the applicability of the values in attributes DEFINITION, PRODUCT and SEQUENCE for grouping on function. The test cases 8-14 are run on the data entries used in test case 4, since test case 4 had the best results among the test cases run on GO terms. The test cases 15-20 are performed on the whole data source, i.e. in total 190 data entries. The test cases include experiments with different thresholds, complex rules combining similarity functions and an experiment with a different grouping algorithm (test case 11). Similarly as for grouping on function, we tested the applicability of DEFINITION, PRODUCT, GO ANNOTATION and SEQUENCE for grouping on isozymes. Table 2 contains the grouping rules applied on the data entries analyzed in test case 4 and the grouping rules applied on all data entries in the data source.

3.4 Pairwise Grouping and Grouping

To perform the actual grouping of the data entries based on a given rule, first pairwise grouping between the data entries and then, grouping of similar data entries are performed. In our experiments, all pairs of data entries in the data source are compared to each other and are identified as similar or not. To organize the similar data entries into groups we experimented with two approaches: cliques and connected components. *Cliques* require that all data entries in a group are similar to each other. In this approach, the generated groups may overlap as all similarity relationships are taken into account. *Connected components* collect all data entries that are directly or transitively similar to each other into a single group. As a result, the approach generates disjoint groups. From the discussed test cases only test case 11 uses cliques. For the other test cases we used connected components.

3.5 Evaluation

To evaluate the results of the test cases we used external quality measures described in [Str02], purity, F-measure, entropy and mutual information. These measures are defined as follows. Let n be the total number of analyzed data entries, n^g the number of generated groups and n^c the number of given classes. Let n_i^g denote the number of entries in group i and n_j^c denote the number of entries in class j. Let n_{ij} represent the number of entries that are common to group i and class j. For each group i and class j, the precision is defined as $p_{ij} = \frac{n_{ij}}{n_i^g}$ and the recall as $r_{ij} = \frac{n_{ij}}{n_j^c}$.

Purity evaluates the average precision of the groups with respect to their best matching classes. For each group i purity is defined as $p_i = \max_j\{p_{ij}\}$. The purity for the whole grouping is defined as

$$p = \sum_{i=1}^{n^g} \frac{n_i^g}{n} p_i$$

F-measure is the average F-measure of the classes with respect to their best matching groups. The measure combines precision and recall into a single value. For each combination of group i and class j the F-measure is $F_{ij} = \frac{2 \cdot p_{ij} \cdot r_{ij}}{p_{ij} + r_{ij}}$. The F-measure for class j is defined as $F_j = \max_i\{F_{ij}\}$ and the F-measure for the whole grouping is defined by

$$F = \sum_{j=1}^{n^c} \frac{n_j^c}{n} F_j$$

Normalized entropy analyzes how on average the data entries in each group distribute among the classes. $E_i = -\sum_{j=1}^{n^c} p_{ij} \log_{n^c} p_{ij}$ is the normalized entropy for group i and the total normalized entropy for the whole grouping is

$$E = \sum_{i=1}^{n^g} \frac{n_i^g}{n} E_i = -\frac{1}{n} \sum_{i=1}^{n^g} \sum_{j=1}^{n^c} n_{ij} \log_{n^c} p_{ij}$$

Mutual information is the average measure of correspondence between each group and class. The mutual information is calculated as

$$MI = \frac{2}{n} \sum_{i=1}^{n^g} \sum_{j=1}^{n^c} n_{ij} \log_{n^g \cdot n^c} \left(\frac{n_{ij} \cdot n}{n_i^g \cdot n_j^c} \right)$$

The evaluation results for each test case are shown in tables 1 and 2.

3.6 Analysis

In this subsection we take a closer look at the grouping and evaluation results for our test cases. We compare different test cases and discuss issues that have an impact on the results. Further, using 3 examples we discuss interesting cases in some more details.

Best Test Cases. The test cases described in the tables 1 and 2 reveal that grouping on GO ANNOTATIONS combining GO_{ann} and GO_{ec} is best suited for grouping the data entries on function (test cases 4, 8 and 9) and that grouping on SEQUENCE is best suited for grouping on isozymes (test cases 24 and 27). For the test cases 4, 8 and 9, the grouping results were only imprecise in the distribution of data entries of one class between two groups (see example 2 below). The same grouping results for the test cases 4, 8 and 9 could be caused by the type of the compared GO annotations and the type of the used grouping approaches. In the case of grouping on isozymes, grouping on sequence performed reasonably well both on the fragment of the data source and on the whole data source.

Grouping on GO Annotation. Test cases 1 and 2 show that the removal of the component and process terms from GO_{ann} increases the quality of the grouping on function. For instance, each group in test case 2 includes entries from a single class (p=1). However, in some cases valuable information may be removed (see example 2 below). This suggests that a method that assigns different weights to different types of GO terms may improve the results.

From the analysis of test cases 2-8 we conclude that *spkw2go* mappings are not suitable for grouping on function. SWISS-PROT keywords are quite general and are mapped to high level GO terms. For instance, some SWISS-PROT data entries contain 'Glycolysis' as a keyword, while all the data entries in the data source relate to 'Glycolysis'. Therefore, some data entries where grouped together even though they differed in more specific functions, i.e. belonged to different classes. For instance, test case 3 generated 2 groups containing data entries from several classes. In contrast to *spkw2go*, GO terms obtained through *ec2go* mappings were specific enough. This is because EC numbers precisely identify the function of the described sequence. For instance, EC:2.7.1.11 maps to the GO term '6-phosphofructokinase activity', which is a very specific function.

Test cases 5 and 6 show that only using the most specific GO terms in the data entries, does not have an impact on the grouping result. This depends partly on the available GO annotations, which for some data entries match exactly. It also depends on the approach to compare sets of GO terms.

Grouping on Definition and Product. DEFINITION- and PRODUCT-based groupings perform worse than SEQUENCE-based groupings both for function and isozymes. The large number of generated groups in test cases 15-18 shows that the data values in these fields vary a lot. For instance, about 3 times more groups are generated than there are classes in the case of grouping on function for the threshold 0.85. Also, PRODUCT values are not available for some data entries. From the current test cases no definite conclusions can be made about the suitability of these types of data for grouping tasks. Further studies are needed.

Grouping on Sequence. Based on test cases 14 and 20, we can conclude that grouping on sequences is too specific to be used for grouping data entries on function. There are nearly twice as many groups as there are classes. For instance, the data entries of the *Pyruvate kinase* class were distributed between a group covering muscle-type sequences and a group covering liver-/red blood cell-type sequences. This example confirms the observation that sequence similarity based grouping is better suited for grouping on isozymes.

Impact of Threshold. Test cases 4, 9 and 10 where grouping is performed on GO_{ann} and GO_{ec} with thresholds 0.95, 1 and 0.85, show that the best quality results are returned when all GO terms of one data entry appear among GO terms of the other data entry. For a slightly lower threshold, the quality of the results drops fast. The test cases 15-18 show that PRODUCT performs better for the lower threshold, while for DEFINITION-based grouping, although less groups are generated, the decrease of the threshold produces lower quality results. In general, experiments with different thresholds allow us to explore the correlation of data entries at different levels of similarity.

Complex Rules. Test cases 7, 8 and 19 illustrate the use of complex rules that gave us a better understanding of the data and enabled an increase of the quality of the grouping results. The negative impact of GO_{sw} is shown by test case 7 that performs much worse than test case 8. The grouping rule combining DEFINITION and PRODUCT (test case 19) resulted in increased quality of the results in comparison with test cases 15-18, which perform grouping on a single attribute.

Impact of Grouping Algorithm. Test cases 4 and 11 illustrate the impact of the different grouping approaches, in our case, connected components and cliques. As cliques put stronger requirements on the grouped data entries and allow overlapping groups (see example 3 below), a larger number of groups are generated and a lower F-measure is obtained. Based on the measures, however, we cannot make a decisive claim about which of the two grouping approaches performs better. The nature of the approaches is very different. They complement each other and give different ways of presenting the results. For instance, by comparing the results of the grouping approaches, subgroups of data entries that are interconnected in a stronger way to each other than to the rest of the entries in the group, can be located. Such subgroups could be generated for several reasons, such as the fact that the described sequences may slightly differ in functionality or that the data entries may have incomplete information.

In the remainder of this section we investigate some of the results in more details.

Example 1. A Group Covers Several Classes. In this example we take a closer look at group 2 of test case 1. The group includes data entries from four classes: *Phosphofructokinase* (class 3), *Pyruvate dehydrogenase* (class 11), *Nuclear receptor subfamily 1* (class 16) and *Pyruvate dehydrogenase kinase* (class 27). The data entries belonging to these classes together with their GO annotations are given in table 3.

A combination of reasons caused the data entries to be organized into the same group. 1) The algorithm that compares sets of GO terms considers the

Table 3. Test case 1. Data entries in group 2.

Accession #	Class #	GO ANNOTATIONS
Q01813	3	**GO:0005945**, GO:0003872, GO:0006096
NP_000280	3	**GO:0005945**, GO:0005524, GO:0016301 GO:0000166, GO:0016740, GO:0000287 GO:0003872, GO:0006096, GO:0006006 GO:0005977, GO:0006110
NP_002618	3	**GO:0005945**, GO:0005524, GO:0016301 GO:0000166, GO:0016740, GO:0000287 GO:0003872, GO:0006096
P11177	11	*GO:0004739*, GO:0006099
P29803	11	*GO:0004739*
P10515	11	**GO:0005967**, GO:0004742, GO:0006085
NP_000275	11	**GO:0005739**, GO:0016491, *GO:0004739* GO:0016624, GO:0006096, GO:0008152 GO:0006084
P08559	11	**GO:0005739**
NP_005114	16	**GO:0005634**, GO:0046872, GO:0003707 GO:0003700, GO:0003713, GO:0003714 GO:0006350, GO:0008203, GO:0007165 GO:0008206, GO:0006355
NP_002603	27	**GO:0005739**, GO:0005524, GO:0004672 GO:0004740, GO:0006006, GO:0006468

data entries Q01813, NP_000280, NP_002618, P10515, NP_000275, NP_005114 and NP_002603 to be similar to P08559 since the only GO term in the data entry P08559, GO:0005739 - the component term "Mitochondrion", has high similarity with other component terms in the other data entries (marked by bold in table 3). Similarly, the data entries P11177 and NP_000275 are similar to P29803 because of the included GO:0004739. The comparison algorithm ignores the fact that some data entries have more GO terms assigned than the others. 2) The test case uses a grouping approach that assumes that similarity between data entries is transitive. As a result, two groups of similar data entries identified previously are connected by NP_000275 into a single group of similar data entries. 3) The fact that some GO terms in the annotation were general and that some of the data entries contained very few GO terms also caused the classes to be grouped into a single group.

Example 2. A Class Distributed Among Several Groups. In test case 4 only class 11, describing *Pyruvate dehydrogenase complex*, is not matched perfectly by a group. The data entries are divided into two groups: P11177, NP_000275, P08559 and P29803 are grouped together, while P10515 appears in a separate group. The grouping result can be explained by the fact that class 11 describes an enzyme complex that consists of multiple copies of the three types of enzymes E_1, E_2 and E_3. The goal of the whole enzyme complex is to build the molecule acetyl-CoA from Pyruvate and CoA, but different types of enzymes belonging to the complex vary in their function. For instance, E_1 has the major catalytic function, namely the pyruvate dehydrogenase function [BTS02]. In our case, P11177, NP_000275, P08559 and P29803 describe E_1, while P10515 describes E_2. The difference between functions is also reflected in the available GO annotations: P11177, NP_000275, P08559 and P29803 are annotated with "pyruvate dehydrogenase (acetyl-transferring) activity", while P10515 has "dihydrolipoyllysine-residue acetyltransferase activity". These two terms are far from each other in GO. Test case 1 is the only one that organizes all the data entries into a single group. This illustrates how the knowledge from the component and process GO ontologies may positively contribute to the grouping on function. For the other test cases, where only GO terms from the function ontology were used, the available GO annotations are too specific to identify the whole enzyme complex.

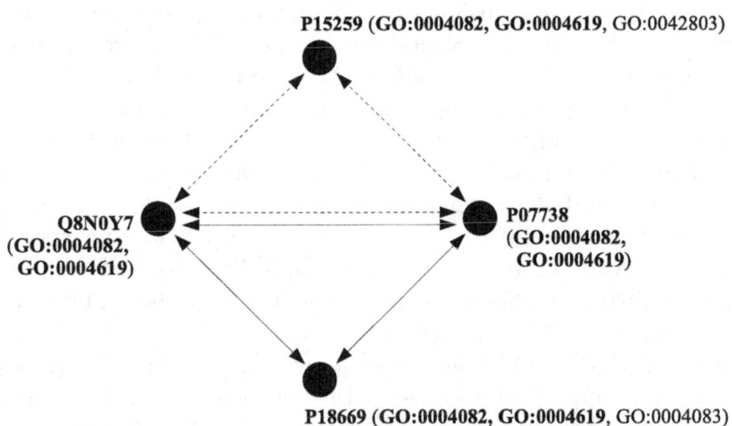

Fig. 2. Clique-based grouping. Class 8.

Example 3. Clique-Based Grouping. In test case 11 we used a clique-based algorithm for grouping similar data entries, which requires that all data entries in a group are similar to each other. The grouping result had a few cases where data entries in a class are distributed between 2 groups. For instance, the data entries of class 8, which describes *Phosphoglycerate mutase*, were distributed between 2 overlapping groups: one group contained P07738, Q8N0Y7 and P15259, while the other group included P07738, Q8N0Y7 and P18669 (figure 2). P15259 and P18669 are not found to be similar as the GO annotations differ from each other by GO:0042803 and GO:0004083, respectively. As a result, the data entries are moved to separate groups. P07738 and Q8N0Y7 are included in both of the groups as they are found to be similar to P15259 and P18669. We have checked that lowering the threshold, e.g. to 0.8, would combine all four data entries to a single group. This example illustrates the high impact of the different aspects in the grouping procedures on the grouping result; in this case the threshold, the approach for comparing sets of GO terms and the approach for grouping similar data entries.

4 Conclusion

In this paper we motivated the need for environments that support the development and evaluation of similarity-based grouping procedures. We proposed a method that identifies the main components and steps that are important for such environments. Further, we illustrated the method by analyzing test cases for grouping of protein data entries with respect to their function and with respect to what classes of isozymes they belong to. The test cases illustrate the complexity of similarity-based grouping tasks. The choices made at the different steps in a grouping procedure have a large impact on the quality of the grouping results.

The test cases gave us also insights in different issues as well as interesting topics for future work. For instance, for the analyzed data source grouping on GO ANNOTATIONS combining GO_{ann} and GO_{ec} is best suited for grouping function and grouping on SEQUENCE is best suited for grouping on isozymes. Further studies are needed to investigate how other attributes can be useful for grouping tasks in life sciences. When grouping based on GO annotations, it is important to be aware of the fact that the annotations may be incomplete. In this paper we illustrated the possibility of partially compensating the lacking information by using mappings available at the GO Consortium. We observed that different mappings can be useful to different degrees. For instance, the mappings translating EC-NUMBER into GO terms (*ec2go*) gave good results, while mappings translating KEYWORDS in GO terms (*spkw2go*) were too general. When working with GO terms it is important to distinguish between general and more specific terms as they contribute differently to our knowledge. A number of test cases showed the importance for deeper studies to develop suitable methods to compare sets of GO terms and to explore their impact on the results.

The analysis and evaluation of the test cases was a time-consuming process. Tools are needed to support the different steps in our method. For instance, we need support at different levels of detail for the generation and analysis of the grouping results, the visualization and analysis of related data entries, and the analysis of the influence of external knowledge.

Acknowledgements

This research work was funded by CUGS (the Swedish national graduate school in computer science) and CENIIT (Center for Industrial Information Technology). The first and third authors are also members of the EU Network of Excellence REWERSE (Sixth Framework Programme project 506779, working group on a Semantic Web for Bioinformatics).

References

[AGMML90] Altschul SF, Gish W, Miller W, Myers EW, Lipman DJ (1990) Basic local alignment search tool. *Journal of Molecular Biology*, 215:403-410.

[BTS02] Berg JM, Tymoczko JL, Stryer L (2002) *Biochemistry*. W.H. Freeman and Company, New York.

[BBBDN05] Bilke A, Bleiholder J, Böhm C, Draba K, Naumann F (2005) Automatic Data Fusion with HumMer. Demo at *VLDB Conference*, pp 1251-1254.

[CSC05] Couto FM, Silva MJ, Coutinho P (2005) Semantic similarity over the gene ontology: family correlation and selecting disjunctive ancestors. *Conference on Information and Knowledge Management*, pp 343-344.

[DS05] Doms A, Schroeder M (2005) GoPubMed: Exploring PubMed with the GeneOntology. *Nucleic Acids Research*, 33:W783-W786.

[GH04] Gabaldon T, Huynen MA (2004) Prediction of protein function and pathways in the genome era. *Cellular and molecular life sciences : CMLS*, 61(7-8):930-944.

[GO00] The Gene Ontology Consortium (2000) Gene Ontology: tool for the unification of biology. *Nature Genetics*, 25(1):25-29. http://www.geneontology.org/.

[HGPWW04] Herbert KG, Gehani NH, Piel WH, Wang J, Wu CH (2004) BIO-AJAX: An Extensible Framework for Biological Data Cleaning. *SIGMOD Record*, 33(2):51-57.

[JAligner] Java implementation of the Smith-Waterman algorithm for biological sequence alignment. http://jaligner.sourceforge.net/

[KLKTB04] Koh JLY, Lee ML, Khan AM, Tan PTJ, Brusic V (2004) Duplicate Detection in Biological Data using Association Rule Mining. *ECML/PKDD Workshop on Data Mining and Text Mining for Bioinformatics*, pp 31-37.

[Lev66] Levenshtein VI (1966) Binary codes capable of correcting deletions, insertions, and reversals. *Soviet Physics Doklady*, 10(8):707-710.

[LSBG03] Lord PW, Stevens R, Brass A, Goble CA (2003) Investigating semantic similarity measures across the Gene Ontology: the relationship between sequence and annotation. *Bioinformatics*, 19(10):1275-1283.

[PPF95] Pfeifer U, Poersch T, Fuhr N (1995) Searching Proper Names in Databases. *Conference on Hypertext - Information Retrieval - Multimedia*, pp 259-275.

[SS02] Shamir R, Sharan R (2002) Algorithmic Approaches to Clustering Gene Expression Data. Chapter in *Current Topics in Computational Biology*, Jiang T, Smith T, Xu Y, Zhang MQ editors, MIT Press, pp 269-299.

[SFSZ05] Speer N, Fröhlich H, Spieth C, Zell A (2005) Functional Distances for Genes Based on GO Feature Maps and their Application to Clustering. *IEEE Symposium on Computational Intelligence in Bioinformatics and Computational Biology (CIBCB)*, pp 142-149.

[SKK00] Steinbach M, Karypis G, Kumar V (2000) A comparison of document clustering techniques. *KDD Workshop on Text Mining*.

[Str02] Strehl A (2002) *Relationship-based Clustering and Cluster Ensembles for High-dimensional Data Mining*. PhD thesis, University of Texas at Austin.

On Querying OBO Ontologies Using a DAG Pattern Query Language*

Amarnath Gupta and Simone Santini

San Diego Supercomputer Center
University of California San Diego, La Jolla, CA 92093, USA
{ssantini, gupta}@sdsc.edu

Abstract. The Open Biomedical Ontologies (OBO) is a consortium that serves as a repository of ontologies that are structured like directed acyclic graphs. In this paper we present a language DQL for querying a database of directed acyclic graphs. The query language has a comprehension style syntax and contains a pattern specification sub-language DPL. DPL can be viewed as an extension of tree-pattern query language like XPath. The language allows extraction of nodes, paths and subgraphs from DAGs, and permits construction of result structures by composing them. We show that using such a language on OBO ontologies (such as the gene ontology), we can express more complex and scientifically valuable queries.

1 Introduction

Query languages and query evaluation techniques for the retrieval and manipulation of graph-structured data have been investigated since the late 80s [1,2], through the era of object-oriented data models [3,4,5] up to the more recent general interest in semistructured data [6,7,8] and ontologies represented in RDF [9]. Graph-structured data appear naturally in many modern applications, especially in biological information systems [10], chemical structure analysis [11], and social network analysis. In these application domains, a surprisingly large fragment of graph-structured data turn out to be directed and acyclic. Specifically in the domain of biomedical and biological ontologies, the majority of the ontological structures are designed to be directed acyclic graphs (DAGs). The Open Biomedical Ontologies (http://obo.sourceforge.net/) is an umbrella consortium that serves as a repository of many different but often inter-related ontologies, where the nodes of the graphs represent terms used in the vocabulary of a specialized biological domain, and the edges between nodes are typically labelled by the strings "isa", "part-of" or "develops-from". Furthermore, given the multiplicity and categories of ontologies emerging today, new needs are developing to query across ontologies and composing ontologies together. As the ontologies grow and become more complex, searching through them will require a more complex query mechanism that natively operates on graphs, especially DAGs.

* Supported in part by NSF ITR Grant EIA-0205061, and the NLADR grant from NSF.

U. Leser, F. Naumann, and B. Eckman (Eds.): DILS 2006, LNBI 4075, pp. 152–167, 2006.

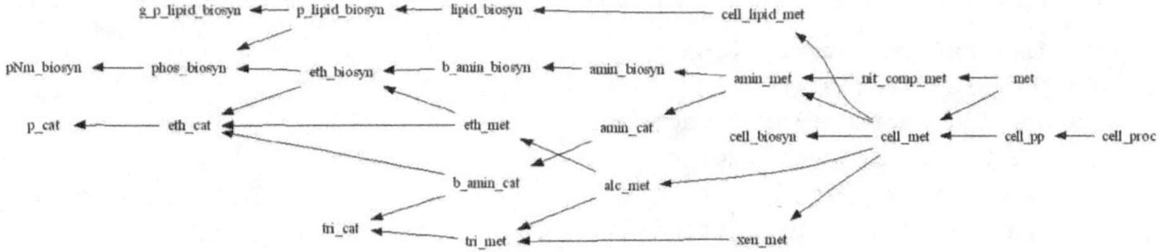

Fig. 1. A simplified fragment of the Biological Process component of Gene Ontology. The names of the nodes have been abbreviated for clarity.

Despite this need, most of the systems available to life scientists are mostly operated with visual interfaces allow only simple operations like keyword based node search, descendant enumeration, shortest path finding and neighborhood operations on graphs. This paper is an early step toward searching repositories of large ontological structures using a DAG query language, and similar in its intent as [12].

Example 1. As a motivational example, consider the well known Gene Ontology (GO) (`www.geneontology.org`) that consists of three DAG-structured components called biological processes (BP), molecular functions (MF) and subcellular components (SC). In Figure 1, a fragment of the BP DAG is shown. Here, an edge represents an *superclass* relation, such that $n_1 \rightarrow n_2$ means that the process n_2 is a specialization of the process n_1. Nodes in this graph represent tuples of a relation N which, in our simplified example, has three attributes *id*, *name* and *definition.* To make the node names simpler, just consider that a node with the substring "_met" is a metabolism process, a node with "_cat" is a catabolism process and a node with "_biosyn" is a biosynthesis process. Given this example DAG, a number of different types of queries can be asked:

1. Which biosynthesis processes under lipid biosynthesis are also classified as amine biosynthesis? (Q1)
2. How does phosphatidylethanolamine biosynthesis (phos_biosyn in Fig. 1) derive from cellular metabolism (cell_met)? (Q2)
3. Is there a case where a xenobiotic process (e.g., xen_met) is a subprocess of at least two forms of cellular metabolism? (Q3)
4. construct a reduced data graph by deleting all metabolism nodes except *met*, and connecting the non-deleted parent(s) of a deleted node n to its non-deleted children. (Q4)

Consider the first query. Since the graph represents a classification structure (i.e., an is-a graph) we interpret the expression "A classified as B" to mean "A reachable from B" in this DAG. Thus, this query can be expressed as the pattern query

reachable_from(X, lipid_biosyn) $\wedge$ reachable_from(X, amin_biosyn) $\wedge$ substr('biosyn', X) (Q1′)

```
root
    conditional phenotypes (cp)
    cell cycle defects (ccd)
    mating and sporulation defects (msd)
        mating efficiency (me)
        sporulation efficiency (se)
        inappropriate sporulation (is)
            KAR4
            RIM1 ***
            ABP1
            ...
        other mating and sporulation defects (omsd)
    ...
    cell morphology and organelle nutrients (cmon)
        flocculence (fl)
        budding mutants (bm)
            bud localization (bl)
            multibudded cells (mbc)
            pseudohyphae formation (phf)
                GDH3
                TEC1
                RIM1 ***
            ...
        ...
    stress response defects (srd)
    ...
```

Fig. 2. A fragment of the Yeast Phenotype Classification. Some genes (leaf level) like RIM1 have multiple parents. The edge from the parent term of a gene to the gene is "produced-by-mutating".

where the last predicate is a syntactic way to state that X is a biosynthesis process; the query would return the set of two nodes (*phos_biosyn, pNm_biosyn*). Notice that since X must be reachable from lipid_biosyn as well as amin_biosyn, the query expresses a DAG-pattern, notionally akin to tree pattern queries expressed by XPath queries. Of course, our data model is much simpler than XML models, in that we only consider child and descendant relationships and completely disregard order among the children of a node.

Example 2. Next consider the yeast phenotype classification (YPC) scheme (available at `http://mips.gsf.de/genre/proj/yeast/searchCatalogFirstAction.do?db=CYGD`) – the non-leaf nodes of this scheme represent phenotype terms, while the leaf nodes represent genes. Since one gene can be responsible for multiple phenotypes, the structure of the YPC (Figure 2) is a DAG when leaves are considered (such as the gene RIM1 in the figure) and is an is-a tree over the rest of the nodes. For the purpose of this discussion, we will ignore all the descriptors associated with the genes and the YPC terms – we will only use the fact that almost all yeast genes have references to GO-ids. Hence, the YPC structure is joinable with the BP ontology through the GO-ids. This enables us to ask scientific queries like: "merge all paths P_1 reachable from the node named *tanscription*, with all paths P_2 reachable from the yeast phenotype *sporulation defect* such that X is a node in P_1, Y is a node in P_2 and $X.id = Y.GOid$ by creating an edge labeled e from X to Y". Biologically, query creates an association between biological processes and relevant phenotypes. Computationally, the query first retrieves P_1 and P_2 through a join query and then merges them through a construction.

The intent of this paper is to present a query language and its corresponding algebra for the retrieval and manipulation of DAG-structured data to achieve the capabilities described above. The query language will have a sublanguage to express query patterns, a formal way to manage collections of intermediate result graphs, and operations to manipulate and construct graphs.

2 The DAG Data Model

As mentioned, the DAGs we consider have nodes that represent tuples from some relation N. With no loss of generality, we can assume that the *id* attribute of nodes is globally unique, so that nodes are represented by their ids. We only consider DAGs with unlabeled edges; we also assume that the children of any node of the data DAG are unordered. Henceforth, unless explicitly mentioned, we use the term *graph* (correspondingly, sub-graph) to mean this class of DAGs and its substructures.

In this paper, our focus is to introduce a language to manipulate the structure of the graph (we use the term graph to refer to DAGs from this point on); the retrieval and manipulation of the node content is performed using standard relational algebra. To manipulate the graph structure, on the other hand, we need to have a type system with basic types and type constructors. For example, to define a set of paths, we need to have the type set(list(skolem)) where skolem is the data type of all ids, and is a data type for which no operations are defined except value equality. In our model all nodes are typed. If α is the type of a node then we will use the notation $\nu[\alpha]$ for a generic collection monoid of type ν with elements of type α (e.g., set(GO-node-type)). For this monoid three functions are defined:

i) $\text{nil}_\nu : \nu(\alpha)$ (the empty collection);
ii) $\mathcal{U}_\nu : \alpha \to \nu(\alpha)$ (the singleton);
iii) $\_\boxed{\nu}\_ : \nu(\alpha) \times \nu(\alpha) \to \nu(\alpha)$ (the *join*, like the union operator for a set monoid).

We also define the projection operator $p_i : \alpha_1 \times \cdots \times \alpha_n \to \alpha_i$ and the record construction operator $(\_, \_, \ldots, \_) : \alpha_1 \times \cdots \alpha_n \to \langle \alpha_1 \times \cdots \times \alpha_n \rangle$. To introduce a few terms used through the paper, a node n of a graph g is *terminal* if it has no outgoing edges, and it is *initial* if it has no incoming edges; $\bot(g)$ is the set of all terminal nodes of the graph g, and $\top(g)$ is the set of initial nodes. Since our graphs are acyclic, for each g, neither $\bot(g)$ nor $\top(g)$ are empty.

3 Our Query Language *DQL*

We start by observing that the query language presented here assumes that the data is in the form of a DAG and not a graph containing cycles. While it is easy to show that the pattern language cannot express a cycle, we do not ask how the queries would behave if the underlying data had cycles.

3.1 An Informal Introduction

We introduce the query language using the graph G_1 shown in Figure 3 as the reference. First, we focus on the pattern language DP. The pattern $(v = 1)$ matches a set of nodes for which the value of the variable v is 1. The pattern *true* matches all nodes of the DAG. We use the symbol $-$ to denote an edge from the node to the left of $-$ to the node to the right of it. Thus, the pattern $(v = 1) - (v = 2)$ matches the edges $[1, 1] \to [3, 2]$ and $[2, 1] \to [4, 2]$. DP allows the use of the Kleene star to refer to 0 or more occurrences of the subpattern within its scope. The pattern $(v = 1)[-(v = 2)] * -(v = 1)$ matches the graphs have a node with $v = 1$ is followed by a chain of any number of nodes with $v = 2$, which is then followed by another node with $v = 1$ (not the same node as the first since the graph is acyclic). The edge chains matching the pattern are $[1, 1] \to [3, 2] \to [7, 1]$, $[1, 1] \to [3, 2] \to [2, 1]$, $[1, 1] \to [3, 2] \to [4, 2] \to [8, 1]$, and so on. Now, let us associate variables x and y to two elements of the pattern. The augmented pattern becomes $y : (v = 1)[-(v = 2)] * -x : (v = 1)$. Although here the variables are only associated with nodes, in general, variables can associated with any subpattern, such as an edge chain or a subgraph, as illustrated later in the paper. The variable association implicitly produces matches for the variables in addition to the match for the whole pattern. In this example, the pattern produces the y, x tuples $\{([1, 1], [2, 1]), ([1, 1], [7, 1]), ([1, 1], [8, 1]), ([2, 1], [8, 1])\}$ if we eliminate duplicates.

As the final element in this section, we would like to produce a graph for each xy pair by constructing an edge from each instance of x to its corresponding y. This operation of graph creation requires us to produce a set of edges,

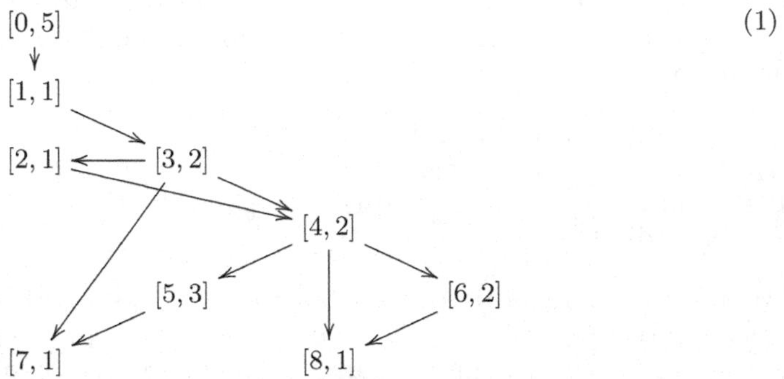

Fig. 3. Our running example. Each node has an id (the first number) and an attribute called v, whose value is shown in the second number.

which the pattern language cannot express because the pattern language only performs matching. To accomplish the graph construction, we place the pattern in a monoid comprehension framework, and express it as:

$$\cup[\{x - y | g \vdash y : (v = 1)[-(v = 2)] * -x : (v = 1) \leftarrow G_1\}]$$

which is read as: Let g be that substructure of G_1 that satisfies the specified pattern π. Using the variables x and y of g construct the edge $x - y$ for each instance x, y satisfying π, and form a set union of these edges.

3.2 Formal Description of Pattern Language *DP*

To formalize the ideas described in the previous subsection, we observe that pattern π in the pattern language *DP* is generated by the following rules:

i) A predicate C in which the free variables are the names of the components of the node data type is a pattern; in particular t (the value "true") is a pattern;
ii) if π_1 and π_2 are patterns, then $\pi_1 - \pi_2$ is a pattern;
iii) if π and π' are patterns, then $\pi'[-\pi]*$ and $[\pi-] * \pi'$ are patterns;
iv) if $\pi_1, \ldots, \pi_n$ are patterns, and ν is a patterns then $\{\pi_1-, \ldots, \pi_n-\}\nu$, and $\nu\{-\pi_1, \ldots, -\pi_n\}$ are patterns;
v) if $\pi_1, \ldots, \pi_n$ are patterns, then $\pi_1 | \ldots | \pi_n$ is a pattern;
vi) if π is a pattern and v a variable name, then $v : \pi$ is a pattern;
vii) nothing else is a pattern.

These cases are illustrated in Figure 4.

The grammar of the language is

$$\begin{array}{rl} <\pi> ::= & C | <\pi> - <\pi> \\ & | <\pi>[-<\pi>]* \\ & | [<\pi>-] * <\pi> \\ & | <\pi>\text{'\{'} - <\pi>\{, -<\pi>\}^{*}\text{'\}'} \\ & | \text{'\{'}<\pi> - \{, <\pi>-\}^{*}\text{'\}'}<\pi> \\ & | <\pi>|<\pi>\{|<\pi>\}^{*} \\ & | (<\pi>)|<\mathit{literal}> : <\pi> \end{array}$$

Note that the brackets { and } have been placed in quotes when they appear as terminals to avoid confusion with the repetition operator of the grammar. Condition have higher precedence that the structural operators, and – has precedence over |. Parentheses can be used whenever necessary. We use the shortcut $\# \equiv [-t] * -$ (or, equivalently, $-[t-]*$), making the symbol # the notional equivalent of // in XPath.

It is important to point out a few distinctive aspects of this DAG pattern language.

In the informal example, we stated that – represents an edge between two *nodes*. In this section, we generalize this notion to represent a "connection" between two subDAGs, one satisfying pattern π_1 and another satisfying pattern π_2

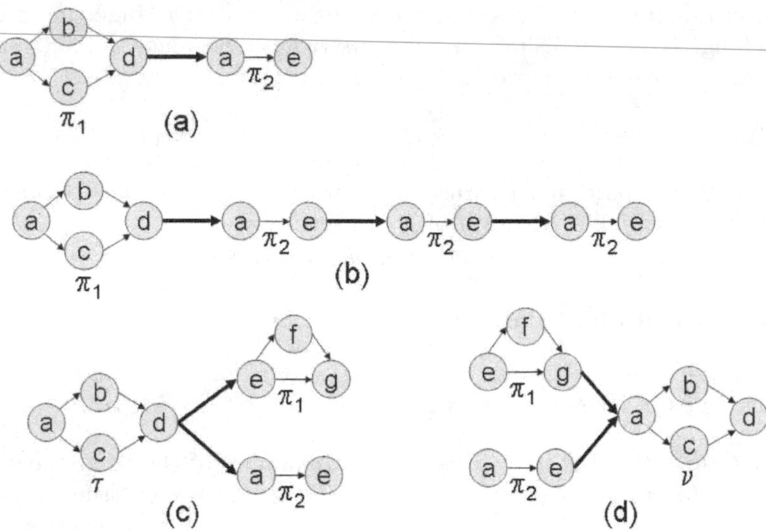

Fig. 4. Examples DAGs corresponding to the different patterns described in the text. (a) $\pi_1 - \pi_2$, (b) $\pi_1[-\pi_2](2, 4)$, (c) $\tau\{-\pi_1, -\pi_2\}$ and (d) $\{\pi_1-, \pi_2-\}\nu$.

(construction rule ii above). To this end, we define a *stitch* relationship ($|\rangle$), which generalizes the *child* relationship for tree-structured data. given two graphs g_1 and g_2, let $g_1|\rangle g_2$ be the graph obtained by connecting all terminal nodes of g_1 to all initial nodes of g_2. Thus the semantics of rule ii is that if combined pattern $\pi \equiv \pi_1 - \pi_2$, and the graph g matches π then there are two disjoint sub-graphs of g, namely g_1 and g_2 such that g_1 matches π_1, g_2 matches π_2, and $g = g_1|\rangle g_2$.

Next, we use the $|\rangle$ operation to generalize the twig structure of tree pattern languages. For DAGs we need both the *split* structure of trees denoting branches emanating from a node, as well as a *merge* structure denoting edges converging on to a node. In this vein, the expression within the $\{...\}$ (rule iv) is a branching where the patterns on the different branches are required to be distinct. Thus the whole pattern represents a fork ($\tau\{...\}$) or merge ($\{...\}\nu$) pattern, or a combination. Formally, if $\pi \equiv \tau - \{\pi_1-, \ldots, \pi_n-\}\nu$, then g matches π if there are disjoint sub-graphs of g $g', g_1, \ldots, g_n, g''$ such that g' matches τ, for each i g_i matches π_i, g'' matches ν, and, for all i $g'|\rangle g_i$ and $g_i|\rangle g''$ are sub-graphs of g.

Finally, the language offers a syntax to specify the number of recurrences of a pattern. We use the shortcuts $\pi^n \equiv \overbrace{\pi - \pi - \cdots - \pi}^{n}$, $\pi'[-\pi](m, n) \equiv \pi' - \pi^m | \pi' - \pi^{m+1} | \cdots | \pi' - \pi^n$ ($n > m$, the shortcut $[\pi-](m, n)\pi'$ is defined analogously), $\pi'[-\pi](m, \infty) \equiv \pi' - \pi^m[\pi]*$ (the shortcut $[\pi-](m, \infty)\pi'$ is defined analogously.

3.3 The DQL Language

The use of monoids and the comprehension syntax is common in query languages that allow complex types [13]. For example, the query Q1 presented in Example 1 can now be expressed as:

$\cup[\{x|g \vdash \{substr(name, \text{“lipid\_biosyn”}), substr(name, \text{“amin\_biosyn”})\}\#x : substr(name, \text{“biosyn”}) \leftarrow GO\}]$

Q2 and Q3 can be expressed with queries having a similar form. For example, Q3 can be expressed as:

$$\cup[\{z|g \vdash (name = \text{“cell\_met”})\#\{x,y\}\#z, (name = \text{“xen\_met”})\#z \leftarrow GO\}]$$

This is a conjunctive query the result variable z must satisfy two patterns. Note the x and y are implicitly existentially qualified, and by the semantics of the $\{\ldots\}$ construct the same node cannot instantiate both variables. Query Q2 (and all queries that ask "*what is the relationship between* nodes satisfying condition C_1 and those satisfying condition C_2?") is an example of a graph-returning query. It is simply formulated as:

$$\cup[\{x|g \vdash x : ((name = \text{“cell\_met”})\#(name = \text{“xen\_met”})) \leftarrow GO\}]$$

where the scope for the result variable x is the entire subgraph satisfying the given pattern.

Now we turn our attention from the pattern language DP to the monoid comprehension structure in which it is embedded. Since most of the monoids we need are standard for sets, lists, and arithmetic and if-then-else constructs, we do not describe them in detail here. However, in addition to the collection and the simple monoids of the comprehension calculus, graphs come with their own monoids, each one defined on the set of graphs, and characterized by a join operation. The most important are:

merge: puts together two graphs by identifying nodes with the same id;
gmax: given two graphs g_1, g_2, $g = \text{gmax}(g_1, g_2)$ is the smallest graph for which $g_1, g_2 \subseteq g$;
gmin: the largest graph contained in two graphs.

All these operators can be easily extended to take a set of graphs as input. We omit the proof that these operations are associative, as required by the definition of monoid. In the previous example, the query

$$merge\{x - y|g \vdash y : (v = 1)[-(v = 2)] * -x : (v = 1)\} \quad (2)$$

would return as a result the graph

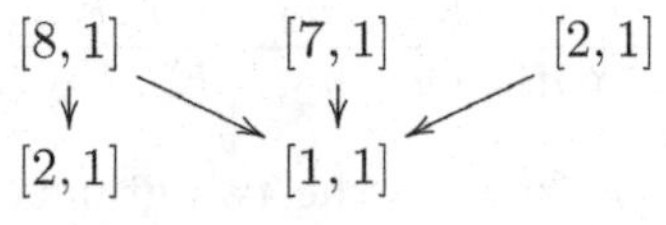

$$\begin{array}{ccc} [8,1] & [7,1] & [2,1] \\ \downarrow \searrow & \downarrow & \swarrow \\ [2,1] & [1,1] & \end{array} \quad (3)$$

As we have seen in some examples above, any graph that matches a fragment of the pattern can be assigned to a variable. This gives us the possibility of assembling a result out of portions of the graphs in the data base. Consider the constructive query presented as Q4 in Section 1. Abstractly, the query can be modeled as:

Given three conditions on nodes A (nodes with attribute name containing "_met"), B, C (these conditions are empty in example Q4), remove from the graph all the nodes which satisfy condition A; every time one of these nodes has a parent that satisfies B and a child that satisfies C, join the parent and the child.

Consider first this query:

$$merge\{x - y | g \vdash x : ([t-] * B) - A - y : (C[-t]*)\} \tag{4}$$

Here x represents the subgraph "up to" the B-satisfying nodes and y represents the subgraph beyond the C-satisfying nodes. The edge-construction between these nodes effectively deletes the nodes satisfying A from the output. This query performs the required job for the portion of the graph that contains nodes that satisfy the conditions on A, B, C. However, other portions of the graph do not match any pattern, and hence, will be lost. The solution in this case is to use the negation of A to match all paths that do not contain A, and then merge the graphs thus obtained:

$$\begin{aligned} &merge(merge\{x - y | g \vdash x : ([t-] * B) - A - y : (C[-t]*)\},\\ &merge\{z | g \vdash z : ([\neg A-] * \neg A)\}) \end{aligned} \tag{5}$$

4 Translating into an Algebra

The algebra in which the patterns are translated can be divided in two parts: on one hand there are the operations that deal with the values of each nodes, on the other there are the *structural* operations that manipulate the graph structure. The first part is fairly standard (e.g., textbook operations for relational systems, [13] for object-valued data, [3] for tree-valued data and so on). In this section, we will concentrate mainly on the second. The graph operations for a graph with nodes of type α work on three data types: the data type of the nodes themselves (α), that of *paths* of nodes (equivalent to lists of nodes, i.e. $[\alpha]$), and that of graphs ($\Gamma(\alpha)$), with the sub-typing relations $\alpha < [\alpha] < \Gamma(\alpha)$.

There are three graph manipulation operators in the algebra:

path: the call path(g, n_1, n_2, h, k) return the set of paths between the nodes n_1 and n_2 in the graph g such that the length of the path is between h and k; the typing of this function is

$$\frac{g : \Gamma(\alpha) \quad n_1, n_2 : \alpha \quad h, k : \text{int}}{\text{path}(g, n_1, n_2, h, k) : \{[\alpha]\}} \tag{6}$$

merge: the call merge(g_1, g_2) merges the two graphs g_1 and g_2 by identifying the nodes with equal value; the operator requires that the two graphs have at least one node that can be identified: it returns null for disconnected graphs; its typing is

$$\frac{g_1, g_2 : \Gamma(\alpha)}{\text{merge}(g_1, g_2) : \Gamma(\alpha)} \tag{7}$$

σ: the call $\sigma(g, P)$ returns the set of all nodes of the graph g that satisfy the predicate P; its typing is

$$\frac{g : \Gamma(\alpha) \quad P : \alpha \rightarrow \mathbf{2}}{\sigma(g, P) : \{\alpha\}} \tag{8}$$

The proof of the following property is quite obvious, and we don't report it here:

Theorem 1. *The algebra* $(path, merge, \sigma)$ *is minimal: none of its operators can be expressed as a combination of the others.*

In addition to the graph operators there are two structural operators: *apply* and *chain*.

apply: The operator apply$[\omega](A, f)$ applies the function f to all the elements of the structure A, and collects the results in a structure of type ω. It typing is:

$$\frac{A : \nu(\alpha) \quad f : \alpha \rightarrow \beta \cup \{\bot\}}{\text{apply}[\omega](A, f) : \omega(\beta)} \tag{9}$$

Formally, define the modified singleton for ω as

$$s'_\omega(x) = \begin{cases} s_\omega(x) & \text{if } x \neq \bot \\ 0_\omega & \text{if } x = \bot \end{cases} \tag{10}$$

then, if $A = a_1 \boxed{\nu} \cdots \boxed{\nu} a_n$ one has

$$\text{apply}[\omega](A, f) = s'_\omega(f(a_1)) \boxed{\omega} \cdots \boxed{\omega} s'_\omega(f(a_n)) \tag{11}$$

chain: given a set of paths S, a graph g that contains them, and two integers h, k, chain(g, S, h, k) builds all the chains that can be built out of paths in S taking each path between h and k times. Its typing is:

$$\frac{S : \{[\alpha]\} \quad g : \Gamma(\alpha) \quad h, k : \text{int}}{\text{chain}[\omega](g, S, h, k) : \{[\alpha]\}} \tag{12}$$

Consider now a pattern π for which a translation is sought in the previous algebra. Formally, the planning algorithm is a function plan(π, g, U) where π is the pattern for which a plan is sought, g is the variable name for the input graph, and U is the variable name for the set of *environments* which is the collection of instantiated pattern variables produced at any stage of the plan. The value of the function plan is a list of algebra functions and variable assignments. We give a couple of simple examples of plans, before going into the details of the algorithm. Here, as elsewhere, $u_1, u_2, \ldots$, and $p_{11}, p_{12}, \ldots, p_{ij}, \ldots$ are unique variable names generated by the planning algorithm.

plan($z : C$, g, e) =
    $u_1 = \sigma$(g, C);
    e = apply[set](u_1,
                fun x => (z $\mapsto$ x)
)

Note that we write a list of (in this case) two elements as a;b rather than [a,b] for ease of notation, and that we use the ML-style notation "fun x => v" for $\lambda x.v$. The value "(z $\mapsto$ x)" is the environment constructor: it creates an environment in which the only assignment is that of the value x to the variable z.

plan($z : C_1 - C_2$, g, e) =
    $u_1 = \sigma(\mathrm{g}, C_1)$;
    $u_2 = \sigma(\mathrm{g}, C_2)$;
    p_{12} = apply[set](u_1
            fun x_1 => apply[set](u_2,
                        fun x_2 => path(x_1, x_2);
            );
    e = apply[set](u_1,
            fun x_3 => (z $\mapsto$ x_3)
)

We illustrate the algorithm through an example. Consider the pattern

$$y : (C_1[-t] * C_2[-t](5,7) - x : (C_3[-C_4 - C_5] * -C_6) - C_7) \tag{13}$$

where $C_1, \ldots, C_7$ are suitable conditions on the nodes and t stands for the value *true*. The first rewriting consists in isolating the portions that are assigned to a variable (except for the variable that contains the whole pattern; this is necessary because, in the final algorithm we will have to create not only the sub-graphs that match the whole pattern, but also the sub-graphs that match the individual variables). We represent this rewritten pattern as follows:

$$\begin{array}{ccc} C_1\ [-t]*\ C_2\ [-t](5,7) & x & -\ C_7 \\ & | & \\ C_3 & [-C_4 - C_5]* - C_6 & \end{array} \tag{14}$$

Then we replace all the patterns with $[-t]$ or $[t-]$ with the path symbols #, $-$, or (a,b), which indicates a path of length between a and b:

$$\begin{array}{ccc} C_1\ \#\ C_2\ (5,7) & x & -\ C_7 \\ & | & \\ C_3 & [-C_4 - C_5]* - C_6 & \end{array} \tag{15}$$

Then we expand the "star" elements:

$$\begin{array}{ccccc} C_1\ \#\ C_2\ (5,7) & & x & - & C_7 \\ & & | & & \\ C_3 & - & * & - & C_6 \\ & & | & & \\ & C_4 & - C_5 & & \end{array} \tag{16}$$

The planning algorithm operates on this representation.

First, each repeated pattern is eliminated: For each pattern $[-\pi](n,m)$, the planning algorithm is called recursively to generate a plan for $x : \pi$, where x is a new variable, and then the function chin is used to generate the set of structures that match the repeated pattern. In other words, we have, for a path $[-\pi](n,m)$, the fragment

plan($x_1 : \pi$, g, u_1);
u_2 = apply[set](u_1
fun x_2 => $u_1(x_2)$ (Transform the set of environments into a set of graphs)
);
p_{45} = chain(g, u_2, n, m);

In the representation, the star operator is replaced by the name of the path set that contains the graphs that satisfy the pattern:

$$\begin{array}{ccccc} C_1 \# C_2 \ (5,7) & x & - C_7 \\ & | & \\ C_3 \ - & p_{45} & - C_6 \end{array} \tag{17}$$

Note that for the purpose of the paper, the name p_{45} to the variable that holds the path has been given for ease of exposition, since these are paths that go from nodes for which condition C_4 holds to nodes for which condition C_5 holds. The same convention will be followed in the paper for all variable names; the actual names used by the algorithm may, of course, vary.

Now the instructions are generated to replace each condition C_i with the set of nodes that satisfy it

$$U_1 = \sigma(g, C_1)$$
$$\vdots$$
$$U_7 = \sigma(g, C_7)$$

and the sets of paths that join contiguous nodes in the traversal of the structure are generated, with the conditions established for that path:

p_{12} = apply[set](U_1, fun x => apply[set](U_2, fun y => path(x, y, 0, infty))
p_{23} = apply[set](U_2, fun x => apply[set](U_3, fun y => path(x, y, 5, 7))
p_{34} = apply[set](U_3, fun x => apply[set](U_4, fun y => path(x, y, 1, 1))
p_{56} = apply[set](U_5, fun x => apply[set](U_6, fun y => path(x, y, 1, 1))
p_{67} = apply[set](U_6, fun x => apply[set](U_7, fun y => path(x, y, 1, 1))

The data structure is updated by eliminating the node sets and replacing each path symbol with the set of paths that implement it:

$$\begin{array}{ccc} p_{12} \sim p_{23} \sim & x & \sim p_{67} \\ & | & \\ p_{34} \sim & p_{45} & \sim p_{56} \end{array} \tag{18}$$

The next step of the algorithm is a traversal of this structure where, for each p_{ij} a loop is generated to chain (denoted by the symbol $\sim$) every path in it with the paths of the following set p_{jk}. In addition, the paths that depend on a variable are joined separately, and environments are created in which the paths are assigned to the variable. The path corresponding to each variable is expanded, going from the variables deeper in the structure towards the top. In this case there is only one variable, so there will be a single expansion:

```
p36 = apply[set](p34, fun x34 =>
    apply[set](p45, fun x45 =>
        apply[set](p56, fun x56 => merge(x34, merge(x45, x56)))
    )
)
```

The structure is then updated as follows:

$$p_{12} \sim p_{23} \sim p_{36} \sim p_{67} \tag{19}$$

and an entry is made in a variable table to associate the variable x with the set p_{36}. The operation is repeated until the complete structure has been eliminated. In this case there will be only one more generation:

```
p17 = apply[set](p12, fun x12 =>
    apply[set](p23, fun x23 =>
        apply[set](p36, fun x36 =>
          apply[set](p67, fun x67 => merge(x12, merge(x23, merge(x36, x67))))
        )
    )
)
```

Now all the structures that are necessary to contain the result are contained in the p variables: the final step is the construction of the set of environments; the apply functions loop over all the structure sets associated to output variables x and y:

```
U = apply[set](p17, fun x17 =>
        apply[set](p36, fun x36 => (x ↦ x36) ⊕ (y ↦ x17)
    );
```

The $\oplus$ operator creates the tuples of all x, y pairs that satisfy the plan.

The fundamental correctness result for the algorithm is the following:

Theorem 2. *Let π be a pattern in DP^{-v} the variable-free fragment of DP, g a graph, and U the set of environments created by the execution of the plan plan$(x : \pi, g, U)$, with $U = \{(x \mapsto q_i)\}$ then:*

i) $q_i \subseteq g$;
ii) $q_i \models \pi$.

The proof, not formally presented here, is conceptually very simple: it is based on the fact that all the paths that are generated are between nodes that satisfy the corresponding end-path conditions and therefore each path corresponds to a fragment of the pattern. The semantics of the chain operator guarantees that this is true for repeated patterns as well. The way in which the sub-patterns for variables are expanded guarantees that at the end of the plan each graph that has to be assigned to a variable is present in one of the p_{ij}.

5 Applying DQL to Life Science Problems

We are in the process of constructing a composite ontology for disease specific information by combining relevant substructures from multiple different ontologies and other standard databases. A full description of this on-going work is beyond the scope and page limit of this paper. Here we present a few illustrations of how the features of *DQL* are used in the task.

ICD-10 (`http://www3.who.int/icd/vol1htm2003/fr-icd.htm`) is a taxonomy that categorizes diseases based on the system (e.g., cardiopulmonary) they affect. The pathway ontology (`http://cvs.sourceforge.net/viewcvs.py/obo/obo/ontology/genomic-proteomic/pathway.obo`) relates certain diseases with the molecular pathways they affect. The "biological processes" fragment of the Gene Ontology relates major pathways to component pathways that constitute the major pathways. These component pathways often formally refer to the molecular elements or biological processes that participate in them. Finally, genes are functionally annotated by GO-ids to terms from the gene ontology. Thus it is notionally possible to start with a family of diseases per the ICD-10 classification, ultimately relate them to the biological processes and corresponding genes. In our preliminary experiments to construct such connections, we have successfully created integrated graphs for closely related neurodegenerative disorders (like Alzheimer's disease, Parkinson's disease, Lewy body disease etc.) and identified subgraphs that are common to these diseases. In performing these exercises, we have identified a number of "query patterns" that are very convenient to express with the *DQL*:

- "Find node n_1's reachability graph in G_1 until some node n_2 such that n_2 can be joined with some descendant of n_3 of graph G_2.
- "Find that subgraph of the n_1's reachability graph that reaches n_2 but not any n_3 that is reachable from both n_4 and n_5.
- "Merge two subgraphs found by subqueries S_1 and S_2 such that the merged nodes refer to the same GO-id or UMLS id". UMLS is a large vocabulary from the National Library of Medicine.

6 Related Work

Querying ontologies as graphs is a relatively new area of research. [12] has developed an algorithm to index DAG-structured data to make queries like transitive closure and least common ancestor more efficient. [14] has developed an algorithm to perform pattern matching queries on DAGs, and used in on the Gene Ontology. [15,16] have developed algorithms for DAG searching. However, to our knowledge this is the first attempt to develop a query language for DAG data, and apply it to address an emerging area of life sciences.

In terms of query languages, we mark distinction between *DQL* and schema-based graph query languages like [4,5] in that ours is a pattern language and does not operate in the paradigm of querying against a graph-schema. On the other hand, *DQL* is closely related to [13,17] on the one hand and XML query languages on the other. We view the primary contribution of this work in extending a monoid comprehension framework with a DAG-manipulating pattern language. We contrast our language with Lorel [6], UnQL [8] and StruQL [7] in two ways. 1) Our pattern sublanguage DP is specifically designed for DAGs (and not for general graphs) and although not shown here, can be proven to express serially connected minimal vertex series-parallel graphs (MVSPs) [18]. 2) Our language permits more powerful construction capabilities than these languages. Lorel does not have any graph restructuring operation, UnQL's graph construction operations are simpler than ours. StruQL is closer to our language; but StruQL was designed for web site construction and did not need nesting. DQL allows naturally allows nesting through environments, where at each level of nesting we can have selection, aggregation and construction.

7 Conclusion

In this paper, we have made the case that having the ability to query a repository of ontologies will provide a useful tool to enable new types of analysis that were not possible hitherto. To this end, we have presented the DQL query language and the DAG pattern definition sublanguage DP, a corresponding algebra, and a trace of the query planning process. In this paper, we have taken the narrow view that ontologies are merely DAGs and adopt a closed world assumption. The semantic aspect of ontologies that leads to knowledge representation and logical inference problems have been ignored. This allows us to focus on the formulation of structural queries. Even with structural queries alone, interesting life science problems can be addressed. We have not covered systems design and query evaluation algorithms in this paper.

References

1. Consens, M.P.: Graphlog: Real life recursive queries using graphs. Master's thesis, Dept. of Computer Science, University of Toronto (1989)
2. Agrawal, R., Jagadish, H.V.: Direct algorithms for computing the transitive closure of database relations. In: Proc. 13th Int. Conf. on VLDB. (1987) 255–266
3. Subramanian, B., Zdonik, S.B., Leung, T.W., Vandenberg, S.L.: Ordered types in the aqua data model. In: Proc. of the 4th Int. Workshop on Database Programming Languages (DBPL), London, UK, Springer-Verlag (1994) 115–135
4. Gyssens, M., Paredaens, J., den Bussche, J.V., van Gucht, D.: A graph-oriented object database model. IEEE Transactions on Knowledge and Data Engineering **6** (1994) 572–586
5. Poulovassilis, A., Levene, M.: A nested-graph model for the representation and manipulation of complex objects. ACM Trans. Inf. Syst. **12** (1994) 35–68
6. McHugh, J., Abiteboul, S., Goldman, R., Quass, D., Widom, J.: Lore: a database management system for semistructured data. SIGMOD Rec. **26** (1997) 54–66
7. Fernandez, M.F., Florescu, D., Levy, A.Y., Suciu, D.: Declarative specification of web sites with strudel. VLDB Journal **9** (2000) 38–55
8. Buneman, P., Fernandez, M., Suciu, D.: Unql: a query language and algebra for semistructured data based on structural recursion. The VLDB Journal **9** (2000) 76–110

9. Seaborne, A.: SPARQL query language for RDF. W3C Working Draft 21 (2005)
10. Zimnyi, E., dit Gabouje, S.S.: Semantic visualization of biochemical databases. In: Semantics of a Networked World: Semantics for Grid Databases, LNCS 3226. (2004)
11. Yan, X., Yu, P.S., Han, J.: Substructure similarity search in graph databases. In: Proc. ACM SIGMOD International Conference on Management of Data, New York, NY, USA, ACM Press (2005) 766–777
12. Tri"sl, S., Leser, U.: Querying ontologies in relational database systems. In: DILS '05: Proc. 2nd International Conference on Data Integration in Life Sciences. (2005)
13. Fegaras, L., Maier, D.: Towards an effective calculus for object query languages. In: ACM SIGMOD International Conference on Management of Data, San Jose, CA, ACM (1995) 47–58
14. Chen, L., Gupta, A., Kurul, M.E.: Stack-based algorithms for pattern matching on dags. In: Proc. 31st Int. Conf. on Very Large Databases (VLDB), Stockholm. (2005) 493–504
15. Vagena, Z., Moro, M.M., Tsotras, V.J.: Twig query processing over graph-structured xml data. In: WebDB '04: Proc. 7th International Workshop on the Web and Databases. (2004) 43–48
16. Wang, H., He, H., Yang, J., Yu, P., Yu, J.X.: Dual labeling: Answering graph reachability queries in constant time. In: ICDE '06: Proc. 22nd International Conference on Data Engineering. (2006 (to appear))
17. Fegaras, L., Elmasri, R.: Query engines for web-accessible xml data. In: Proceedings of the 27th Int. Conf. on Very Large Data Bases (VLDB), San Francisco, CA, USA, Morgan Kaufmann Publishers Inc. (2001) 251–260
18. Bang-Jensen, J., Gutin, G.: Digraphs: Theory, Algorithms and Applications. Springer-Verlag, London (2001)

Using Term Lists and Inverted Files to Improve Search Speed for Metabolic Pathway Databases

Greeshma Neglur[1], Robert L. Grossman[2], Natalia Maltsev[3], and Clement Yu[4]

[1] Laboratory for Advanced Computing,
University of Illinois at Chicago, Chicago, IL 60607, USA
neglur@lac.uic.edu
[2] Laboratory for Advanced Computing,
University of Illinois at Chicago, Chicago, IL 60607, USA
grossman@uic.edu
[3] Math and Computer Science Division,
Argonne National Laboratory, Argonne, IL 60439, USA
maltsev@mcs.anl.gov
[4] Department of Computer Science,
University of Illinois at Chicago, Chicago, IL 60607, USA
yu@cs.uic.edu

Abstract. This paper describes a technique for efficiently searching metabolic pathways similar to a given query pathway, from a pathway database. Metabolic pathways can be converted into labeled directed graphs where the nodes represent chemical compounds. Similarity between two graphs can be computed using a metric based on Maximal Common Subgraph (MCS). By maintaining an inverted file that indexes all pathways in a database on their edges, our algorithm finds and ranks all pathways similar to the user input query pathway in time, which is linear in the total number of occurrences of the edges in common with the query in the entire database.

1 Introduction

Understanding of the complex architecture of metabolic networks provides insights into the fundamental design principles underlying the structure and function of living organisms. Common ancestry leads to the similarity of many molecular functions observed in all domains of life (Eukaryotes, Prokaryotes and Archaeabacteria). However, differences in organisms' physiology and lifestyle result in divergent evolution and emergence of variants of metabolic organization and phenotypic features. Large amount of metabolic and proteomic data available in public databases now allows for systematic exploration of adaptive mechanisms that led to the diversification of biological systems and the emergence of metabolic pathways characteristic of particular taxonomic or phenotypic groups of organisms. Such evolutionary and comparative analysis of metabolic pathways represents one of the essential problems in life sciences and is essential for progress in medicine, biotechnology and bioremediation. Metabolic pathways corresponding to various metabolic processes in an organism may be represented as a labeled directed graph.

U. Leser, F. Naumann, and B. Eckman (Eds.): DILS 2006, LNBI 4075, pp. 168 – 184, 2006.

The basic elements of metabolic pathways are chemical reactions that include compounds (e.g., substrates, products) and enzymes. Hence, computing similarity between pathways involves matching the constituent reactions (that include substrates and enzymes) and the connectivity between them. Brute force methods of matching the structures of substrates, enzymes and the pathway itself involve graph isomorphism tests at three levels and turn out to be computationally very expensive. The problem is further complicated if we are trying to query a database consisting of thousands of pathways with an average pathway size of 20+ nodes. Hence, efficient techniques for querying pathway databases are essential.

Our technique is able to search and retrieve pathways from a database similar to a query pathway in *time linear in the total number of occurrences of the pathway edges that are in common with the query* in the entire database. We do this by employing a simple indexing technique that uses terms defined from pathway edges and inverted files containing these terms.

2 Related Work

Many metabolic pathway similarity computing algorithms [8, 9] are based on abstracting pathways as enzyme graphs, i.e., directed labeled graphs where nodes are labeled with enzyme EC [12] (Enzyme Commission) numbers and a directed edge from one node to another implies that the product of the former node is the substrate of the latter. EC numbers provide a hierarchical classification of enzymes based on the reactions they catalyze. The classification tree consists of 4 levels with a root. Each enzyme is assigned a string consisting of 4 numbers, each of which corresponds to a level, for example: 1.2.3.4. In [8] the algorithm takes as input a sequence of EC numbers representing the enzyme graphs, aligns one sequence to another and attempts to find all EC numbers with the same 4-level hierarchical numbers, scores the similarities and cuts the sequences by removing the identical EC numbers and each pair of sub-sequences is initialized to begin a new round of 3-level hierarchical EC number match and so on. Another polynomial time algorithm described in [9] uses Approximate Labeled Subgraph Homeomorphism. A disadvantage of the enzyme graph representation is that it does not incorporate the similarity verification (i.e., in terms of structural similarity or chemical formula/sequence similarity) of substrates and products[1] in the pathway graphs.

Another technique [11] overcomes this disadvantage by combining sequence information of substrates and enzymes with graph topology of the underlying pathway. Several algorithms that efficiently perform pairwise pathway comparison are known [8, 9, 11, 17, 19]. One of the popular techniques outlined in [18] is called PathBLAST, which performs pairwise protein-protein interaction network alignment to detect linear paths and clusters [19] that are conserved between different species. This approach incorporates a refined probabilistic model for protein interaction data and also includes an automatic system for laying out and visualizing the resulting conserved subnetworks. This method is useful in evolutionary analysis by comparing the same pathway from different organisms, but may not scale efficiently to search

[1] Products in a pathway are chemical compounds.

large databases as it performs pair wise comparison of the pathways and does not describe any graph indexing techniques.

The problem of graph indexing is a critical problem in the field of computational biology. Several efficient graph indexing techniques [20, 21, 22, 23, and 24] for semi structured/XML databases and complex graph databases have been proposed. DataGuide [20] describes efficient index techniques for path expressions and Apex [21] considers the adaptivity of index structure to fit the query load. Another popular XML indexing technique called HOPI [22] provides support for path expression search with wildcards. However, these techniques are optimal and more suitable for path expressions and tree-structured data than for arbitrary graph queries.

There has been a variety of prior work that has used paths to index graphs, including Shasha et al. [23] and the Daylight System [25]. More recently, Han et. al. [24] have used labels attached to frequent subgraphs to index graphs.

3 Background

Definition. A metabolic pathway is a *series of 2 or more interconnected enzyme-mediated (or spontaneous) chemical reactions* that take place in a cell. A chemical reaction consists of one or more substrates (chemical compounds) transforming into one or more products (chemical compounds) via an enzyme (represented with an Enzyme Commission Number or ECN). The basic structure of a bio-chemical pathway is shown in Figure 1.

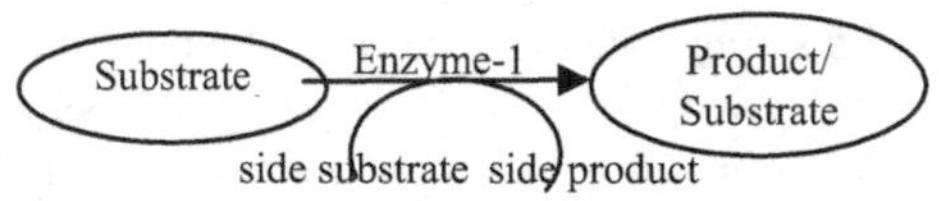

Fig. 1. Basic structure of metabolic pathway

Based on their functionality, pathways are classified into the following:

1. *Metabolic pathways*: consist of a series of chemical reactions occurring in an organism for energy production, synthesis of carbohydrates, etc. For example, photosynthesis.
2. *Signaling pathways*: consists of chemical reactions for information transmission and processing.
3. *Protein Interaction networks*: used to record pairs of proteins, which are experimentally observed to interact with each other.
4. *Gene regulatory networks*: orchestrate the level of expression for each gene in the genome by controlling whether and how vigorously that gene will be transcribed into RNA.

We will be chiefly dealing with metabolic pathways as most metabolic pathways in different organisms have been identified and a large collection of them can be found in various pathway databases. Also, the concepts developed here can be extended to index protein interaction networks, signaling pathways, gene regulatory networks and other directed or undirected graphs.

Metabolic Pathway Databases. There are several metabolic pathway databases [1], each of which stores thousands of pathways for different organisms. Differences between the databases are in their source of information, classification of pathways for organisms, level of detail, graph representations, etc. Metabolic pathway databases include: KEGG [2], ENZYME [3], BRENDA [4] and EcoCyc/BioCyc/MetaCyc [5].

In this paper, we primarily focus on MetaCyc. The MetaCyc database consists of pathways, reactions, enzymes, substrates and citation to source. It also contains Super-pathways: i.e., groups of pathways linked by common substrates. Pathways are represented as directed graphs with nodes for each enzyme and substrate; graph edges connect substrates. The BioCyc database also has a web-interface to retrieve pathways, given a single chemical compound name, Enzyme name or EC number or pathway name.

4 Terms and Definitions

4.1 Uniquely Labeled Graph

By a uniquely labeled directed graph, we mean a directed graph in which there is a unique label attached to each node in the graph. Figure 2 is a uniquely labeled directed graph. As we will see below in Section 4.5.2, there are a number of reasons that it is difficult to associate uniquely labeled graphs with pathways.

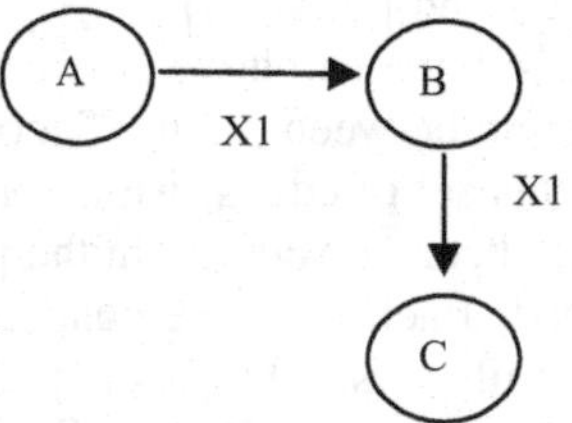

Fig. 2. Uniquely labeled directed graph

4.2 From Directed to Undirected Graphs

Graphs representing metabolic pathways are generally directed. To compute the common subgraph between the query and the pathways in the database, we first enumerate all the directed edges in common between the query and each of the indexed pathways. Next we form the corresponding undirected graph from the set of common edges for each, as in Figure 3. Hence, we are not losing the directionality information, it is already being considered in the index.

A directed graph is said to be weakly connected if its undirected version is connected, i.e., there is a path from each vertex to any other vertex in its undirected version.

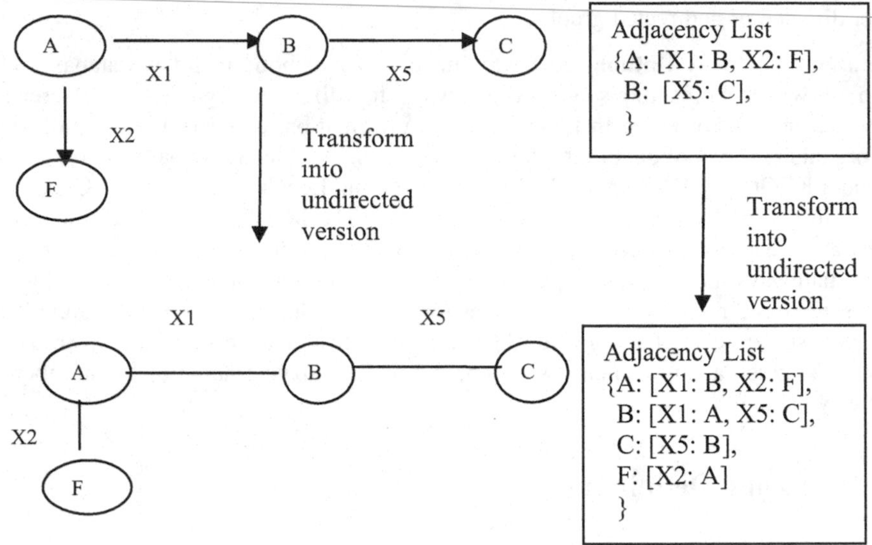

Fig. 3. Weakly connected directed graph

4.3 Common Subgraph

A common subgraph between two directed graphs Q and P represents a subgraph in common between Q and P. For example, the common subgraphs between P and Q in Figure 4 are {A:X1:B, A:X2:F} and {C:X3:D}.

We compute the common subgraphs between two pathways by first enumerating all directed edges in common between them. Second, to obtain the component subgraphs in common from this set of edges, we ignore the directions of these edges (as the directions have already been considered in the previous step) and compute the connected components formed. Each of these connected components is a common subgraph between the two pathways. For example: The set of edges (A:X1:B, A:X2:F, C:X3:D) in common between P and Q form two common subgraphs {A:X1:B, A:X2:F} and {C:X3:D}.

4.4 Maximal Common Subgraph

A common subgraph that has the maximum number of edges is called a Maximal Common Subgraph. In Figure 4, out of the two subgraphs in common {A:X1:B, A:X2:F} is maximal.

4.5 Metabolic Pathways as Uniquely Labeled Directed Graphs

4.5.1 Motivation

Metabolic pathways are representations of reactions. In general, there is no unique way to describe a reaction. For example, different researchers may expand or contract the definition of a reaction by incorporating more compounds or fewer compounds respectively. For this reason, graphs that abstract reactions are not unique. In contrast, graphs that represent chemical structures are unique. Even so, by coding pathways as uniquely labeled graphs, we can speed up many common pathway queries, which is the point of view we take in this paper.

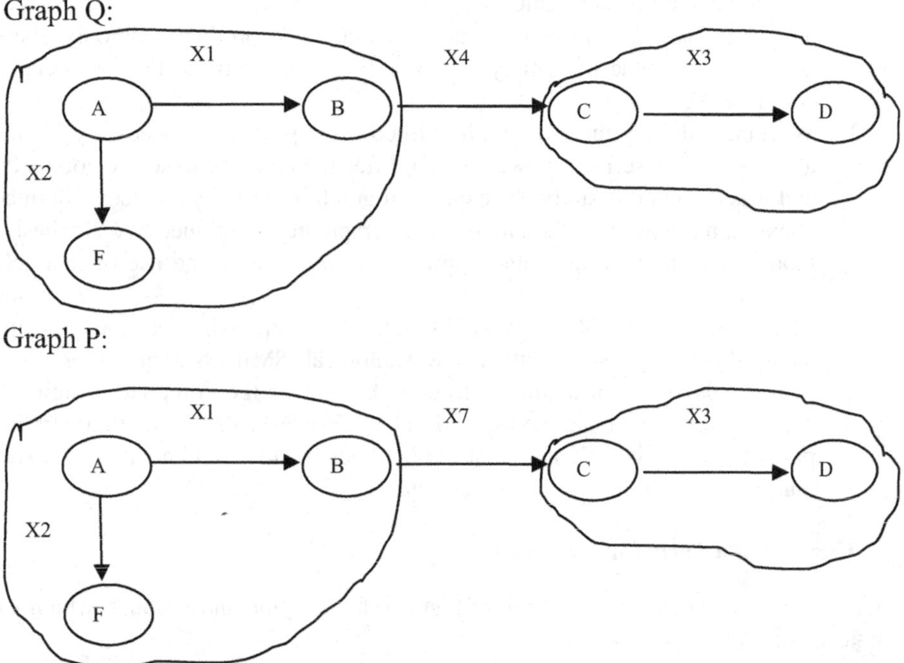

Fig. 4. The common subgraphs between graph P and graph Q are {A:X1:B, A:X2:F} and {C:X3:D}

4.5.2 The Labeled Graph Associated with a Pathway

We associate a labeled graph to a pathway as follows:

1) The nodes of the graph are the chemical compounds in the pathway.
2) Two nodes are connected by an edge when there is a chemical reaction in the pathway transforming one node into the other.
3) The edge is labeled with the enzyme using the EC number of the enzyme.

For the study described in this paper, we perform the following two additional steps:

4) We label the nodes with Canonical SMILES string of the chemical compound associated with the node. The Canonical SMILES string can be obtained from the PubChem database [6].
5) We identify all nodes whose labels are the same. That is, if 1) – 4) above define the graph G, then the graph associated with the pathway is the graph $G' = G / \sim$, where $\sim$ is the equivalence relation defined as follows: $u \sim v$ in case the nodes u and v in G have the same label defined in Step 4.

Figure 5 illustrates Steps 1) – 4) above. The names of the compounds in the nodes have been skipped for brevity. Instead, just letters A, B, C are used to denote different compound names.

Some remarks about this construction:

1. Although Canonical SMILES are not unique [16], for the purposes of the work here, they are adequate.
2. This construction ignores side substrates and side products, although these are easy to include by simply incorporating them into the label associated with an edge.
3. Working with the quotient graph defined in Step 5) is not necessary for the algorithm we describe below, but simplifies the algorithm and the code a bit and was used in this study. The quotient graph is uniquely labeled as defined above in the sense that the labels of the graph are all unique. We emphasize though that the algorithm and approach do not require the use of uniquely labeled graphs.
4. Some pathways in MetaCyc could not be converted into a labeled graph using these steps since either the Canonical SMILES strings for some compounds were not available from PubChem or the complete 4-digit EC numbers were not available from MetaCyc. However there are relatively few pathways like this. We note that the methods in [15] could be used to assign unique labels for chemical compounds.

4.6 Concept of a Term for a Pathway

A term is an ordered-triplet consisting of a substrate, enzyme and product, which we denote as follows:

substrate:enzyme:product. (term)

Note that a term represents an edge in the uniquely labeled graph of the pathway, or one or more edges in the labeled graph associated with a pathway. Because of this, we use “term” and “edge” synonymously below. For example: C=C1CN:2.4.8.1: C2C3=SC is a term (refer to the above pathway in Figure 5). Terms will be used below to build an index structure, which is an inverted file.

4.7 Pathway Vector

Given a labeled graph and an ordering of terms (for example, a lexigraphical ordering), one can associate a vector with the pathway. The component of the vector associated with a term is simply the number of times the term occurs in the graph. Note that if the graph is uniquely labeled, then the pathway vector is a binary vector in the sense that each component is either 0 or 1.

As we will see below, treating pathways as vectors of terms and using inverted files provides a natural index. For the pathways in Figure 4 the vectors are as below, the top row represents the terms of the pathway graphs.

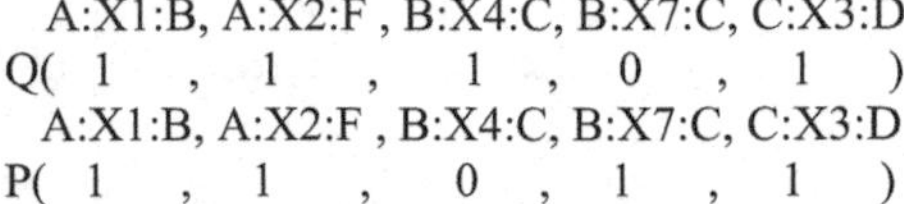

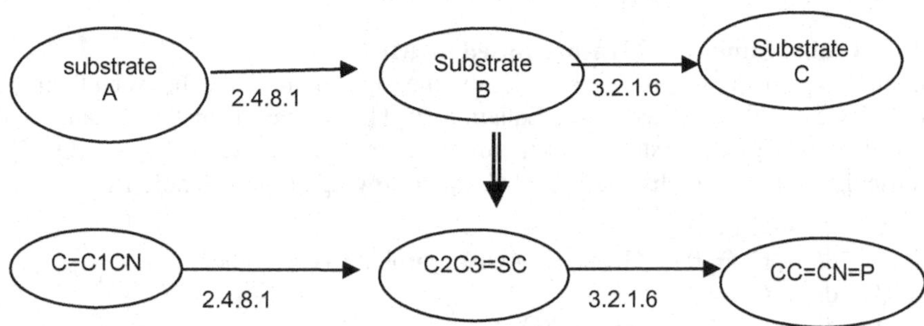

Fig. 5. Transformation of a pathway into uniquely labeled directed graph representation

4.8 Pathway Adjacency List

To capture the structure of the pathway, we can represent it using an adjacency list, which for each node 'p' in an undirected graph, stores a list consisting of nodes that are incident at 'p', together with the label of the corresponding edges. For the undirected pathway in Figure 3 the adjacency list is to the left in Figure 6. The adjacency list for the node with label 'C' consists of a single adjacent node 'B' via an edge with label 'X5'. The adjacency list representation can be converted into a vector and vice versa. Similarly, for a directed graph the adjacency list for each node 'p' is a set of nodes that are pointed to by 'p'. For example, the adjacency list for the directed pathway in Figure 3 is to the right in Figure 6.

```
Adjacency List
{A : [ X1:B,  X2:F],
 B : [ X1:A, X5:C],
 C : [X5:B],
 F : [X2:A]
 }
```

```
Adjacency List
{A : [ X1:B,  X2:F],
 B :  [ X5:C],
 }
```

Fig. 6. Left: Adjacency list for undirected graph in Fig 3. Right: Adjacency list for the directed graph in Fig 3.

4.9 Similarity Functions

4.9.1 Cosine Similarity Function

Given two pathway vectors 'Q' and 'G' the measure of cosine similarity between them can be computed as in Equation 1. The cosine similarity is a measure of the number of terms/edges in common between the pathway vector 'Q' and vector 'G' [Salton and McGrill 1983].

$$F(Q,G) = \frac{\sum_{i}^{n} q_i G_i}{\sqrt{\sum q_i^2}\sqrt{\sum G_i^2}} \quad \ldots\ldots (1)$$

where $Q = (q_1, q_2, \ldots\ldots, q_n)$ and $G = (G_1, G_2, \ldots\ldots, G_n)$

4.9.2 Global Similarity Measure Based on MCS

Given two pathway graphs 'Q' and 'G' the measure of similarity between them based on MCS can be computed as in Equation 2. This is based on the distance metric described in [7], which states that the distance metric d(Q, G) = 1 - Sim(Q, G) is a metric, i.e., for any graphs G_1, G_2 and G_3 the following properties hold true:

1. $0 \le d(G_1, G_2) \le 1$,
2. $d(G_1, G_2) = 0 \Leftrightarrow G_1$ and G_2 are isomorphic to each other,
3. $d(G_1, G_2) = d(G_2, G_1)$,
4. $d(G_1, G_3) \le d(G_1, G_2) + d(G_2, G_3)$.

$$Sim(Q, G) = \frac{|mcs(Q, G)|}{Max(|Q|, |G|)} \quad \ldots\ldots (2)$$

Where mcs(Q, G) is the Maximal Common Subgraph between Q and G and |G| is the size of the graph in terms of number of edges (E) in the graph.

5 Algorithms

The following sections describe in detail the algorithms employed for preprocessing the database, building an index structure for the database, and for searching and displaying pathways similar to the user input query pathway in descending order of similarity.

5.1 Database Preprocessing

The database preprocessing consists of the following two steps:

- Transforming the pathways stored in the database to a labeled graph.
- Building of the index structure for the database, which indexes each term or edge consisting of a triplet of the form substrate:enzyme:product

5.2 Database Index Structure

We construct an indexed structure for the database that indexes pathways based on terms/edges present in them. Hence for each term/edge in the index structure, there will be a list of pathways that contain that term/edge, i.e., a list of pathways that contain the corresponding edge. Each pathway has a unique identifier that is stored in the index structure. The size of the pathway is computed and stored as well. An example is given in Figure 7.

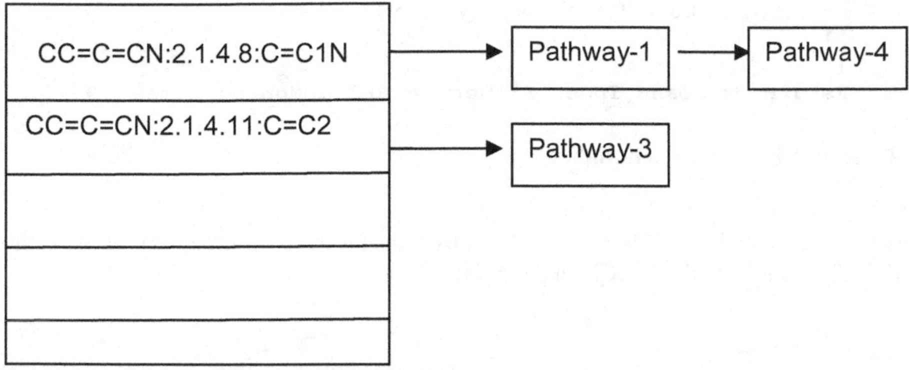

Fig. 7. Example index structure

5.3 Single Pass Algorithm for Indexing the Database

Here is the pseudo-code of the single pass algorithm for generating an index structure for a database of pathways to facilitate pathway queries:

```
//database index structure indexes terms/edges and stores all
//the pathways that have the term as a list pointed to by the
//term.
Define Database_Index_Structure as a hash table

//temporary data structure which records the unique canonical
//string for a given compound name obtained from the PubChem
//database. This facilitates a quick local check to see if a
//unique string for the compound was obtained from
//PubChem previously and reduces the number of calls to the
//PubChem database.
Define Compound_Names as a hash table

  For each pathway in the DB do
{
  Pathway_Vector = ( )
  Begin a Breadth First Traversal
    For each edge(substrate:enzyme:product) encountered in
  the pathway traversal do
  {
    Term = ''
    For each substrate in the edge do
          If compound does not have unique string in
          Compound_Names Then {
                  Connect to PubChem, obtain unique string
                  Insert entry into Compound_Names
      }
      Append substrate unique string to the Term
      Append the enzyme EC number to the Term
      Append product unique string to the Term
      Insert term into Database_Index_Structure along
            with Pathway id
```

```
            Append term to Pathway_Vector
        }
    }
    Return Database_Index_Structure and Compound_Names
```

5.4 Search Engine Functionality

5.4.1

Given a query pathway; for each of the substrates/products in the query retrieve the SMILES string from the PubChem database.

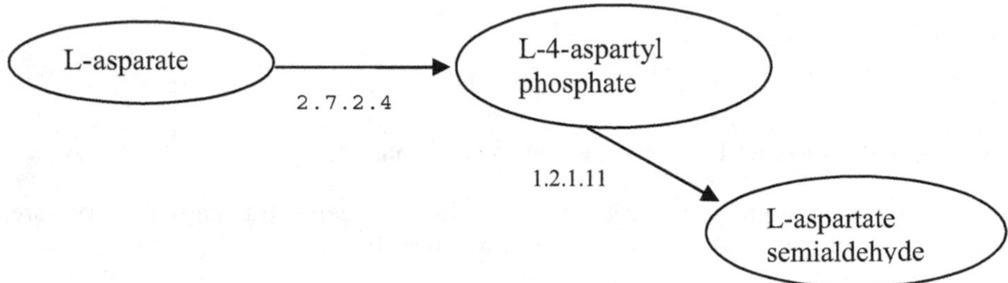

Fig. 8. Example user query to be retrieved from MetaCyc database

For the query in Figure 8: We submit the names L-asparate, L-4-aspartyl phosphate and L-asparate semialdehyde to PubChem to obtain strings:

(1) L-asparate – C(C(C(=O)O)N)C(=O)O
(2) L-4-aspartyl phosphate – C(C(C(=O)O)N)C(=O)OP(=O)(O)O
(3) L-asparate semialdehyde - C(C=O)C(C(=O)O)N

5.4.2

Submit the new query graph represented in the form of a labeled graph (as shown in Figure 9 for the query graph) to the search engine, which will retrieve all the similar pathway graphs from the given database, e.g., MetaCyc, and display them in descending order of the similarity measure calculated using a similarity metric chosen by the user, i.e., either based on maximal common subgraphs or cosine similarity. Details of the algorithm are given in the next section.

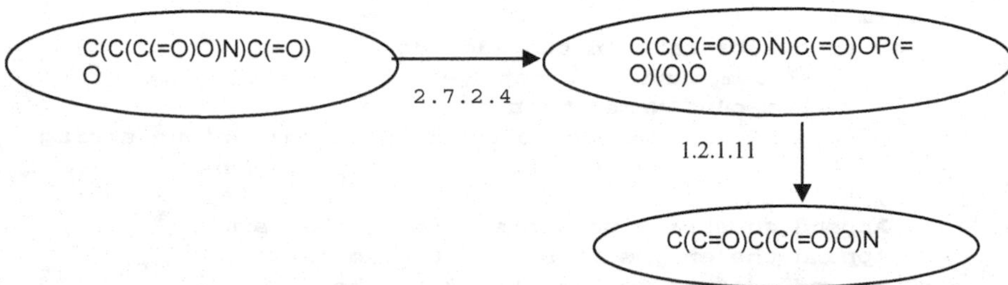

Fig. 9. Labeled graph representation for graph in Fig. 8

Q = { A:5.3.1.9:B, B:10.10.10.10:C, C:2.7.2.3:D, D:5.4.2.1:E, E:4.2.1.11:F, F:2.7.1.40:G }

P1 = { A:5.3.1.9:B, B:2.7.1.11:H, H:4.1.2.13:I, H:4.1.2.13:J, J:5.3.1.1:I, I:1.2.1.12:C, C:2.7.2.3:D, D:5.4.2.1:E, E:4.2.1.11:F , F:2.7.1.40:G }
P2 = { K:2.7.1.147:A, A:5.3.1.9:B, B:2.7.1.146:H, H:4.1.2.13:I, H:4.1.2.13:J, I:X:D, J:5.3.1.1:I, G:2.6.1.2:L, G:1.2.7.1:M , M:6.2.1.13:N, D:5.4.2.1:E, E:4.2.1.11:F, F:2.7.1.40:G }

Fig. 10. Example query graph 'Q' and example database with two pathways 'P1' and 'P2'

5.5 Algorithm for Computing Similarity Based on MCS and Cosine Similarity

The words edges and terms are used synonymously.

Overview. The important tasks performed by the algorithm are as follows:

Step 1.1 For each edge given in the query pathway; find all the database pathways that have the edge.

Step 1.2 For each pathway obtained in Step 1.1; find all the common edges between the pathway and the query graph.

Step 1.3 Represent the common edges in the form of adjacency lists (refer to section 4.8 for details) for the undirected graph (this is required to obtain the maximal common subgraph defined as a connected component). The common edges for the graphs 'P1' and 'P2' of Figure 10 are shown in Figure 11.

P1 = { A:5.3.1.9:B, C:2.7.2.3:D, D:5.4.2.1:E, E:4.2.1.11:F, F:2.7.1.40:G} = 5 common edges
P2 = { A:5.3.1.9:B, D:5.4.2.1:E, E:4.2.1.11:F , F:2.7.1.40:G} = 4 common edges
An adjacency list representation would retain the graph structure as follows :

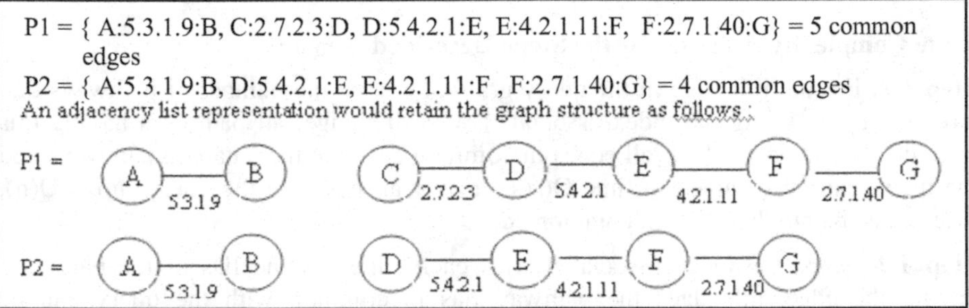

Fig. 11. Common edges after Step 1 for the example in Fig 10

Step 2. For each pathway with common edges found above, perform a simple Depth First Traversal (DFT) on the undirected graph obtained in Step 1. The connected components (trees) obtained in the Depth First Traversal forest will represent the common subgraphs between Q and the pathway. The common subgraphs/connected components obtained from Figure 11 are shown in Figure 12.

Step 3. From the above computed common subgraphs, find a maximal subgraph and use it to compute the similarity measure based on Eq 2 (MCS similarity). Use the set of common edges to compute the similarity measure based on Eq 1 (Cosine

Similarity). Rank the pathways in descending order of similarity based on the similarity measure chosen by the user.

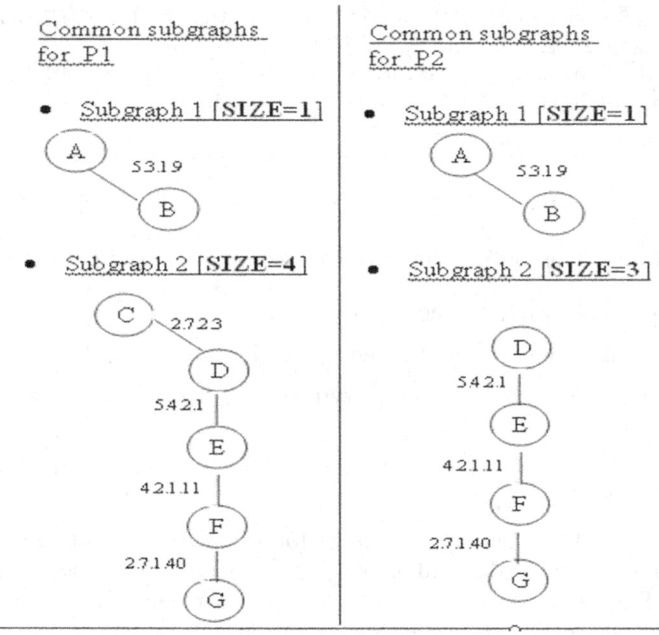

Fig. 12. Common subgraphs found for 'P1' and 'P2'

Time Complexity Analysis for the Steps Described Above:

Step 1.1: For the i'th edge in the query graph, let n_i be the number of pathways that have the edge. Using the index associated with this edge, all pathways having this edge can be obtained. Thus, all edges in common between the database pathways and the query can be obtained in time O((sum over all edges in the query) n_i) = O(n), where n is the number of such common edges.

Step 1.2: Since the index associated with each edge has the IDs of the pathways having the edge, all edges the pathway has in common with the query can be assembled in linear time. Thus, this sub-step also takes O(n) time.

Step 1.3: This is the same as Step 1.2, except that the representation of the edges is in the form of adjacency lists. This takes O(n) time.

Step 2: The depth-first algorithm to find all connected components based on adjacency lists can be obtained in linear time, i.e., O(n) [14].

Step 3: During the computation of the connected components in Step 2, we keep track of the sizes of the connected components for each database pathway that has common edges with the query. Every time that an edge is added to a connected component, its size is increased by 1. Thus, the time to compute the sizes of the connected components is bounded by O(n).

The sizes of the database pathways are pre-computed in Algorithm 4.3 and stored. Thus, the time to compute the two similarity measures is bounded by O(n + Q) = O(n), assuming that each edge in the query Q occurs in some database pathway.

Hence, the search time/retrieval time given a query pathway graph is linear in the total number of edges (n) in common with the query in the entire database.

6 Experimental Results

We performed some preliminary experimental studies and verified that the performance is approximately linear. We have implemented the algorithms in python and have provided a PHP [13] web interface to the search engine that can be accessed at [10]. We ported the search engine onto a web server on an Intel® Xeon™ CPU 2.4GHz and 1GB RAM. We analyzed pathway query graphs of different sizes, and plotted the algorithm's performance as shown in Figure 13 for the MetaCyc database.

No. of input edges	Total No. of common edges in the database **(X axis)**	No. of output pathways	Retrieval time in secs **(Y axis)**
1	2	2	0.00075
1	6	6	0.00088
1	13	13	0.00124
3	28	16	0.00181
3	40	17	0.00241
6	55	21	0.00332
7	61	26	0.00372
8	66	28	0.0041
9	72	34	0.0046
10	84	27	0.0057

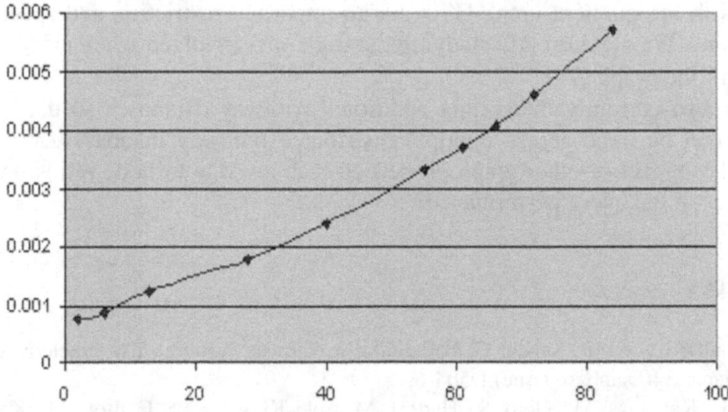

Fig. 13. Bottom: X-axis: total no. of edges in common with the query in the entire database, Y-axis: retrieval time in secs. Top: tabular description of performance.

The Metacyc dataset we used has 547 pathways, 4955 enzymatic reactions, 1940 enzymes and 3551 chemical compounds. The total number of unique edges in the database that are computed and stored in the index is 2294. The total number of edges in the entire database is 41561 and the average pathway size is 78 edges. The maximum length of the list of pathways corresponding to an edge in the index is 41.

The time to compute the index was 11.77 seconds. Since this was a very small database the queries of varying sizes were selected at random to target different sets of pathways in the database. Also, queries consisting of most occurring edges and least occurring edges were formed.

We are working on extending our algorithm to larger databases such as KEGG [2] and EMP[26] to test its scalability and performance. We are also currently working on performing a comparative analysis of our algorithm with other popular graph indexing techniques like GraphGrep [23] or GIndex [24].

7 Summary and Conclusion

Finding similar pathways has important applications in drug discovery and in the study of evolution. For this reason, it is important to develop efficient techniques to compute pathway similarity. In this paper, we have introduced an algorithm for retrieving similar pathways that uses an inverted file for the pathway database and indexes all the pathways in the database based on their edges. In this way, we are able to find and rank all the pathways similar to the user input query pathway in linear time.

We have implemented this algorithm using data from the MetaCyc database and provided a web interface. We have also performed a preliminary experimental analysis showing that the queries are indeed linear as expected.

The study in this paper assumed that the graph associated with a pathway was uniquely labeled. This is easily accomplished by working with the quotient graph G,′ defined above, instead of the graph G. Essentially the same algorithm described in Section 4 works for the graph G – the only difference is that the components of the pathway vector are not restricted to 0 or 1, but can be any positive integer. In practice, working with the quotient graph G' is not an important restriction and speeds up the computations. We are currently studying the trade-offs involved when using the graph G instead of the graph G'.

We are also currently integrating additional pathway databases so that similarity searching can be done across multiple distributed pathway databases. We are also studying the impact of other graph properties such as edge to node ratios, number of cliques, etc. on the query performance.

References

1. Bader GD, Cary MP, Sander C. Pathguide: a pathway resource list. Nucleic Acids Res. 2006 Jan 1;34(Database issue):D504-6.
2. KEGG - Kanehisa, M., Goto, S., Hattori, M.,Aoki-Kinoshita, K.F., Itoh, M., Kawashima, S.,Katayama, T., Araki, M., and Hirakawa, M.; From genomics to chemical genomics: new developments in KEGG. Nucleic Acids Res. 34, D354-357 (2006).
3. Bairoch A. The ENZYME database in 2000 Nucleic Acids Res 28:304-305(2000).

4. BRENDA, enzyme data and metabolic information Schomburg, I., Chang, A., Schomburg, D. Nucleic Acids Res. (2002) 30, 7-9
5. MetaCyc - Cynthia J. Krieger, Peifen Zhang, Lukas A. Mueller, Alfred Wang, Suzanne Paley, Martha Arnaud, John Pick, Seung Y. Rhee, and Peter D. Karp (2004) MetaCyc: A Multiorganism Database of Metabolic Pathways and Enzymes Nucleic Acids Research, 32(1):D438-42.
6. PubChem database : http://pubchem.ncbi.nlm.nih.gov/
7. Horst Bunke , Kim Shearer, A graph distance metric based on the maximal common subgraph, Pattern Recognition Letters, v.19 n.3-4, p.255-259, March 1998
8. Ming Chen, Ralf Hofestaedt, PathAligner: Metabolic Pathway Retrieval and Alignment, *Applied Bioinformatics*, 2004, 3(4): 241-252.
9. Ron Pinter et al. - Tree-based Comparison of Metabolic Pathways
10. Metabolic Pathway Search Engine - http://data.dataspaceweb.net/pathways/Search.php
11. Forst CV, Schulten K. Evolution of metabolisms: a new method for the comparison of metabolic pathways using genomics information.. J Comput Biol. 1999 Fall-Winter;6(3-4):343-60
12. EC-Published in Enzyme Nomenclature 1992 [Academic Press, San Diego, California, ISBN 0-12-227164-5 (hardback), 0-12-227165-3 (paperback)] with Supplement 1 (1993), Supplement 2 (1994), Supplement 3 (1995), Supplement 4 (1997) and Supplement 5 (in Eur. J.Biochem. 1994, 223, 1-5; Eur. J. Biochem. 1995, 232, 1-6; Eur. J. Biochem. 1996, 237, 1-5; Eur. J. Biochem. 1997, 250; 1-6, and Eur. J. Biochem. 1999, 264, 610-650; respectively) [Copyright IUBMB].
13. Rasmus Lerdorf, Kevin Tatroe. Programming PHP. Published: 05/04/2002 ISBN: 1565926102
14. Thomas H. Cormen, Charles E. Leiserson, Ronald L. Rivest and Clifford Stein. Introduction to Algorithms, Second Edition. Section 22.3 Depth First Search.
15. Robert L. Grossman, Pavan Kasturi, Donald Hamelberg, Bing Liu, An Empirical Study of the Universal Chemical Key Algorithm for Assigning Unique Keys to Chemical Compounds, Journal of Bioinformatics and Computational Biology, 2004, Volume 2, Number 1, 2004, pages 155-171.
16. Greeshma Neglur, Robert L. Grossman, Bing Liu: Assigning Unique Keys to Chemical Compounds for Data Integration: Some Interesting Counter Examples. DILS 2005: 145-157.
17. Kelley, B. P., Sharan, R., Karp, R., Sittler, E. T., Root, D. E., Stockwell, B. R., and Ideker, T. Conserved pathways within bacteria and yeast as revealed by global protein network alignment. Proc Natl Acad Sci U S A 100, 11394-9 (2003).
18. Kelley, B. P., Yuan, B., Lewitter, F., Sharan, R., Stockwell, B. R., and Ideker T. PathBLAST: a tool for alignment of protein interaction networks. Nucleic Acids Res.32(Web Server issue):W83-8. 2004.
19. Sharan, R., Suthram, S., Kelley, R. M., Kuhn, T., McCuine, S., Uetz, P., Sittler, T., Karp, R. M., and Ideker, T. Conserved patterns of protein interaction in multiple species. Proc Natl Acad Sci U S A. 8:102(6) 1974-79 (2005).
20. R. Goldman and J. Widom. Dataguides:enabling query formulation and optimization in semistructured databases. In Proceedings of VLDB, pages 436--445, 1997.
21. C.-W. Chung, J.-K. Min, and K. Shim. Apex: an adaptive path index for XML data. In SIGMOD, 121--132, 2002.
22. Ralf Schenkel, Anja Theobald, Gerhard Weikum, "Efficient Creation and Incremental Maintenance of the HOPI Index for Complex XML Document Collections," icde, pp. 360-371, 21st International Conference on Data Engineering (ICDE'05), 2005.

23. D. Shasha, J. T. L. Wang, and R. Giugno. Algorithmics and applications of tree and graph searching. In Symposium on Principles of Database Systems, pages 39--52, 2002.
24. X. Yan, P. S. Yu, and J. Han. Graph indexing: A frequent structure based approach. In Proceedings of SIGMOD 2004.
25. C.A. James, D. Weininger, and J. Delany. Daylight theory manual daylight version 4.82. Daylight Chemical Information Systems, Inc, 2003.
26. E. Selkov, S. Basmanova, T. Gaasterland, I. Goryanin, Y. Gretchkin, N. Maltsev, V. Nenashev, R. Overbeek, E. Panyushkina, L. Pronevitch, E. Selkov, Jr, and I. Yunus. The metabolic pathway collection from EMP: the enzymes and metabolic pathways database: Nucleic Acids Res. 1996 January 1; 24(1): 26–28.

Arevir: A Secure Platform for Designing Personalized Antiretroviral Therapies Against HIV

Kirsten Roomp[1], Niko Beerenwinkel[2], Tobias Sing[1], Eugen Schülter[3], Joachim Büch[1], Saleta Sierra-Aragon[4], Martin Däumer[4], Daniel Hoffmann[3], Rolf Kaiser[4], Thomas Lengauer[1], and Joachim Selbig[5]

[1] Max Planck Institute for Informatics, Stuhlsatzenhausweg 85, 66123 Saarbrücken, Germany
{roomp, tobias.sing, buech, lengauer}@mpi-sb.mpg.de

[2] Department of Mathematics, University of California, Berkeley, USA
niko@math.berkeley.edu

[3] Center of Advanced European Studies and Research (caesar), Friedensplatz 16, 53175 Bonn, Germany
{eugen.schuelter, daniel.hoffmann}@caesar.de

[4] Institute of Virology, University of Cologne, Fürst-Pückler-Str. 56, 50935 Köln, Germany
saleta@arcor.de, martin.daeumer@medizin.uni-koeln.de, rolf.kaiser@uk-koeln.de

[5] University of Potsdam and Max Planck Institute for Molecular Plant Physiology, Am Mühlenberg 1, 14476 Golm-Potsdam, Germany
selbig@mpimp-golm.mpg.de

Abstract. Despite the availability of antiretroviral combination therapies, success in drug treatment of HIV-infected patients is limited. One reason for therapy failure is the development of drug-resistant genetic variants. In principle, the viral genomic sequence provides resistance information and could thus guide the selection of an optimal drug combination. In practice however, the benefit of this procedure is impaired by (1) the difficulty in inferring the clinically relevant information from the genotype of the virus and (2) the restricted availability of this information. We have developed a secure platform for collaborative research aimed at optimizing anti-HIV therapies, called *Arevir*. A relational database schema was designed and implemented together with a web-based user interface. Our system provides a basis for monitoring patients, decision-support, and computational analyses. Thus, it merges clinical, diagnostic and bioinformatics efforts to exploit genomic and patient therapy data in clinical practice.

1 Introduction

1.1 Antiretroviral Therapy

There are currently 25 licensed, antiretroviral agents (including 5 fixed-dose combinations) available in industrialized countries for the treatment of HIV-infected patients. The majority of these drugs targets one of the two viral enzymes, the protease or the reverse transcriptase (RT). Additionally, a new class of drugs called *entry inhibitors* is

U. Leser, F. Naumann, and B. Eckman (Eds.): DILS 2006, LNBI 4075, pp. 185 – 194, 2006.

under development, with a subclass called fusion inhibitors. One such drug has been approved so far, which targets the gp41 protein located on the surface of HIV [1]. However, despite the introduction of combination therapy (called HAART – highly active antiretroviral therapy, usually consisting of three or more drugs) eradication of the virus from the patient's body cannot be achieved by current regimens [2]. Therefore, treatment strategies aim at maximal suppression of the viral load, i.e., the number of free virus particles per mL of the patient's blood serum. Besides the strong side effects of the inhibitors [3], the long-term effectiveness of HAART is also limited by the development of drug-resistant genetic variants [4]. Consequently, HIV resistance testing becomes increasingly important in the management of infected patients.

Resistance testing can be performed either by measuring viral activity in the presence or absence of a drug (phenotypic resistance testing [5]), or by scanning the viral genome for resistance-associated mutations (genotypic resistance testing [6]). It has been shown that patients can benefit from both genotypic and phenotypic testing [7]. Genotyping is faster and cheaper, whereas phenotypic results are easier to interpret. Direct sequencing produces genomic data of about 1200 base pairs of the HIV *pol* gene, which codes for protease and RT. This sequence carries the information about susceptibility or resistance of the patient's virus to each of the available drugs. However, it is challenging to infer resistance from the sequence and the optimal way of interpreting the genotype with respect to clinical outcome is not known.

1.2 Public Databases and Related Work

The HIV Sequence Database in Los Alamos provides a public repository for annotated HIV sequence data and is centered on sequence analysis. The HIV Drug Resistance Database in Stanford, formerly called the HIV RT/Protease Sequence Database, collects and analyzes sequences associated with the development of viral resistance. It is focused on sequences coding for the molecular targets of anti-HIV therapy. It includes drug susceptibility data and clinical histories [8]. The public website contains also a tool, *HIVdb*, for the interpretation of genotypic resistance. This system predicts resistance to a drug by scoring observed mutations in the drug's target protein. Mutation scores are derived manually by human experts based on reviewing links between mutations and resistance phenotypes described in the literature.

Another approach to interpreting genotypic resistance tests that avoids this bias lies in the systematic analysis of large sets of matched genotype-phenotype pairs. Statistical and machine learning methods have been applied successfully to derive models that predict phenotypic resistance from the genotype [9-13]. However, the predicted phenotype is only a first step towards understanding the clinical impact of resistance mutations. Selecting an optimal drug combination has become more difficult, because resistance testing adds complex genomic information to the decision-making process. Furthermore, the growing number of available drugs implies an exponentially growing number of possible drug combinations. Extending the data mining approach to related sequences, therapy histories and clinical outcomes promises to identify the determinants of the clinical resistance phenotype. However, this approach presumes large sets of curated and structured data.

Further related work can be found at the HIV Resistance Response Database Initiative (www.hivrdi.org), the Forum for Collaborative HIV Research (www.hivforum.org) and EuResist (www.euresist.org).

2 The Arevir Database

2.1 Database Rationale

Clinical information systems can support decision-making by providing relevant and valid information at the right time and place [14]. These requirements are generally not met with systems used in the management of HIV-infected patients. Firstly, resistance testing is usually performed in specialized virologic laboratories. Secondly, clinical data management systems – if at all existent in electronic form – are not prepared to handle and interpret genotypic data. Thus, test results remain separated from the electronic patient record. This situation is unsatisfying not only with regard to routine clinical decision-making, but also to research on optimizing therapies by means of incorporating genomic data.

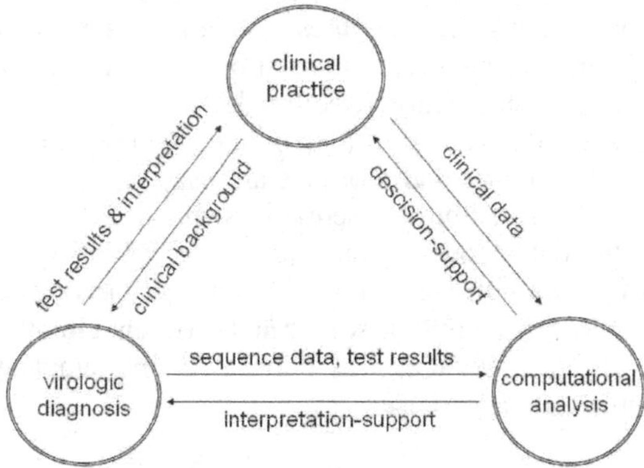

Fig. 1. Supported information flow in the *Arevir* system

We therefore made a concerted effort using the help of clinicians, virologists and bioinformaticians to take maximal advantage of genomic information in clinical practice. We focused on the design and implementation of a data management system that provides the logical and technical basis for therapy optimization. The system is used for monitoring patients, supporting medical decisions, augmenting the quality of resistance tests and the development of new computational tools that support decision-making (Fig. 1). Thus, it is a platform for interdisciplinary cyclic health care quality improvement [15].

Specific design goals for the system were (1) to integrate all relevant patient data, including sequence data resulting from resistance tests; (2) to provide secure access to these data for all authorized healthcare professionals involved, irrespective of their location; (3) to support the interpretation of genotypic resistance tests and the identification of remaining therapeutic options; and (4) to accumulate data sets of sufficient quantity and quality to yield computational tools for the optimization of individual therapies.

2.2 Database Design

The development of *Arevir* involved the design and implementation of a relational database accessible to all project partners across different platforms, ensuring the security and integrity of patient data. In managing HIV-infected patients different types of data arise, including personal patient data, therapy histories, numerous virologic, immunologic and other clinical test results derived from patient samples, and sequence data from genotypic resistance tests. Our database schema captures these data types in different modules, each consisting of a few tables. The key relations between these modules are shown in the entity relationship diagram in Fig. 2.

There is an important relationship between sequences and therapies via the drug targets, which defines a critical feature of the database. The compounds making up a combination therapy target specific viral proteins. In turn, DNA segments coding for these proteins are sequenced in order to gain information on the level of resistance that has been developed by the virus. Thus, given the values of clinical markers the data model allows for inferring the outcomes of therapy types versus mutational patterns within the drug targets. Indeed, these relationships have been used to make the first steps towards therapy optimization (Section 3.4).

The core database schema consists of 36 related tables that were normalized to third normal form. Further tables and modules are used for storing computational results such as alignments, derived protein sequences, and annotations.

We implemented the data model in the open source relational database management system (RDBMS), MySQL version 5.0.16. MySQL provides a client/server system consisting of a multi-threaded SQL server and different client programs, libraries, and programming interfaces. MySQL was chosen for implementation, because it is considered fast, reliable and easy to use.

2.3 Database Content

The current implementation of the system was intended for use on a national level within Germany. Currently, collaborators from 17 clinical centers, three virologic labs and two information technology institutes are participating. As of March 2006, the database contains data from over 5,720 patients, including 9,685 therapies, 5,290 viral genomic sequences and 48,502 clinical test results. Virtually all components of the system are scalable to larger settings. However, since data quality is a key factor and has been identified as a major challenge, emphasis lies on well-defined data sets and close cooperation.

2.4 System Configuration

Electronic patient records require special protection when stored or transferred across networks [16, 17]. In the present case, eavesdropping or theft of information may uncover the fact of someone being HIV-positive or his or her detailed medical history. Fraudulent manipulation of data might lead to wrong interpretations and therapeutical decisions. Thus, we took strong technical measures to protect data from unauthorized access and corruption. The software modules necessary to run the database that have been described so far are the RDBMS client and server, and the web client (browser) and server including the CGI scripts. Security demands on the client side are limited to ensuring that the display is not visible to others. SSH connections are configured with a time-out function, which leads to the interruption of the connection after a certain time without data traffic.

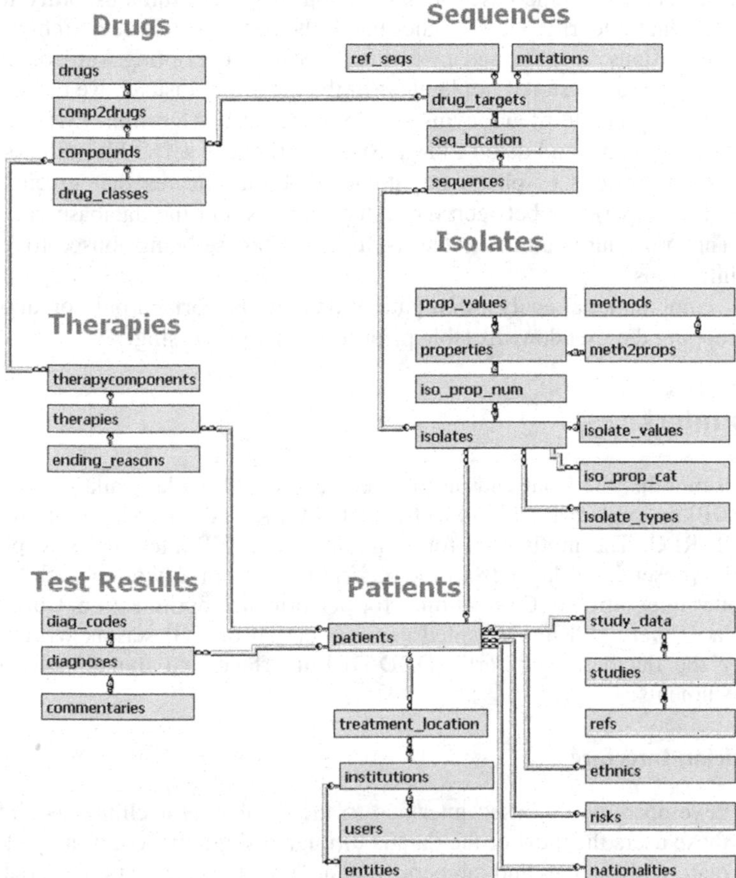

Fig. 2. Simplified entity-relationship diagram of the six main database modules comprising a total of 36 tables

Access to the database is restricted to our project partners via both machine names (TCP/IP addresses) and user names. The only authentication method accepted by the SSH daemon is the public key system [18, 19]. Private keys are protected by pass phrases on the client machines. To establish the connection, a local port is forwarded to the server over the secure channel or tunnel. A second authentication step, the login onto the website, is performed using the basic authentication offered by the Apache web server. The login name must match an entry in the appropriate database table. Access to the database engine is managed by MySQL's built-in privilege system, but since data from different institutions is stored in the same tables, access control is further refined within CGI scripts using the 'user-belongs to-institution' relation on the SQL level. Access to the database is via the clinician's interface (Section 3.1).

2.5 Patient Identifiers

The strict security measures described above allow the data to be accessed over the web by identifying a patient by its name and date of birth. Unlike using anonymous patient identifiers, this name-based identification method assures usability in clinics and promotes data integrity. On the other hand, the restrictive system architecture entails some limitations on speed and ease of use, notably on printing web contents.

Patient names are not stored in plaintext in the database. Instead, we use a one-way hash function to generate pseudonyms. The Secure Hash Algorithm (SHA-1) [20] is applied to patient name and date of birth. SHA-1 produces a 160-bit hash code. Storing pseudonyms instead of plaintext patient names implicates that given the hash function only comparisons between requested patients and the database contents are possible. This procedure minimizes the risk of the database being abused for uncovering HIV-infections.

Finally, computational analyses on patient data are performed only on anonymous data by dropping the pseudonyms table prior to further processing.

3 Web Interfaces

In the next months, the front-end on the web server will made available in PHP, the back-end will be implemented in object-oriented C++. These two levels communicate using XML-RPC. The motivation for switching to a PHP-interface is to physically separate the presentation layer from the algorithm layer in a two-tier architecture, as well as better integration of C++ routines for performance optimization. Currently, the web-interfaces were still implemented as a collection of CGI scripts written in Perl that query the database via Perl's DBD/DBI interface and dynamically generate HTML documents.

3.1 Clinician Interface

We have developed a web-based interface to the database for clinicians and virologists. For these users the view of the data is through a single patient or a single patient sample. Treating physicians and lab personnel are able to access an integrated view of all relevant data for one patient, including personal data, therapy data, clinical data, resistance data, and data on cell tropism (genotypic and phenotypic data). Additionally, clinicians are able to enter and update patient data using the interface.

A graphical representation of the time course of infection facilitates the efficient perception of information. Important clinical parameters, such as viral load and CD4+ cell count, are plotted versus the history of medications. Hyperlinks provide direct access to more detailed information, such as resistance test results.

3.2 geno2pheno[resistance]

The first tool for the interpretation of sequence data that was developed is called geno2pheno[resistance] [12, 21]. This tool uses statistical learning techniques for the prediction of phenotypic drug resistance from genotypes. Based on a set of approximately 800 matched genotype-phenotype pairs from *Arevir*, regression models were constructed. As the range of resistance factors varies considerably between different drugs, two scoring functions are derived from different sets of predicted phenotypes. First, predicted values are compared to those of samples derived from treatment-naive patients and the relative deviance is reported via a z-score. Second, the estimation of the probability density of predicted phenotypes gives rise to an intrinsic definition of a susceptible and a resistant subpopulation. For a predicted phenotype, we calculate the probability of membership in the resistant subpopulation. Both scores provide standardized measures of resistance based on the genotype that are comparable between drugs.

Physicians can therefore not only evaluate a genotypic resistance test result in the context of the patient's medical history and current immunological status, but also have access to a phenotypic interpretation of the genotype. Currently, resistance mutations in the protease and RT genes are used by the system.

A version of geno2pheno[resistance] that is decoupled from the *Arevir* database can be accessed on the Internet at www.geno2pheno.org.

3.3 geno2pheno[coreceptor]

A new subclass of entry inhibitor drugs known as coreceptor anatgonists tries to prevent cell entry of HIV by binding to one of the two chemokine receptors (CCR5 and CXCR4), which are used by the virus as coreceptors for entering the cell. One of these drugs has been approved so far, with several more having entered phase III clinical trials. Drug treatment is complicated by the possibility of inducing a coreceptor switch where the virus, under drug pressure, adapts and begins to use the other coreceptor for which no drug is being given. Thus, treatment should be accompanied by frequent monitoring of viral coreceptor usage. As in the case of resistance testing, in routine diagnostics the only feasible way to accomplish this is by experimentally determining the genotype. Additionally, switching, particularly from CCR5 to CXCR4 virus (also naturally occurring in 50% of patients), is associated with progression towards AIDS.

In light of these considerations, we have developed geno2pheno[coreceptor] for predicting viral coreceptor usage from the third variable loop of the HIV envelope protein gp120 [22, 23]. Alignments of the envelope V3 loop region were obtained using ClustalW and a fixed reference alignment. Six statistical learning methods operating on the entire V3 loop were evaluated on a set of 1110 matched genotype-phenotype pairs from different subtypes using cross-validation. In a receiver operating characteristic (ROC) analysis, classifiers based on support vector machines (SVMs) showed significantly higher area under the ROC curve than other methods (p-value less than 0.001) and dominated all other methods in terms of sensitivity at practically important specificity rates. The tool is able to deal with data from both clonal and population-based sequencing. We have recently shown that predictive performance can be further improved by incorporating information on the host CCR5 genotype and immunological status, making use of both genomic and immunological datasets contained within *Arevir*.

geno2pheno[coreceptor] can also be accessed in a stand-alone version at www.geno2pheno.org

3.4 THEO

The growing number of drugs increases the number of available combination therapies which can be given to patients. There are no objective selection criteria for an optimal regimen, thus a computer-aided solution is called for. Our approach does not only incorporate the genotype itself as a predictor, but also the dynamics of viral evolutionary escape from the considered regimen [24].

We formulated the problem as a binary classification task (therapy failure vs. therapy success) based on the measured virological response. Treatment change episodes (comprising matched genotype, therapy, and outcome data) were extracted from the *Arevir* database yielding 552 therapy failures and 224 successes. For integrating evolutionary information, we compared an explicit search through sequence space [25] to an alternative based on estimating the probability of escape [26]. In the latter approach, we computed the genetic barrier using a probabilistic evolutionary model, in which resistance mutations accumulate along different mutational pathways. Across several statistical learning methods, the genetic barrier-based approach consistently outperformed explicit search through sequence space, improving accuracy of therapy outcome prediction from above 20% to 14.3% (±2.5%).

THEO, a prototypical implementation of the optimal approach, is freely available at www.geno2pheno.org.

4 Legal Aspects

For enrollment in *Arevir*, patients need to consent explicitly to providing their data and can revoke their agreement at any time. They are informed in detail about project goals and technical realizations. Thus, patients decide on their participation being fully aware of the potential benefit versus the risks of exposure. The described security concept, complemented by organizational measures such as a physically secured server machine and well-defined responsibilities for it, has been examined and approved by federal state data security officials.

5 Conclusion

We have presented a web-based data management system for collaborative research on HIV of direct clinical relevance. The system has the goal of optimizing antiretroviral therapies in view of viral sequence data. Our focus is on providing a basis for patient management, evidence-based decision-support and research at the same time. These seemingly diverse tasks can be unified in a natural way into one system on the basis of a common data model. The system design supports a cyclic optimization process that involves clinical practice, diagnostics and computational analysis. This approach may be seen as a real-life example of incorporating bioinformatics methods into clinical practice. Meeting this challenge required the design, implementation and validation of elaborate diagnostic and decision-support tools that operate on complex data in a clinical setting. The presented data model proves its flexibility in admitting new clinical parameters, and new drugs with new target molecules.

Acknowledgements

We are grateful to all participants in the *Arevir* project, Hauke Walter, Klaus Korn, Thomas Berg, Patrick Braun, Gerd Fätkenheuer, Mark Oette, Jürgen Rockstroh, and Bernd Kupfer. Deutsche Forschungsgemeinschaft (DFG) has funded much of the *Arevir* research in the context of the Priority Program on Informatics Methods for the Analysis and Interpretation of Large Genomic Datasets and of the Bioinformatics Center Saar. N.B. is funded by DFG under grant No. BE 3217/1-1.

References

1. Stanic, A., Schneider, T.K.: Overview of Antiretroviral Agents in 2005. Journal of Pharmacy Practice **18** (2005) 228-246
2. Marcello, A.: Latency: the hidden HIV-1 challenge. Retrovirology **3** (2006) 7
3. Powderly, W.G.: Long-term exposure to lifelong therapies. J Acquir Immune Defic Syndr **29 Suppl 1** (2002) S28-40
4. Perrin, L., Telenti, A.: HIV treatment failure: testing for HIV resistance in clinical practice. Science **280** (1998) 1871-1873
5. Walter, H., Schmidt, B., Korn, K., Vandamme, A.M., Harrer, T., Uberla, K.: Rapid, phenotypic HIV-1 drug sensitivity assay for protease and reverse transcriptase inhibitors. J Clin Virol **13** (1999) 71-80
6. Shafer, R.W., Kantor, R., J. Gonzales, M.J.: The Genetic Basis of HIV-1 Resistance to Reverse Transcriptase and Protease Inhibitors. AIDS Rev **2** (2000) 211-228
7. DeGruttola, V., Dix, L., D'Aquila, R., Holder, D., Phillips, A., Ait-Khaled, M., Baxter, J., Clevenbergh, P., Hammer, S., Harrigan, R., Katzenstein, D., Lanier, R., Miller, M., Para, M., Yerly, S., Zolopa, A., Murray, J., Patick, A., Miller, V., Castillo, S., Pedneault, L., Mellors, J.: The relation between baseline HIV drug resistance and response to antiretroviral therapy: re-analysis of retrospective and prospective studies using a standardized data analysis plan. Antivir Ther **5** (2000) 41-48
8. Kuiken, C., Korber, B., Shafer, R.W.: HIV sequence databases. AIDS Rev **5** (2003) 52-61
9. Wang, D., Bloor, S., Larder, B.A.: The application of neural networks in predicting phenotypic resistance from genotypes for HIV-1 protease inhibitors. Antivir Ther (2000) 51-52
10. Sevin, A.D., DeGruttola, V., Nijhuis, M., Schapiro, J.M., Foulkes, A.S., Para, M.F., Boucher, C.A.: Methods for investigation of the relationship between drug-susceptibility phenotype and human immunodeficiency virus type 1 genotype with applications to AIDS clinical trials group 333. J Infect Dis **182** (2000) 59-67
11. Beerenwinkel, N., Schmidt, B., Walter, H., Kaiser, R., Lengauer, T., Hoffmann, D., Korn, K., Selbig, J.: Geno2pheno: Interpreting Genotypic HIV Drug Resistance Tests. IEEE Intelligent Systems in Biology (2001) 35-41
12. Beerenwinkel, N., Schmidt, B., Walter, H., Kaiser, R., Lengauer, T., Hoffmann, D., Korn, K., Selbig, J.: Diversity and complexity of HIV-1 drug resistance: a bioinformatics approach to predicting phenotype from genotype. Proc Natl Acad Sci U S A **99** (2002) 8271-8276
13. Cordes, F., Kaiser, R., Selbig, J.: Bioinformatics approach to predicting HIV drug resistance. Expert Rev Mol Diagn **6** (2006) 207-215
14. Tierney, W.M.: Improving clinical decisions and outcomes with information: a review. Int J Med Inform **62** (2001) 1-9

15. Marshall, W.W., Haley, R.W.: Use of a secure Internet Web site for collaborative medical research. Jama **284** (2000) 1843-1849
16. Schoenberg, R., Safran, C.: Internet based repository of medical records that retains patient confidentiality. Bmj **321** (2000) 1199-1203
17. Mandl, K.D., Szolovits, P., Kohane, I.S.: Public standards and patients' control: how to keep electronic medical records accessible but private. Bmj **322** (2001) 283-287
18. Diffie, W., Hellman, M.: New directions in cryptography. IEEE Transactions on Information Theory (1976) 472-492
19. Rivest, R.L., Shamir, A., Adleman, L.: A method for obtaining digital signatures and public-key cryptosystems. Communications of the ACM (1978) 120-126
20. Secure Hash Standard. FIPS PUB. Federal Information Processing Standards (1995)
21. Beerenwinkel, N., Daumer, M., Oette, M., Korn, K., Hoffmann, D., Kaiser, R., Lengauer, T., Selbig, J., Walter, H.: Geno2pheno: Estimating phenotypic drug resistance from HIV-1 genotypes. Nucleic Acids Res **31** (2003) 3850-3855
22. Sing, T., Sander, O., Beerenwinkel, N., Lengauer, T.: ROCR: visualizing classifier performance in R. Bioinformatics **21** (2005) 3940-3941
23. Sirois, S., Sing, T., Chou, K.C.: HIV-1 gp120 V3 loop for structure-based drug design. Curr Protein Pept Sci **6** (2005) 413-422
24. Beerenwinkel, N., Sing, T., Lengauer, T., Rahnenfuhrer, J., Roomp, K., Savenkov, I., Fischer, R., Hoffmann, D., Selbig, J., Korn, K., Walter, H., Berg, T., Braun, P., Fatkenheuer, G., Oette, M., Rockstroh, J., Kupfer, B., Kaiser, R., Daumer, M.: Computational methods for the design of effective therapies against drug resistant HIV strains. Bioinformatics **21** (2005) 3943-3950
25. Beerenwinkel, N., Lengauer, T., Daumer, M., Kaiser, R., Walter, H., Korn, K., Hoffmann, D., Selbig, J.: Methods for optimizing antiviral combination therapies. Bioinformatics **19 Suppl 1** (2003) i16-25
26. Beerenwinkel, N., Daumer, M., Sing, T., Rahnenfuhrer, J., Lengauer, T., Selbig, J., Hoffmann, D., Kaiser, R.: Estimating HIV evolutionary pathways and the genetic barrier to drug resistance. J Infect Dis **191** (2005) 1953-1960

The Distributed Annotation System for Integration of Biological Data

Andreas Prlić[1], Ewan Birney[2], Tony Cox[1], Thomas A. Down[1], Rob Finn[1], Stefan Gräf[2], David Jackson[1], Andreas Kähäri[2], Eugene Kulesha[1], Roger Pettett[1], James Smith[1], Jim Stalker[1], and Tim J.P. Hubbard[1]

[1]The Wellcome Trust Sanger Institute, Wellcome Trust Genome Campus, Hinxton, Cambridge, CB10 1SA, UK
ap3@sanger.ac.uk,
http://das.sanger.ac.uk/registry/
[2]EMBL - European Bioinformatics Institute, Hinxton, UK

Abstract. The Distributed Annotation System (DAS) is a protocol for sharing of biological data which allows for dynamical data integration. It has become widely used in both the genome and protein bioinformatics communities. Here we provide an overview of the available DAS infrastructure and present our latest developments, including a registration server that facilitates service discovery by DAS clients while automatically monitoring service availability. Currently there are 108 registered DAS servers, provided by 24 institutions in 10 countries.

1 Introduction

Annotation of biological data, such as genome and protein sequences, is one of the central tasks in biological research. This is done by different means, for example manually, computationally and experimentally. There are a number of centralized resources available that are working on the integration of these data. They are facing the problems of how to manage the vast amount of data that is available, the need for frequent updates and releases, and how to exchange data with other institutions and users.

The Distributed Annotation System (DAS) is a protocol that addresses these issues and facilitates the sharing of biological data [1]. It is based on the idea that annotation data is not aggregated into large centralized databases, but instead is spread over multiple sites, generally maintained by the original data creators. DAS is frequently used for

1. integration of personal data into bioinformatics resources,
2. integration of the annotations from external sources into local applications,
3. access to most recent data versions without the need for local installations,

DAS is a web service protocol built upon well established open technologies (HTTP and XML), with some similarities to SOAP-based services. Where SOAP services use XML requests and responses for the transport of information, DAS

U. Leser, F. Naumann, and B. Eckman (Eds.): DILS 2006, LNBI 4075, pp. 195–203, 2006.

provides a data model, a query model, and a transport. The returned XML documents contain objects like *sequence* or *feature*. All data are provided by DAS servers and it is up to a DAS client to retrieve the annotations from multiple servers and to integrate these into a visualization that is presented to the user (see Fig. 1). For a detailed description of the DAS protocol see `http://www.biodas.org/documents/spec.html`.

The DAS protocol was originally designed to serve annotation for genomes. Resources like the *Ensembl* genome browser utilize this protocol to visualize new or personal data in the context of other annotations [2]. Different web pages, *"views"*, provide access to annotation data for e.g. chromosomes, transcripts, genes, or proteins. Each of these views acts as a DAS client. A management interface allows users to configure a list of DAS servers from which annotation should be retrieved. Once a new server has been added in the configuration, Ensembl establishes the contact to the server, fetches the data, and displays it together with other annotations. In this setup the Ensembl web server acts as a data-proxy and the users can access all data via their web browsers.

Over the last few years DAS has also been used to share annotations of proteins. We recently presented *SPICE*, a browser of protein structures, sequences, and their annotations, which is built on DAS [3]. SPICE is a Java application that installs and runs locally using the Java Web-Start technology. It can be launched by simply following a link on a web page. SPICE provides an integrated view of protein sequence and structure and can project annotations from one coordinate system onto another. This, for example, allows it to display protein sequence annotations with respect to their position on the protein structure. SPICE is integrated with Ensembl (see Fig. 2).

Dasty is another protein DAS client [4]. It is a Java application with a Macromedia Flash front-end, and all DAS communication is done via a dedicated server. Other DAS clients that can be easily integrated into web pages are ProView (`http://www.sanger.ac.uk/proview/`) or the CBS DAS Viewer [5].

DAS has been widely adopted in the bioinformatics community, because it is simple to use and simple to set up. Both DAS servers and client software are available with implementations in multiple languages: In Perl there is support for setting up a DAS server using ProServer (`http://www.sanger.ac.uk/proserver/`) or LDAS (`http://www.biodas.org/servers/LDAS.html`), while users who prefer Java can use Dazzle (`http://www.derkholm.net/thomas/dazzle/`). Client libraries are also available in Perl, e.g. the Bio::DasLite library (`http://search.cpan.org/~rpettett/Bio-DasLite/`), and in Java (`http://www.biojava.org/`, `http://www.spice-3d.org/dasobert/`), making integration of DAS support into new and existing bioinformatics tools easy.

Several collaborations are providing support for DAS. The BioSapiens Network of Excellence (`http://www.biosapiens.info/`) is providing a large number of DAS sources, which are listed at the BioSapiens Information Resource (`http://www.biosapiens.info/page.php?page=biosapiensdir`). BioSapiens also provides a Portal that can query UniProt and provides access to several DAS clients (`http://www.biosapiens.info/page.php?page=das_portal`). Another

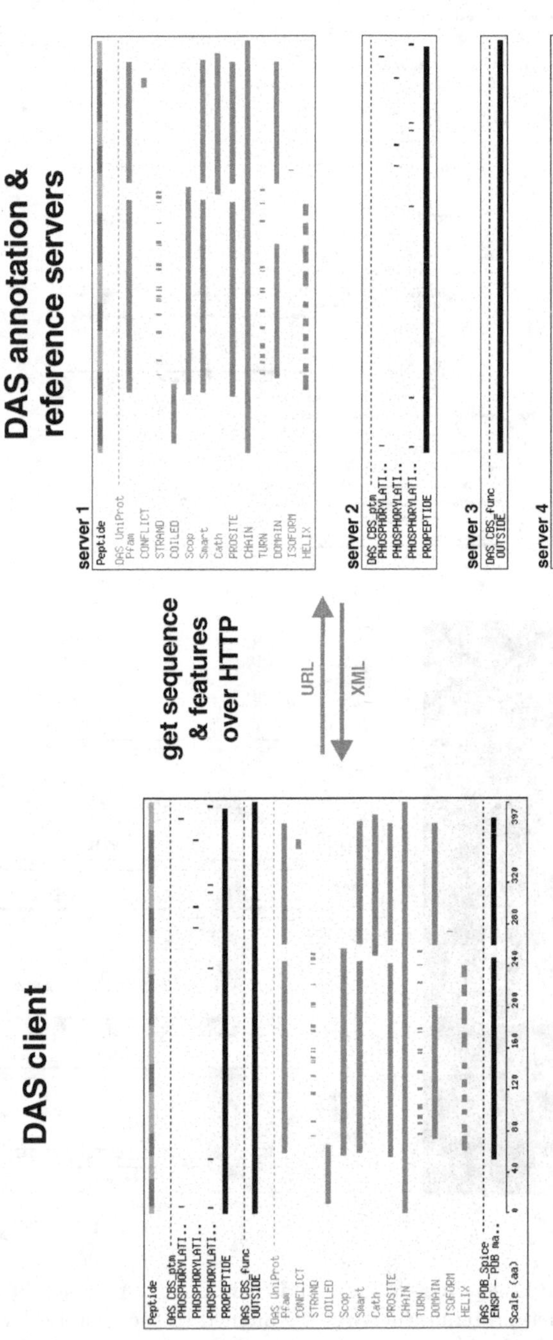

Fig. 1. A DAS client retrieves data from several DAS servers. In this schematic example the UniProt DAS server provides the reference sequence and some annotations. Other DAS servers are available that provide additional annotations for the same sequence. The Ensembl ProtView integrates the data into a common display.

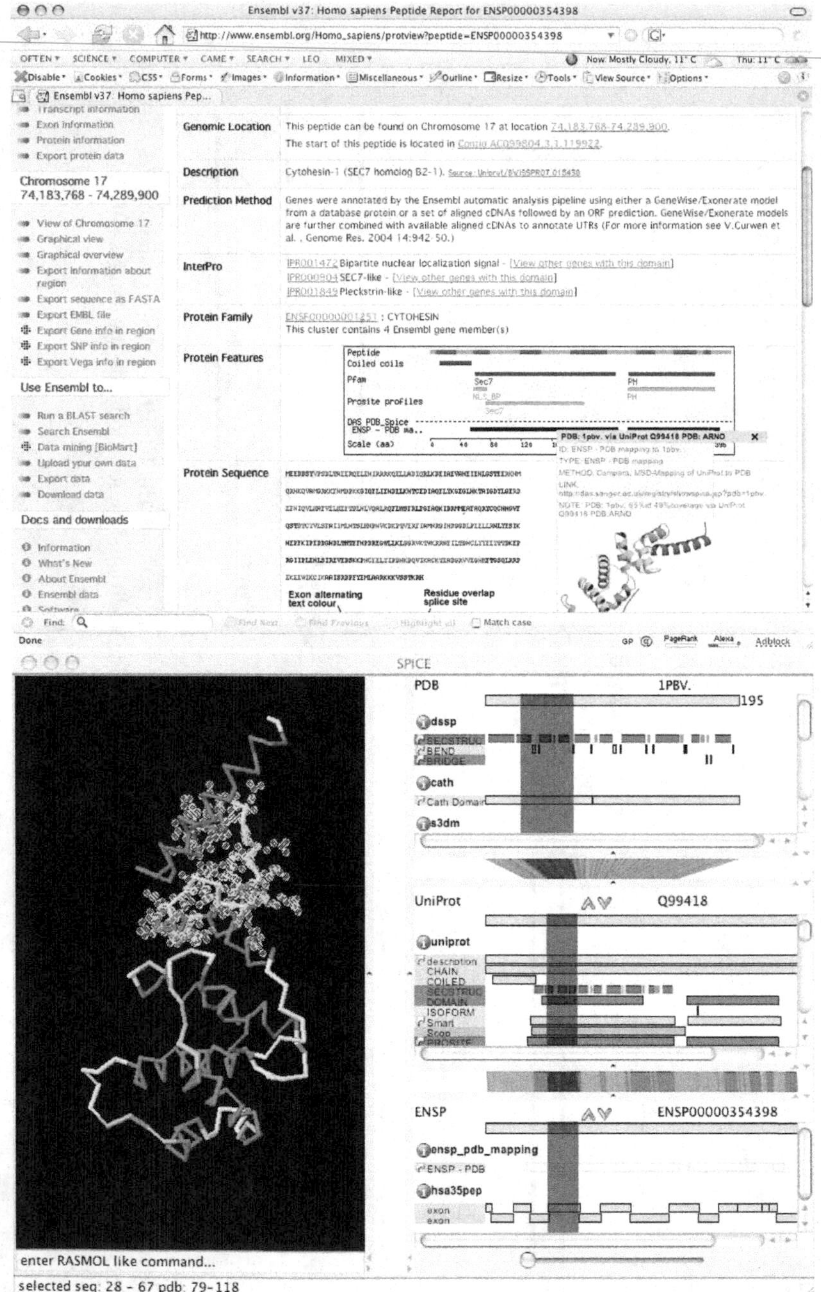

Fig. 2. A screenshot of the Ensembl ProtView and SPICE. A DAS track is available in Ensembl that shows which protein structures are aligned to the protein. By clicking on the DAS track the SPICE browser can be launched. SPICE shows the protein structure, the UniProt sequence, and the Ensembl predicted protein (ENSP) sequence. One of the exons is selected and its position can be projected onto the 3D structure.

project that provides support for DAS is the eFamily project (`http://www.efamily.org.uk/`).

1.1 Registration of DAS Servers

DAS servers are divided into two categories. *Reference sources* provide the object to be annotated, e.g. a sequence or the 3D structure. *Annotation sources* provide the features of these objects. A number of different DAS sources have been released over the years providing annotations for different organisms and on different levels. The DAS protocol does not suggest how DAS clients can discover DAS sources that provide annotation. So far this has been done by hard-coding a list into a client, or requiring users to directly enter the URLs used to communicate with individual servers. Also, DAS does not define how to deal with the fact that different annotation servers provide the data for different types of biological objects. To address this we have developed a DAS registration server.

The DAS registration server fulfils several purposes:

1. It allows users (or their client software) to query and retrieve lists of available DAS sources via either a web interface or a XML web service for programmatic access.
2. It is able to direct a user to any of the most common DAS clients and attach the registered DAS server that the user is interested in seeing annotations from.
3. It automatically validates DAS sources to make sure they provide valid DAS-XML.
4. It can notify the administrator of a DAS source if the server has been down for a while.
5. It groups the registered DAS sources according to the coordinate system of the provided data.

There are three components to a Service Orientated Architecture concept: Service provider, service requestor, and a service registry. Connecting these components together are three operations: publish, find, and bind. The original DAS protocol partially implements this architecture, with DAS servers being the service providers and the clients being the requestors. By providing the new registration service that supports discovery, DAS has become a full Service Oriented Architecture.

2 System and Methods

In this section we present various concepts and terminology that is applicable for DAS.

2.1 Coordinate Systems

DAS is used to annotate many different object types: genomes, gene loci, protein sequences, and structures are currently the most common cases. For each type, there are a number of meaningful 'sets' of objects — for example, the chromosome sequences in a particular assembly of the human genome. To allow data integration, clients must be able to find all the DAS servers which annotate a particular set of objects. We call the description of these sets *coordinate systems.* It can also be thought of as a "namespace". The following information is used for their description:

(1) The *authority* (or name). This is the name of the institution that defines the identifiers or accession numbers for a particular set of objects. In case of genome assemblies this field also contains the version of the assembly. For example, UniProt is an authority to assign protein sequence accession codes, while the currently used build for the human genome is NCBI 36.

(2) The *type* of object that is being annotated. This entity refers to the "physical dimension" of the data. Currently supported are *Chromosome, Clone, Contig, Gene_ID, NT_Contig, Protein Sequence, Protein Structure*, and *Scaffold.*

(3) The *organism.* The scientific name of an organism. This field is optional, since some DAS sources provide annotations for more than one organisms.

2.2 DAS Capabilities

The DAS specification (version 1.5, `http://biodas.org/documents/spec.html`) defines a number of commands that can be sent to DAS servers. They are *sequence, features, types, entry_points, dna, stylesheet.* These are supported by the registration server together with the DAS extensions required for protein 3D structure annotations, *structure, alignment*, as described at `http://www.efamily.org.uk/xml/das/documentation/`.

2.3 Validation

There exist a number of different server side implementations to provide data via a DAS source. Frequently used ones include Dazzle (`http://www.derkholm.net/thomas/dazzle/`), ProServer (`http://www.sanger.ac.uk/proserver/`), and LDAS (`http://biodas.org/download/ldas/`) but sometimes individually implemented CGI scripts are used as well. In order to ensure a DAS source communicates in valid DAS-XML it can only be registered if it successfully validates by returning a correct DAS response for each of the capabilities for which it is registered. For this a *test code* is required, which is an accession code that has been annotated and for which features are provided.

Once a DAS source has been registered, the registry software contacts it periodically and attempts to validate it. Successful validation attempts are logged, and a graphical summary of the availability of a DAS source is available via the registry's web interface. If the DAS source can not be validated for more than two days, a *watchdog* can (optionally) inform the server's administrator. If a server is down for a longer period of time, the server administrator can be contacted to inquire about the status of the server. If the server remains unavailable for an extended period, it will be removed from the listing.

2.4 Auto-activation

At the present stage the registry communicates with three DAS clients: Ensembl [2], SPICE [3], and Dasty [4]. Each of these can retrieve a list of available DAS sources from the repository. With appropriate client support, the registry can also communicate back in the reverse direction: DAS sources can be activated in a client by clicking on an icon in the registry web interface. In Ensembl the DAS server can be automatically added to the configuration of a particular view. The registry also provides a *send to friend* mechanism to share auto-activation links by email.

2.5 Implementation

The registration server at its core is a web service, backed by a MySQL database. The service can be accessed at `http://das.sanger.ac.uk/registry/services/das:das_directory?wsdl`. It can be used by different DAS clients to query and retrieve server listings or to validate DAS sources. A HTML front-end is provided which allows manual interaction with the registry (`http://das.sanger.ac.uk/registry/`). These web pages are based also on the web service. The HTML front-end is implemented as a set of JSP pages running on a Resin server.

3 Discussion

DAS is a data integration technology which is widely used in the genome and protein bioinformatics communities. A repository for registering and discovering DAS servers has been missing so far. Here we provide such a service. The DAS registry can interact with DAS clients and auto-activate a DAS source in the DAS client. On the management side the registry ensures that DAS sources follow the specification, and helps administrators to monitor the availability of their DAS sources. If a DAS server has been inactive for a while, we usually contact the administrators in order to query the status of the server. If it has become obsolete it can be removed from the repository.

We are participating in the development of DAS/2, a major update to the core DAS protocol (`http://biodas.org/documents/das2/das2_protocol.html`). DAS/2 will add the ability to search sets of features by identifiers and other properties (for instance, to find a gene given its name), and provides servers with extension mechanisms, allowing DAS features to be annotated with additional structured information (in XML format) as well as textual notes. DAS/2 also specifies an upload mechanism, so advanced clients can write back manually curated annotations to a DAS server.

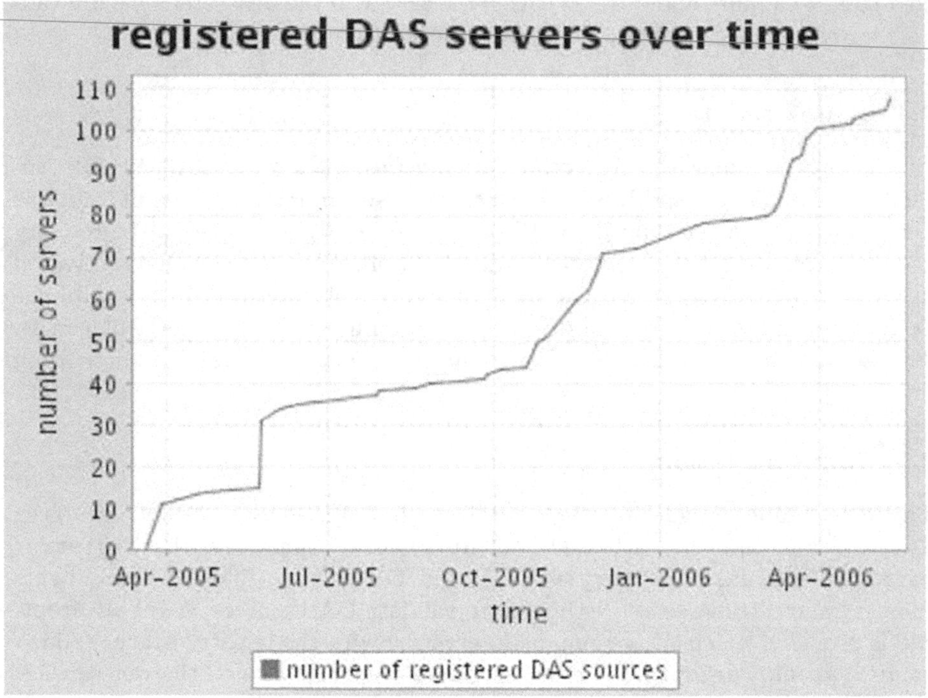

Fig. 3. The number of registered DAS sources over time. Currently there are 108 DAS sources available from 24 institutions in 10 countries.

The registration server currently contains 108 DAS sources provided by 24 institutions in 10 countries. Over the last year the number of registered DAS sources has been constantly growing (see Fig. 3). If this trend continues, at some point additional tools might be required for users to maintain an overview of the provided data. One way to achieve so could be to provide a user rating system, similar to what is known from popular online stores. Ideally such a system would be supported within the DAS clients, so user could rate a DAS source in a client, which would be communicated back to the registration server. DAS clients could sort DAS servers according to their popularity.

Acknowledgments

We want to thank everybody who provides DAS servers and shares the data with the community. The system would not work without you. This work has been supported by the Medical Research Council, The Wellcome Trust, and the BioSapiens Network of Excellence. All source code is available under LGPL from `http://www.derkholm.net/svn/repos/dasregistry/`.

References

1. Dowell, R.D., Jokerst, R.M., Day, A., Eddy, S.R., Stein, L.: The distributed annotation system. BMC Bioinformatics. **2** (2001) 7–7
2. Birney, E., Andrews, D., Caccamo, M., Chen, Y., Clarke, L., Coates, G., Cox, T., Cunningham, F., Curwen, V., Cutts, T., Down, T., Durbin, R., Fernandez-Suarez, X.M., Flicek, P., Gräf, S., Hammond, M., Herrero, J., Howe, K., Iyer, V., Jekosch, K., Kähäri, A., Kasprzyk, A., Keefe, D., Kokocinski, F., Kulesha, E., London, D., Longden, I., Melsopp, C., Meidl, P., Overduin, B., Parker, A., Proctor, G., Prlić, A., Rae, M., Rios, D., Redmond, S., Schuster, M., Sealy, I., Searle, S., Severin, J., Slater, G., Smedley, D., Smith, J., Stabenau, A., Stalker, J., Trevanion, S., Ureta-Vidal, A., Vogel, J., White, S., Woodwark, C., Hubbard, T.J.P.: Ensembl 2006. Nucleic Acids Res **34**(Database issue) (2006) D556–61
3. Prlić, A., Down, T.A., Hubbard, T.J.P.: Adding Some SPICE to DAS. Bioinformatics **21 Suppl 2** (2005) ii40–ii41
4. Jones, P., Vinod, N., Down, T., Hackmann, A., Kahari, A., Kretschmann, E., Quinn, A., Wieser, D., Hermjakob, H., Apweiler, R.: Dasty and UniProt DAS: a perfect pair for protein feature visualization. Bioinformatics **21**(14) (2005) 3198–9
5. Olason, P.I.: Integrating protein annotation resources through the Distributed Annotation System. Nucleic Acids Res **33**(Web Server issue) (2005) W468–70

An Information Management System for Collaboration Within Distributed Working Environment (Systems Paper)

Maria Samsonova, Andrei Pisarev, Konstantin Kozlov, Ekaterina Poustelnikova, and Arthur Tkachenko

St.Petersburg State Polytechnical University, St.Petersburg, 195251 Russia

Abstract. Over a period of several years we apply the systems biology approach to investigate the dynamic regulatory mechanisms controlling the expression of segmentation genes in Drosophila embryo. Due to ongoing data acquisition, development of new processing and analysis methods, as well as modification and improvement of old ones serious problems arose with data and workflows management. Different geographical location of research groups poses additional difficulties. To solve these problems we have developed an information management system using multiagent and REST architectures. This system is easily extendable to deal with new data processing and analysis methods, flexible in specification and modification of these methods, scalable and supports distributed processing and analysis of data.

1 Introduction

Recently the introduction of high-throughput techniques as well as digital recording devices and computers lead to accumulation of large volumes of data. There are hundreds of resources and applications available to a biologist via "command line" applications, databases, flat files, web forms or graphical user interfaces. Publishing of data and providing services via the Internet has a long-lasting tradition in biology. Taking advantage of the broad-bandwidth Internet connections, researches are able to connect remotely to computers to share research data, tools and computing power.

Traditionally a biologist needs access to dozens of data types and services to plan her experiments and analyze results. To obtain such an information a researcher needs to navigate and download data from many computers, process, integrate and analyze the downloaded information manually or to use complex scripts to overcome incompatibilities. This is a very difficult and tedious task, as resources are widely distributed, highly heterogeneous, diverse and autonomous.

The increase in data, the rapid growth in a number of analysis tools and the range of knowledge needed to interpret and use them requires to develop methods for at least partial automation of data and services integration. Among obvious advantages of such an automation are reduction in a number of routine

U. Leser, F. Naumann, and B. Eckman (Eds.): DILS 2006, LNBI 4075, pp. 204–215, 2006.

queries for wet lab biologists, possibility to perform and repeat data analysis multiple times, reduction in research cost, as well as the transparency of code and algorithms.

Currently several service oriented architectures (SOA) are used to integrate heterogeneous resources. CORBA, RMI and DCOM are mainly applied to integrate Intranet applications. SOAs based on Web services concept use specific protocols to access services. Though several widely recognized implementations have been attempted [1,2,3,4] some aspects of these technologies impede their use for creation of a highly integrated global biological data space. These are developing and incomplete standards, introduction of several versions of standards with different policies in the field of patenting by different organizations (W3C, OASIS, Grid, etc.), insufficient solution of safety questions, necessity to modernize the already developed programs. Besides, as we will show in section 3.2, XML-representation of some data types (images, BLOBS, matrices) can decrease the performance of application.

Contrary to all the architectures described the REST (Representational State Transfer) architecture uses URIs to identify resources, and a small, globally defined set of remote HttpMethods to access and manipulate the state of those resources [5]. HTTP is the protocol by which resources are accessed. REST proponents argue that the HTTP's minimal method set and semantics, as well as its ability to extend this method set as required is sufficiently general to model any application domain.

This paper presents the application inspired by REST. It is designed to automate management and analysis of information generated by the consortium of laboratories in USA, Russia and Western Europe. Here we describe the prototype of this system and demonstrate real-life scenarios of data processing and analysis.

2 Materials and Methods

Over a period of several years the consortium of laboratories from St.Petersburg Polytechnical University, the Ioffe Physical-Technical Institute (Russia), Stony Brook University, Los Alamos National laboratory (USA) and University of Amsterdam (the Netherlands) investigates the dynamical regulatory mechanisms which control the expression of segmentation genes in Drosophila embryo [6,7]. To solve this problem the systems biology approach is applied, which encompasses the acquisition of data on a large scale, mathematical modelling and simulations.

Due to ongoing data acquisition, development of new processing and analysis methods, as well as modification and improvement of old ones, serious problems have been encountered with data storage and management of application programs. Additional problems are created by different geographical location of performance sites, as often users need data or programs kept in another laboratory. These problems complicate data analysis and processing and decrease the efficiency of work as a whole. To automate information management and analysis, as well as to integrate different types of data and application programs in all laboratories we start to develop an information management system known as iSIMBioS (integrated Service Infrastructure for Molecular Biology Systems).

2.1 Information Flow

Like all other insects, the body of the fruit fly Drosophila is made up of repeated units called segments. The segment determination happens during the first three hours of the development of fruit fly and is controlled by the network of about sixteen genes [8,9]. The expression of segmentation genes is registered by confocally scanning of fixed embryos stained with fluorescently tagged antibodies. Images of gene expression obtained from these embryos serve as a raw material for quantification of gene expression. The conversion of images into quantitative data is performed in several steps, for each step the specialized methods for image and data processing are developed and implemented [6]. This results in the construction of reference data on expression of segmentation genes at cellular resolution and at each time point. Images and quantitative gene expression data from individual embryos, as well as reference gene expression data are used to study the dynamics of formation of segmentation gene expression domains, precision of development and pattern formation and the mechanisms of segment determination [10].

2.2 System Requirements

iSIMBioS is designed to provide flexible environment for on-line collaboration of investigators from different laboratories via the Internet. As such the requirements to the system can be formulated in the following way: extendability to deal with continuously growing number of images and data volumes, introduction of new processing and analysis methods, integration with third parties tools; flexibility in specification and modification of analysis methods; scalability; support of distributed processing and analysis of data; provision of simultaneous access of multiple users to shared data and methods; no need in programming skills or familiarization how to install special software libraries and program tools for processing and analysis of data; availability of powerful and friendly Web-based user interface, as well as visualization tools; use of heterogeneous software/hardware platforms; provision of access through firewall and proxy servers; support of autonomous task performance upon connection hang up, as well as notification about processing results; provision of continuous work, when new components are added or old one are removed; sufficient response time and readiness characteristics; failure-resistance, if malfunction of hardware or software components happens; preferably based on open source software; portability across software platforms.

The Web-service based SOAs cannot currently support the functionality required. The standardization of these technologies has not finished yet. Some of the standards are on their way to become broadly adopted, while others have overlapping functionality and/or are still immature in terms of software implementation. Incomplete standardization of the Web services' ingredients such as

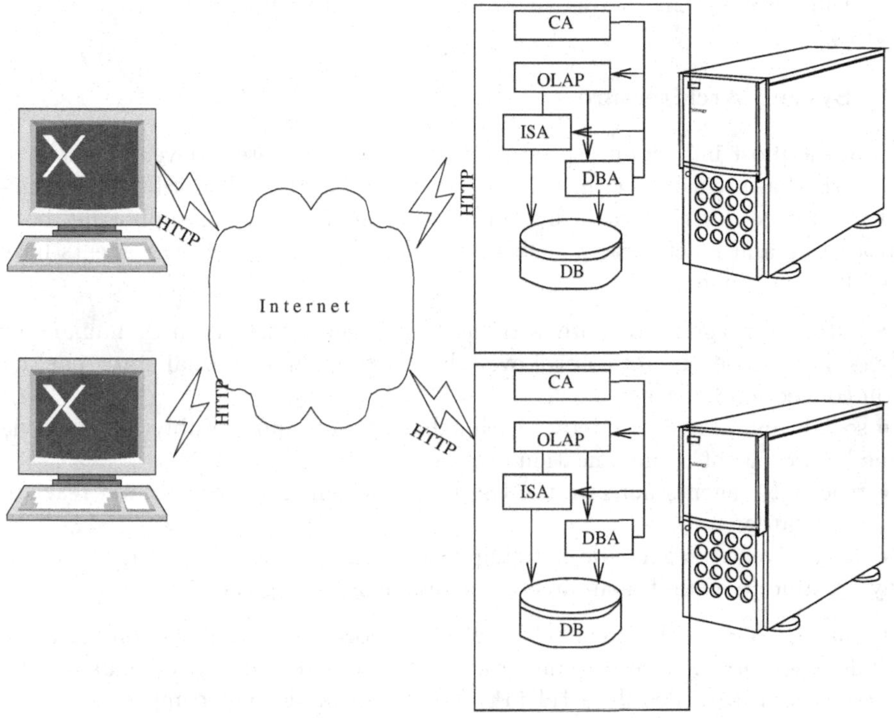

Fig. 1. iSIMBioS architecture

orchestration and choreography makes it problematic to apply these technologies to automate management and analysis of distributed data in distributed environment.

Besides the important consideration in the choice of architectural style was the availability of many in-house programs developed prior to the introduction of Web services. Conversion of this software into Web services requires substantial efforts and resources. At the same time all these tools need to be integrated with XML-RPC and SOAP applications.

Finally the combined approach have been selected based on application of both multiagent architecture and REST architectural style. The important advantage of multiagent systems is in their inherent modularity. Due to modularity these systems are scalable and easy extendable. Moreover multiagent systems with redundant components (databases, agents, application programs, other services) are robust, highly adaptable to functional extensions and have high readiness and reactivity characteristics. REST scales well with large numbers of clients, enables data transfer in streams of unlimited size and type and supports intermediaries (proxies and gateways) as data transformation and caching components. Thus the joint use of multiagent and REST architectural styles enables to satisfy almost all of the requirements to the system behavior. Therefore we decided to built a system in which autonomous agents interact with each other via HTTP protocol and act as adaptors to integrate all types of programs and services.

2.3 System Architecture

One of the most important aspects of the multiagent system development is to define the basic entities within the system – the agents. The iSIMBioS architecture can be described as a hybrid multiagent architecture supporting both deliberative and reactive actions of agents [11]. In this system the agents have the following properties:

• autonomy: agents operate without the direct intervention of humans or others, have some kind of control over their actions and internal state, can act upon connection hang up;

• social ability: agents interact with other agents (and possibly humans) via some kind of agent-communication language;

• reactivity: agents perceive the context in which they operate and react to it appropriately;

• pro-activeness: agents do not simply act in response to their environment, they are able to exhibit goal-directed behavior by taking the initiative.

Figure 1 presents the architecture of iSIMBioS. At present its configuration includes two servers each containing all system components. All agents are designed as multithreaded Java HTTP servers and implement complex scenarios of distributed interactions in heterogeneous environment. The agents exchange messages via HTTP (in public domain implementation) or HTTPS (in secure corporate implementation) protocol and hence can be used in networks with firewall and proxy server.

System Configuration. The information about agents and their functions is stored in a coordination agent (CA) database. To ensure the actuality of information about the system configuration

• each CA agent database stores the list of counteragents and their URLs, list of functions, reference to the monitoring program, load and authorization characteristics;

• each agent registers with CAs reporting its URL, logical names of executed services, as well as the parameters of designed load characteristics;

• each agent notifies CAs about its scheduled sign-off (e.g.,due to decrease of load on a given service or modification);

• all agents update the information about system configuration by notifying CAs about their current load;

• if any agent or service is unavailable its counteragents notify CAs about their failure to establish connection;

• functionality of the system as a whole and each registered service separately is periodically monitored;

• CAs notify registered agents about changes in the configuration of the system by sending HTTP/HTTPS messages;

• system administrator is notified by e-mail if malfunction of the system happens.

Each agent selects a counteragent (or a required service) with regard to its availability and load, a counteragent located on the same server being selected first. This means that the interactions of agents are not static and can be reorganized dynamically. It is the permanent tracking of the actual system configuration and the dynamic reorganization of agent interactions that ensure the capability of the system to reconfigure. This property allows to extend the functionality of the system and to modify it in operation mode increasing the efficiency of the system use.

System Components. In this section we present a detailed description of each system component.

User Interface Agent (UIA). This agent supports user registration, local searches on a client side and uploading of data and images to a server. It also allows to visualize data and images inserted into a database; select application program modules and specify parameters for data analysis and processing; visualize the results of processing and analysis; retrieve images and data from a database, as well as select output formats and download data and images from a database. Besides UIA provides for visualization of uploaded data and image files, as well as for confirmation of user's intention to insert this data into a database.

Database Access Agent (DBA). DBA executes SQL queries to a database via JDBC; formats query result as TXT, HTML or XML files and converts images stored in a database into JPEG format for visualization on a client side.

Image Server Agent (ISA). This agent performs conversion of image formats and image scaling using ImageMagic library. It also executes standard operations on images (e.g. contrast enhancement, intensity filtering, combining of several images, etc.). In addition ISA participates in retrieval of images from a database for processing as well as in visualization of processed images as JPEGs.

OLAP Server. The OLAP server cooperates with DBA and ISA servers to execute complex scenarios for image and data processing and analysis. The logical rules are applied to implement this cooperation. The OLAP server communicates also with the local database via JDBC and the remote database via DBA. In addition this server interacts with registered workflow modules providing for their initialization, function calls and result output. Communication with modules implemented as Web-services is mediated by the adapter agent. This agent performs conversion of messages between SOAP and HTTP using Apache Axis Soap toolkit, which supports the necessary level of functionality, security and robustness. It also formats SOAP query results as a plain text.

Coordination Agent (CA). CA supports agent registration, and notifies agents about current system status if requested. It also monitors the functionality of the system and notifies registered agents about agent failure and other changes in the configuration of the system. The system administrator is notified about these changes via e-mail.

Database. To manage data and images we decide to use IBM DB2 RDBMS, which supports necessary functionality and reliability of the system and requires minimum familiarization efforts.

Workflows and Modules. In general each scenario for image processing or data analysis consists of many steps executed by heterogeneous programs and services. This software, as well as huge amount of data produced by the consortium are distributed over network. To conduct *in silico* experiments a user needs to combine these resources in a specific order thus forming a workflow. A challenge in implementing the workflow is that each program or service (called as workflow module) should be designed with the interface that can talk to the program before and after it in the chain of calculations. This interface needs to be flexible enough to support communication with different modules in different workflows. Our approach provides a very powerful way to implement such an interface. Modules communicate with each other via agents. An agent can insert data into a database, send it directly to the next module and modify configuration files and other auxiliary data, if necessary. Agents interact with workflow modules through different interfaces, XML-RPC, JNI (Java Native Interface), Java Servlet, Java Server Pages, system calls (for command line modules) and SOAP included.

2.4 Security Measures

In corporate implementation iSIMBioS is modified to include the subsystem providing security. To protect confidential user information we use a method which extends standard HTTPS and SSL technologies by implementing additional authorization and encryption (e.g., Blowfish) procedures, as well as supplementary control of IP-address and access rights.

2.5 Control of Functionality

Services of any component based architecture will, at some point, fail. It is therefore the responsibility of a controlling authority, in our case a coordinating agent (CA), to handle such failures.

In iSIMBioS each agent or program module knows which tests it is necessary to perform to check its functionality. When these services register with OLAP the information about tests is collated into XML file. The CA uses this file to test the system functionality periodically or should the failure of any service was detected by one of its counteragents.

The XML file contains the list of test queries for each service, sample results, references to methods used to check the correspondence of a test result to the sample, as well as logical rules to infer the functionality of services. On execution of tests, the CA builds up a fact base containing binary entries about the state of elementary function of each service. Next the productions rules are applied to this base to infer the service functionality. When any service failure was detected the CA notifies the system administrator via e-mail and other registered agents by sending HTTP messages.

3 Implementation

3.1 User Interface

Local Interface. Local user interface supports visual construction of workflows from program modules, workflow execution and visualization of both intermediate and final results (Fig. 2).

The workflow is constructed by joining program modules and represents a directed acyclic graph. In this graph the nodes shown as rectangles are modules, while the edges displayed as arrows are data-dependency links, which specify that the output of one module serves as input to another one. After a program module was visually constructed, its parameters are specified. Program modules forming one workflow may be located on one server or may be distributed among several server machines.

Following construction and debugging a workflow can be saved as a complex program module in Java, Jscript or VBScript. This module can be re-used as an individual workflow or in complex workflows consisting both from elementary and complex program modules. A complex workflow can be executed in parallel threads. This reduces its run-time by N times, where N is a degree of workflow parallelism.

Web Interface. If the iSIMBioS local interface is not installed, a standard Web browser can be used to process and analyze data. A Web form is available to send requests using HTTP GET and POST methods (http://urchin.spbcas.ru/downloads/esimbios/). A user can select files for analysis, send these files to a distant server, visualize the uploaded files and/or processing results. On the server side the program agent WSA (Web-browser to Scenario Agent) operates, which launches the execution of workflow in distributed environment and can act in multiuser mode.

Application Program Interface (API). XML-RPC and SOAP interfaces are implemented to access iSIMBioS from command-line tools located on a client computer. These interfaces are developed with standard libraries and serve to pass a user request for distributed processing and load data into iSIMBioS, as well as to return a result to the client. To implement this RXA (REST/XML-RPC) and RSA (REST/SOAP) adaptor agents were developed converting data from XML format to HTTP GET request and communicating with OLAP and DBA agents.

The other approach to access iSIMBioS from command-line tools is to make use of standard HTTP API present in all program packages for application development. This requires to install and register an OLAP agent which serves to

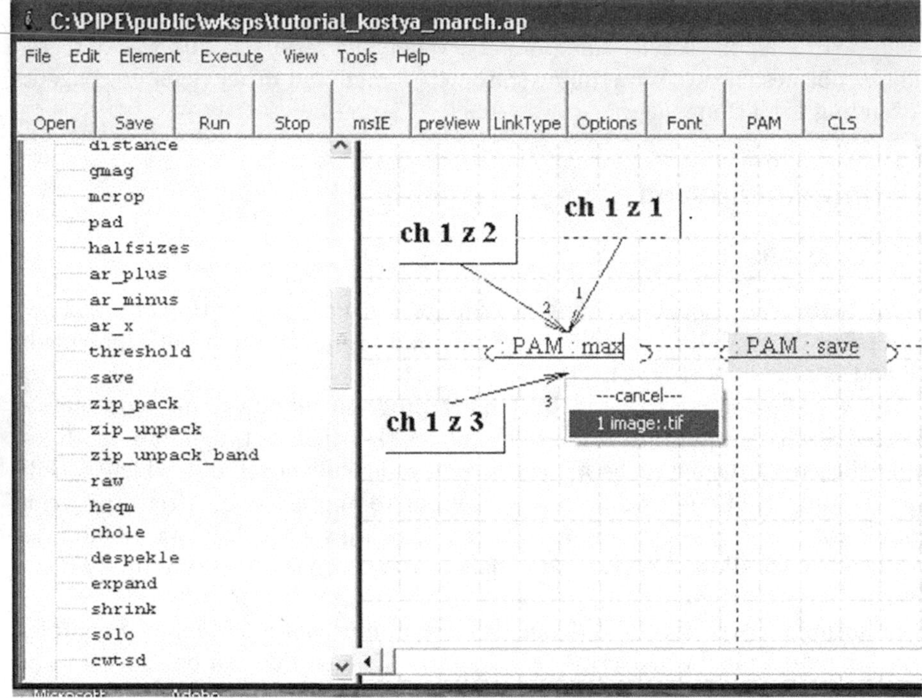

Fig. 2. Local user interface. The main window consists of two frames. The left one displays the list of available workflow modules on each server machine. The right bigger frame shows how to construct the workflow. We illustrate how to select the output port for the workflow module called as max.

pass the HTTP GET request from a command-line application to iSIMBioS and to return the pointer to processing result. The HTTP request contains pointers to data files and program modules located in public $httpd directories of trusted iSIMBioS nodes. Data passing, as well as calls of program and services are mediated by OLAP agents located on these nodes.

3.2 Client Interaction with Services Via SOAP and REST Protocols: Comparative Analysis

The program module wavex was selected to test the efficiency of client interactions with services via SOAP and REST protocols. This module performs the Fast Dyadic Redundant Wavelet Transform (FRDWT) on 2-D array of quantitative gene expression data.

The testing framework was designed as follows. First two clients were designed, SOAP and REST. At the next step both clients were used to send requests to the web service wavex located on a server. Note that SOAP client sends requests directly, while the REST client sends http-GET request via RSA (REST/SOAP) adaptor agent. For both clients the mean response time over a number of simultaneously sent requests was measured. The test was performed using Intel Pentium 3.0 Ghz processor as a client and the Internet connection speed 100 Mbit/s.

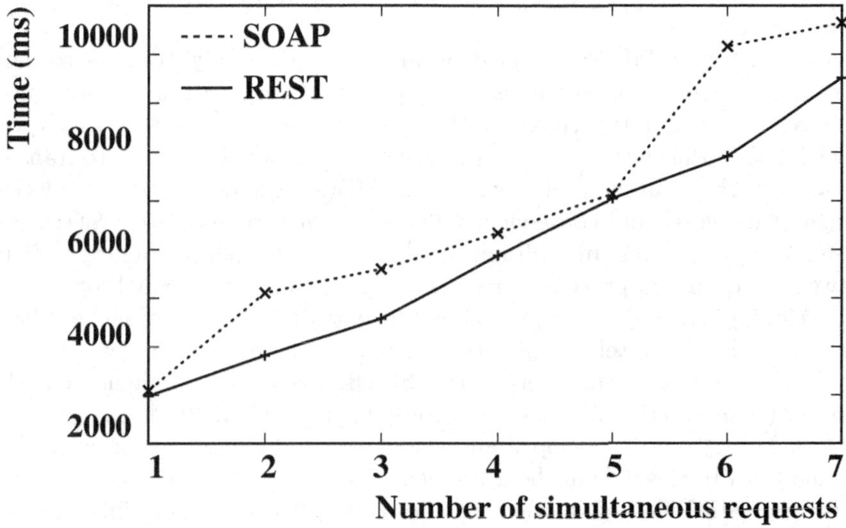

Fig. 3. REST and SOAP client interactions with the wavex web service

Our results show (Fig. 3) that the SOAP client requests are processed on average of 16% more slowly than requests from the REST client. The worse performance of the SOAP client in comparison with that of the REST one is evidently caused by requirement of the SOAP protocol to use XML for the data encapsulation. Conversion into XML format increases data volumes sent to server and thus slows down the performance. We believe that the difference in the SOAP and REST clients' performance will be even more pronounced should slow running client computer or Internet connection be used.

4 Discussion

We have designed iSIMBioS, the system prototype for collaboration within distributed working environment by applying a hybrid approach based on multiagent and REST architectural styles. This prototype is currently tested to automate management and analysis of information on the dynamics of segment determination in fruit fly *Drosophila* by research groups located in different countries.

iSIMBioS is build up from a set of program and services, as well as from an infrastructure used to integrate these components. Program and services can be implemented as XML-RPC, SOAP, RMI, JDBC, CGI and command-line applications. Program agents with different custom interfaces are used to integrate heterogeneous components into workflows. In conformity with REST model all the agents interact via HTTP protocol.

iSIMBioS architecture has two remarkable features. First a workflow can be saved as a complex program module in Java, Jscript or VBScript and re-used in other workflows.

Secondly, when iSIMBioS is called by an application only pointers to data files are transferred. Data passing is mediated by iSIMBioS itself and a pointer to processing result is returned to a user. This is a distinctive feature which sets apart iSIMBioS form the SOAP-based applications, as SOAP requires to transfer data itself [5]. The solution implemented in iSIMBioS results in serious reduction of transport expenses and consequently decreases time required to process and analyze data and images. In addition, it also minimizes requirements to client computers, as data and processing results are solely stored on iSIMBioS nodes.

In developing this prototype we do not concentrate on data storage and have focused on design of system architecture, program components, as well as user and program interfaces. However, when the DB agent was implemented, the technology for interaction of the system components with different DBMS (e.g., IBM DB2, MySQL) was developed and evaluated on different test scenarios for image and data retrieval from the available databases [12].

At present iSIMBioS architecture includes two servers each containing all system components. This functional redundancy provides for robustness of workflow enactments.

There are two features which make the construction of workflow in iSIMBioS fast and convenient. These are possibilities to visualize the intermediate results during workflow enactment and to visually select program components on the iSIMBioS nodes.

iSIMBioS integrates both structured and unstructured data, as well as a variety of data processing and analysis methods implemented with different technologies (e.g. as RMI, JDBC, CGI, command-line applications, SOAP or XML-RPC services) and located on different machines into holistic environment that can be accessed in-real time through a simple and user friendly interface. We believe that this environment may be productively used by biologists to simplify data analysis, hypothesis testing and planning of wet-lab experiments.

Future developments of iSIMBioS will proceed in several directions. We will add new data analysis and processing methods and enhance algorithms for planning sequential and parallel task executions. We are also going to use a DBMS to store and manage all the information. We have already developed the database model which supports the storage of data and programs and is extendable to meet constantly changing user requirements without loss of time on data structure reorganization or software modification. Currently we are implementing program modules executing search, retrieval and saving of all the information. In addition we plan to compare via benchmarks the efficiency of work of iSIMBioS and the analogous application utilizing the SOAP protocol.

Acknowledgements

This work was supported by NIH grant RR07801, NWO-RFFI grant 047.011. 2004.013, contract # 02.467.11.1005 with the Federal agency on science and innovation of the Russian Federation and GAP award RUB1-1578-ST-05. The authors are thankful to Alexander M. Samsonov for valuable discussions and comments.

References

1. Stevens, R.D., Robinson, A.J., Goble, C.A.: mygrid: personalised bioinformatics on the information grid. Bioinformatics **19** (2003) i302–i304
2. Oinn, T., Addis, M., Ferris, J., Marvin, D., Senger, M., Greenwood, M., Carver, T., Glover, K., Pocock, M.R., Wipat, A., Li, P.: Taverna: a tool for the composition and enactment of bioinformatics workflows. Bioinformatics **20** (2004) 3045–3054
3. Wilkinson, M., Schoof, H., Ernst, R., Haase, D.: Biomoby successfully integrates distributed heterogeneous bioinformatics web services. the planet exemplar case. Plant Physiology **138** (2005) 5–17
4. Senger, M., Rice, P., Oinn, T.: Soaplab - a unified sesame door to analysis tools. In: Proceedings of the UK eScience All Hands Meeting. (2003) 509–513
5. Fielding, R.T.: Architectural styles and the design of network-based software architectures. Doctoral dissertation (2000)
6. Myasnikova, E., Samsonova, A., Kozlov, K., Samsonova, M., Reinitz, J.: Registration of the expression patterns of *drosophila* segmentation genes by two independent methods. Bioinformatics **17** (2001) 3–12
7. Jaeger, J., Surkova, S., Blagov, M., Janssens, H., Kosman, D., Kozlov, K.N., Manu, Myasnikova, E., Vanario-Alonso, C.E., Samsonova, M., Sharp, D.H., Reinitz, J.: Dynamic control of positional information in the early *drosophila* embryo. Nature **430** (2004) 368–371
8. Foe, V.E., Alberts, B.M.: Studies of nuclear and cytoplasmic behaviour during the five mitotic cycles that precede gastrulation in *drosophila* embryogenesis. The Journal of Cell Science **61** (1983) 31–70
9. Ingham, P.W.: The molecular genetics of embryonic pattern formation in *drosophila*. Nature **335** (1988) 25–34
10. Jaeger, J., Blagov, M., Kosman, D., Kozlov, K.N., Manu, Myasnikova, E., Surkova, S., Vanario-Alonso, C.E., Samsonova, M., Sharp, D.H., Reinitz, J.: Dynamical analysis of regulatory interactions in the gap gene system of *drosophila melanogaster*. Genetics **167** (2004) 1721–1737
11. Genesereth, M., Ketchpel, S.: Software agents. Communications of the ACM **37(7)** (1994) 48–53
12. Poustelnikova, E., Pisarev, A., Blagov, M., Samsonova, M., Reinitz, J.: A database for management of gene expression data *in situ*. Bioinformatics **20** (2004) 2212–2221

Ontology Analysis on Complexity and Evolution Based on Conceptual Model*

Zhe Yang, Dalu Zhang, and Chuan Ye

Department of Computer Science and Technology, Tongji University,
Postal Code 20 18 04, Shanghai, P.R.C
{Hatasen, Shtjjsjyjx}@hotmail.com,
Daluz@ieee.org

Abstract. With the tremendous development in size, the complexity of ontology increases. Thus ontology evaluation becomes extremely important for developers to determine the fundamental characteristics of ontologies in order to improve the quality, estimate cost and reduce future maintenance. Our research examines the concepts and their hierarchy in ontology conceptual model, the common feature of most ontologies, which reflects the fundamental complexity. We suggest some well-defined metrics of complexity, which mainly examine the quantity, ratio and correlativity of concepts and relationships, to evaluate ontology from the viewpoint of complexity and evolution. In the study, we measured three ontologies in Gene Ontology to verify our metrics. The results indicate that these metrics works well, and the biological process ontology is the most complex one from the view of complexity, and the molecular function ontology is the unsteadiest one from the view of evolution.

1 Introduction

Large standardized ontologies are often developed by several researchers in parallel, such as GO [1]; a number of ontologies grow in the context of peer-to-peer applications [2]; other ontologies are constructed dynamically [3]. Although it becomes important to determine fundamental characteristics of ontologies [4], there are still very few commonly agreed methodologies and metrics to analyze and evaluate ontology complexity and evolution [5, 6]. Thus, metrics are expected to help developers to design ontologies, improve quality, estimate and reduce future maintenance costs in all life cycle.

The rest of this paper is structured as following. Section 2 reviews some related works about ontology, evolution, evaluation and other metrics. Section 3 introduces some formal notations of ontology conceptual model and proposes some complexity metrics. In section 4, the complexity analysis results of Gene Ontology are given to show its evolution trend. Section 6 is summary and outlook for future works.

---

* Supported by National Natural Science Foundation of China under Grant No. 90204010.

U. Leser, F. Naumann, and B. Eckman (Eds.): DILS 2006, LNBI 4075, pp. 216 – 223, 2006.

2 Related Works

The variety of causes and consequences of the ontology changes makes ontology evolution a very complex operation that should be considered as both an organizational and a technical process [7].

In contrast to the researches on ontology evolution and versioning, only little empirical work has focused on evaluation [8]. Ontology metrics is desired in ontology evaluation. Although some metrics have been suggested [9], more work is needed [8]. The most existing metrics are proposed to evaluate the syntactic, semantic, and structure of ontology conceptual model. There are few metrics investigating the ontology complexity and evolution.

Burton et al. assessed the effectiveness of the DAML ontologies [10]. They suggested an ontology auditor metrics suite, and mainly considered the syntactic, semantic, pragmatic and social quality of ontologies.

Literature [11] proposed a set of ontology cohesion metrics to measure the modular relatedness of OWL ontologies. These metrics is focused on the number of classes and depth of inheritance tree of all classes. And it computes cohesion metrics conceptually based on predefined OWL primitives, which explicitly defined tree-based semantic hierarchies in OWL ontologies.

In literature [12], authors use weighted class dependence graphs to represent a class diagrams, and present a structure complexity measure for the UML class diagrams based on entropy distance. It considers complexity of both classes and relationships between the classes. This method can measure the structure complexity of class diagrams objectively.

Idris studied two conceptual integrity metrics based on graph theory is his PhD thesis [13], which are conceptual coherence and conceptual complexity. Conceptual coherence uses average distance between nodes in a graph to measure the interrelatedness of concepts. And conceptual complexity reflects the average number of relationships per node with the average degree across all nodes in a graph.

Chris Mungall researched the increased complexity of Gene Ontology [14]. He measured the average number of paths-to-top of a term and used the path-to-term ratio to measure of complexity in an ontology, which is represented in DAG (directed acyclic graph). However, in his calculation of the total number of terms, the obsolete terms does not be eliminated. While calculating the paths-to-top of terms, the paths of these obsolete terms are not counted. Thus his result is not correct.

3 Ontology Conceptual Model and Complexity Metrics

3.1 Common Formal Notation

We use the following notation to represent some terms in the ontology conceptual model. And small letters are used to identify the notations related to concepts and relationships, while capital letters are used to identify the terminology related to ontology and metrics.

$C = \{c_1, c_2, \cdots, c_m\}$: the set of m concepts defined in an ontology explicitly. In other ontologies, concept may be named as "class" or "term".

$R=\{r_1,r_2,\cdots,r_n\}$: the set of n relationships defined in an ontology explicitly. In other ontologies, relationship may be named as "slot". It only includes those inherited relationships that reflect the hierarchy of concepts, such as "is_a", "part_of", etc.

In ontology conceptual model, concepts hierarchy is typically expressed in DAG (directed acyclic graph) showed in figure 1. Each node represents a concept and each directed arc represents subtype relationship to present the hierarchical structure between concepts in ontology.

$P=\{p_1,p_2,\cdots,p_k\}$: the set of k paths in conceptual model of an ontology. In DAG, path is a distinct trace that can be taken from a specific particular concept to the most general concept in the ontology, which is the concept without any parent.

Different path has its own length, thus the path length is defined as the sum of relationships on a path. So the set of path length in ontology is denoted as $PL=\{pl_1,pl_2,\cdots,pl_k\}$.

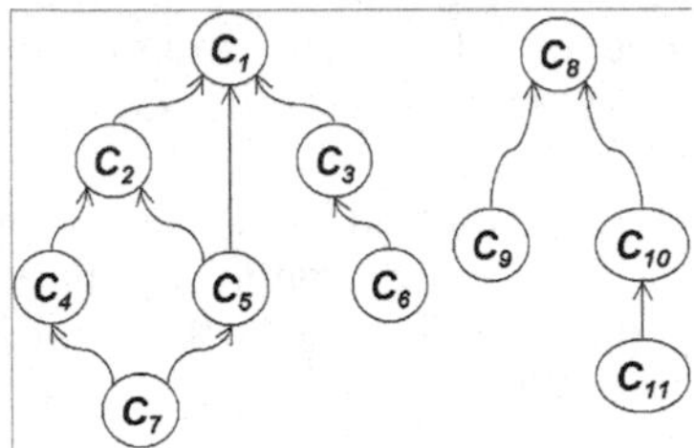

Fig. 1. One ontology conceptual model is represented in directed acyclic graph. There are 11 concepts (m=11); two general concepts (c_1 and c_8) and concept c_7 have three paths to general concept c_1 (c_7-c_5-c_2-c_1, c_7-c_5-c_1 and c_7-c_4-c_2-c_1).

3.2 Complexity Metrics

Concepts: is the sum of concepts in an ontology. $Concepts=|C|=m$.

Relationships: is the sum of relationships in an ontology. $Relationships=|R|=n$.

Paths: is the sum of paths in an ontology. $Paths=|P|=k$.

Λ: is the max path length of ontology. $\Lambda=max(pl_i), 1\le i\le k$.

λ: is the average path length of ontology. $\lambda=\sum_{i=1}^{k} pl_i \Big/ k$.

These two metrics indicate the radius of the ontology in DAG, and the extension of the general concept. They measure the semantic scope covered by the ontology.

μ: the average relationships per concept. $\mu=Relationships/Concepts=n/m$. It indicates the average connectivity degree of a concept.

ρ: the average paths per concept. $\rho=Paths/Concepts=k/m$. For any ontology, ρ must be greater than or equal to 1 (each concept must have a parent except for the general concept). If $\rho=1$, then the ontology is a tree (each concept has a single parent, and thus a single path to the most general concept). Multi-relationship concepts (higher μ ratio) result in higher ρ ratio for an ontology.

σ : is the ratio of max path length to average path length of the ontology, $\sigma = \Lambda / \lambda$. This metric examines the concept aggregation of ontology.

4 Statistics Analyses and Conclusions

We measured the growing complexity of Gene Ontology [1] with the above metrics since we began archiving the ontologies from DEC.2002 to Jun. 2005. GO has three organizing ontologies: BP (biological process), CC (cellular component) and MF (molecular function). The graphs below illustrate the complexity evolution if the three GO ontologies over time.

4.1 Biological Process

In figure 2(a), it is indicated that *Concepts* and *Relationships* increased at a steady but slow rate. The average monthly increase rates are 1.17% and 1.44% respectively. The *Paths* have a rapid growth in quantity. The average monthly increase rate reaches 8.75%. Moreover, before DEC. 2004, *Paths* increased relative steadily in most time. While after that, it increased leapingly, such as at time of DEC. 2004, Jan. 2005 and Apr. 2005. In figure 2(b), the line of μ shows that increment of average relationships per concept is very little; the average monthly increase rate is only 0.26%. While the line of ρ indicates the average paths per concept increased enormously, and the average monthly increase rate is 7.51%. Furthermore, it has the same evolution trend with the line of *Paths* in (a).

If compare (a) and (b) carefully, we can find that the line of μ in (b) also has the same evolution trend with the line of *Relationships* in (a), which leapingly increased over time. This trend is not so obviously in (a) only because of the numerical range on left Y-axis. While in (b), it is magnified.

And if we examine the leap points on two lines in (b) by the time, we can conclude that the μ and ρ increased synchronously. Because path consists of relationships, the increase of μ means the ontology in DAG becomes more complex, the ρ will grow rapidly. This feature can be also observed on the lines of *Relationships* and *Paths* in (a), though not markedly. So it is concluded that the complexity of ontology can be indicated by the metrics of μ and ρ explicitly.

From the line of Λ in (c), it shows an inerratic increase of max path length with two leap points over time. And the λ increased steadily in most time. The average monthly increase rate is 0.48%. This metric reflects the average knowledge range covered by ontology; and its increase indicates the ontology is filling the knowledge extension with concepts and relationships But if magnify the fluctuation on the line, we can find that actually it increased leapingly at some time points, which are the same with lines of ρ in (b) and *Paths* in (a).

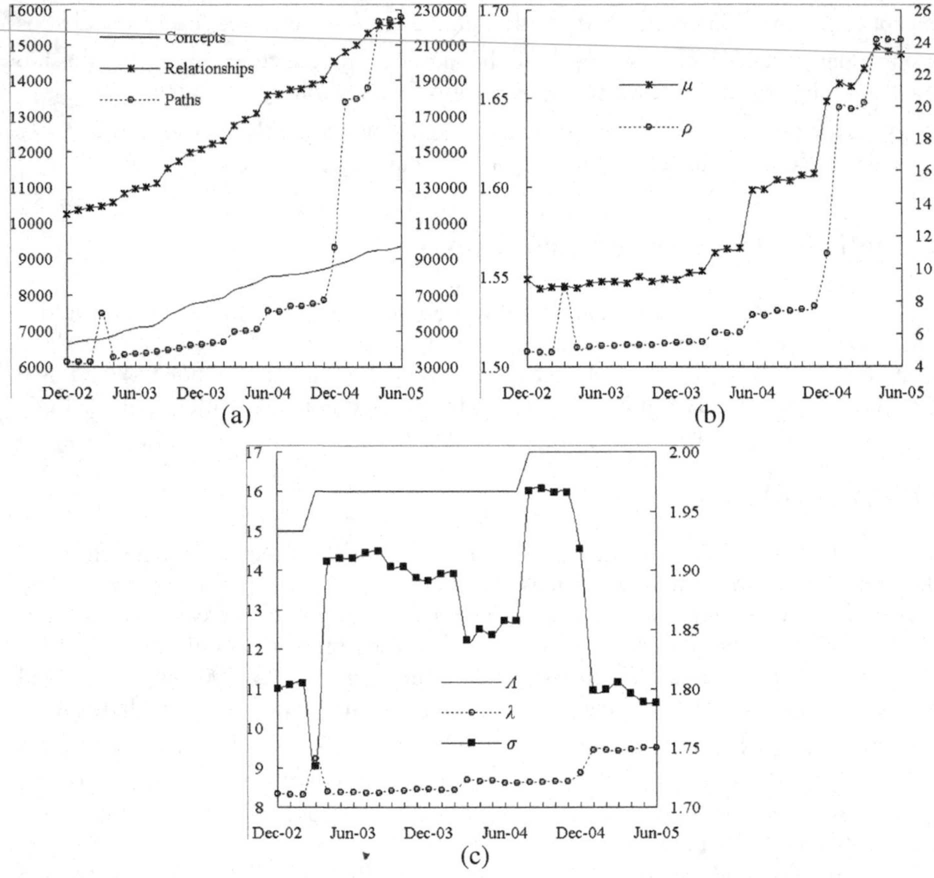

Fig. 2. The complexity evolution of BP ontology. In (a) the left Y-axis shows the increase of concepts and relationships, and the right Y-axis shows the increase of paths. In (b) the left Y-axis shows the increase of μ. The right Y-axis shows the increase of ρ. In (c) the left Y-axis shows the increase of Λ and λ. The right Y-axis indicates the increase of σ ratio.

After examining the line of σ, it is concluded the following results. First, all values of this metric are less than 2, which means most concepts tightly surround the general concept or the core. Professionally, the concept aggregation is high. Second, when the Λ increases, the σ metric will have a leaping increment synchronously. After that, when Λ remains while λ increases, so the σ ratio will decrease until the Λ increases next time. Moreover, the faster the λ increases, the faster the σ decreases. So according to the up or down of the line σ, we can qualitative analysis the variety of Λ and λ with only one metric.

4.2 Cellular Component

In figure 3(a), it is observed that the evolutions of concepts, relationships and paths are almost the same as the figure 2(a). And the monthly average increase rates are

1.28%, 1.92% and 7.24% respectively. The increase of paths is smoother than BP relatively. It has a big rise at Nov. 2004 only.

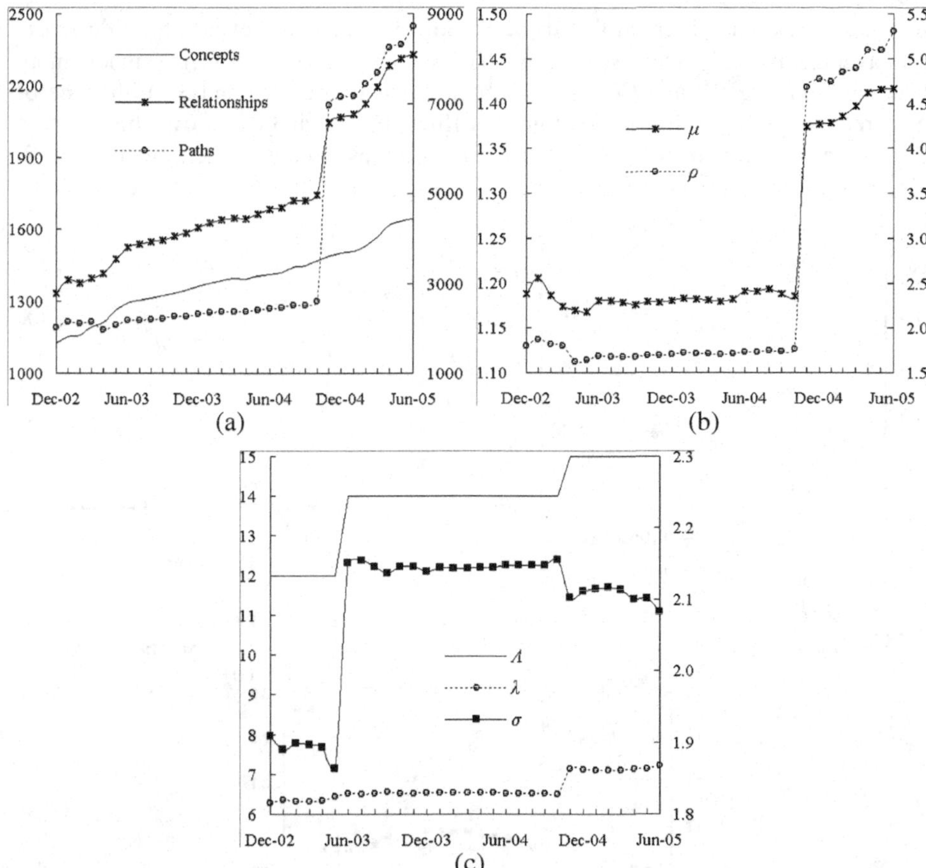

Fig. 3. The complexity evolution of CC ontology. All configurations are the same as figure 2.

Like the figure 3(b), the monthly average increase rates of μ and ρ are 0.63% and 5.88%. As (c) showing, the evolution trends of Λ and λ are similar to BP ontology in figure 2(c). The average monthly increase rate of λ is 0.47%. However, the speed of Λ increase is faster than λ. The biggest difference lies in the values of σ ratio, most of which are over 2. It means the concept organization and aggregation of CC ontology are loose relatively.

4.3 Molecular Function

In figure 4(a), it is observed that concepts and relationships increase steadily except for the third quarter of 2003. However the paths fluctuated a lot over time. The average monthly increase rates are 1.07%, 0.91% and 0.79%. In figure 4(b), fluctuation trend of μ and ρ clearly indicate the complexity of MF is less than the beginning, and it fluctuated wider and rapidly than BP and CC ontologies in some degree. So the average monthly increase rates of μ and ρ are -0.14% and -0.23%. In the figure 4(c), it is observed that the λ is almost changeless, even a little decrease. The average monthly increase rate is -0.02%. The evolution of Λ is much more different from the BP and CC. As we observed in figure 2(c) and 3(c), the line of Λ increases with time. But in this figure, it fluctuates up and down over time. These lead to the same fluctuation trend of σ ratio. And most values of the ratio are over 2, which means the concept organization and aggregation of MF are also loose.

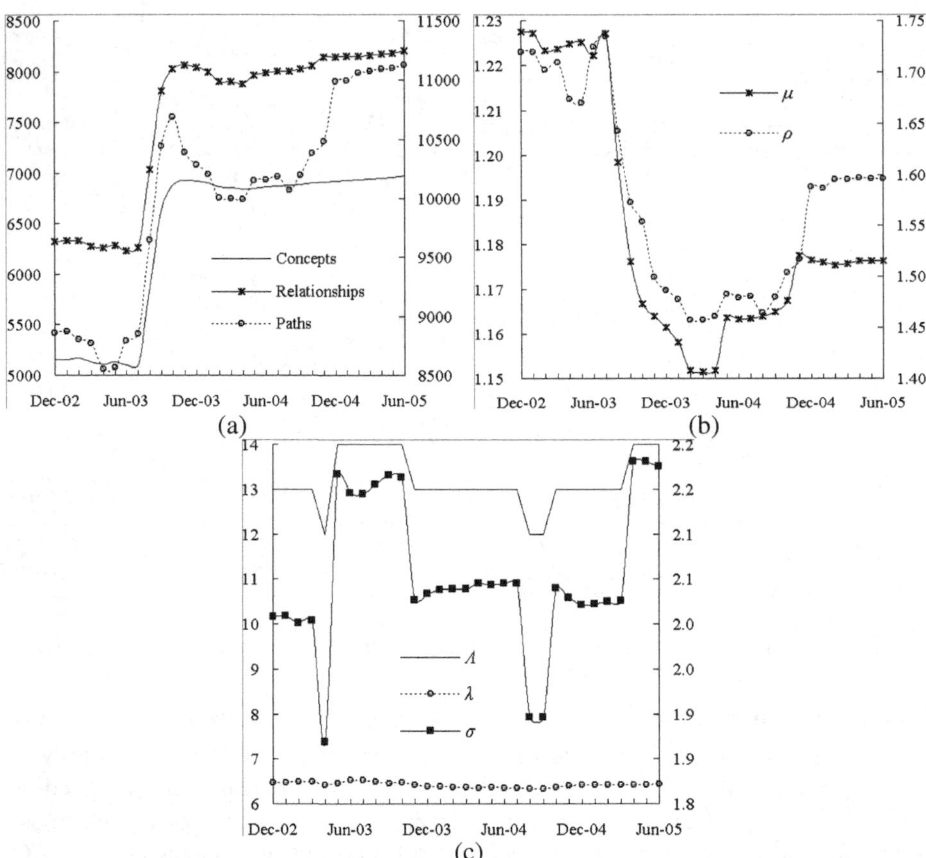

Fig. 4. The complexity evolution of MF ontology. All configurations are the same as figure 2.

According to all graphs, we draw the conclusion that the MF is unsteadier than BP and CC from the view of evolution. And if viewed from the view of complexity, the BP is more complex than CC and MF.

5 Summary and Future Works

In the future, we will continue work on the ontology complexity metrics and other ontology metrics. In this research, our metrics come mainly form the hierarchy in conceptual model. We may add more metrics form other sides of ontology. And upon these metrics, we have had some insights of semantic field of ontology. So, future research may include additional works for the metrics on ontology semantic ability.

References

1. http://www.geneontology.org/
2. W. Nejdl, B. Wolf, C. Qu, S. Decker, M. Sintek, A. Naeve, M. Nilsson, M. Palmr, and T. Risch. Edutella: a p2p networking infrastructure based on rdf. In Proceedings of the eleventh international conference on World Wide Web, pages 604–615. ACM Press, 2002.
3. T. Berners-Lee, J. Hendler, and O. Lassila. The semantic web. Scientific American, 284(5): 34-43, May 2001.
4. A. Das, W. Wu, D. McGuinness, Industrial Strength Ontology Management', The Emerging Semantic Web, IOS Press, 2002.
5. Sirin, E., J. Hendler and B. Parsia, 2003. Semiautomatic composition of web services using semantic description. Web Services: Modeling, Architecture and Infrastructure Workshop in Conjunction with ICEIS2003.
6. Stojanovic L. and Motik B.: Ontology Evolution within Ontology Editors. In Proceedings of 13th International Conference on Knowledge Engineering and Knowledge Management 2002, pp 53-62. Siguenza, Spain, 30th September.
7. S. Staab, H.-P. Schnurr, R. Studer and Y. Sure, Knowledge Processes and Ontologies, IEEE Intelligent Systems. 16(1), Jan./Feb. 2001. Special Issue on Knowledge Management, 2001.
8. Wand, Y. and Weber, R.. Research Commentary: Information Systems and Conceptual Modeling - A Research Agenda. Information Systems Research, 2002, 13 (4), pp. 363-376.
9. Corcho, O. & Gomez-Perez A. A road map to ontology specification languages. In Proceedings of the 12th International Conference on Knowledge Acquisition, Modeling and Management, Juan-les-Pins, France, 2000, pp. 80-96.
10. A. Burton-Jones, V.C. Storey, V. Sugumaran, and P. Ahluwalia. Assessing the effectiveness of the daml ontologies for the semantic web. In Eighth International Conference on Applications of Natural Language to Information Systems, June 2003.
11. Haining Yao, Anthony Mark Orme and Letha Etzkorn. Cohesion Metrics for Ontology Design and Application, Journal of Computer Science 1(1): 107-113, 2005
12. Dazhou Kang, Baowen Xu, Jianjiang Lu, William C. Chu. "A Complexity Measure for Ontology Based on UML," 10th IEEE International Workshop on Future Trends of Distributed Computing Systems (FTDCS'04) pp. 222-228
13. Idris Hsi, "Analyzing the Conceptual Coherence of Computing Applications Through Ontological Excavation", PhD Thesis Proposal, May 13, 2004.
14. Chris. Mungall, Increased complexity in the GO, http://www.fruitfly.org/~cjm/obol/doc/go-complexity.html

Distributed Execution of Workflows in the INB

Ismael Navas-Delgado[1], Antonio J. Pérez[2],
Jose F. Aldana-Montes[1], and Oswaldo Trelles[2]

[1] University of Malaga, Computing Languages and Computer Science Department,
29071, Malaga, Spain
{ismael, jfam}@lcc.uma.es
http://khaos.uma.es
[2] University of Malaga, Computer Architectures Department,
29071, Malaga, Spain
ots@ac.uma.es, aperezp@uma.es
http://www.ac.uma.es

Abstract. Our workflow platform offers a view of the different tools available as a single and uniform pool of services readily available for enhancing query processing. This proposal is based on an architecture for publishing biological data and services, and is designed to be a flexible client for making use of BioMOBY servers, extending them with persistency of the information retrieved for each user. We also present in this paper some biological results, which have been obtained by taking advantage of the proposed workflow execution system. This work has been developed and implemented in the National Institute for Bioinformatics (INB) in Spain (available at http://www.inab.org/MOWServ).

1 Introduction

A Web-based service facilitates access to remote resources promoting the development and availability of highly diverse and specific tools. These new resource capabilities are of special interest in the bioinformatics domain where a variety of databases and services are required in order to produce a more complete view of biological problems. Unfortunately, the common bioinformatics research field becomes hard to operate since it can involve finding appropriate web services by collecting URLs of the useful ones, selecting the most popular or suitable ones, getting familiar with their specific interfaces (e.g. see the very popular sites like NCBI, EBI, ExPASy, etc.), copying and pasting data, manually selecting and combining partial results, and scheduling and pipelining tasks by-hand. Thus, the main component of a bioinformatician's daily work is to carry out a set of simple activities which they usually perform using a diverse set of tools for solving a problem. This implies interacting with different interfaces and storing partial results for use in another tool. This manual interaction with services is costly and error-prone.

To take full advantage of the amount of information available, researchers need to be able to access, link, combine, and query these biological data sets easily and efficiently, and then to integrate the significant number of tools which use these data sources. To address this problem, a growing effort is being made to develop common

U. Leser, F. Naumann, and B. Eckman (Eds.): DILS 2006, LNBI 4075, pp. 224 – 231, 2006.

data-interchange methods, common reference ontologies and automated query engines. Data and service integration has become of particular interest in bioinformatics due to the potential payoff in terms of improved efficiency. Several groups have addressed general solutions for such integration infrastructures: TAMBIS [1], Model-Based Mediation [2], BioDataServer [3], PAT [4], BioBroker [5] and BioMoby [6]. However, these infrastructures have to be inter-related and this process requires human interaction, thereby unnecessarily increasing the time required to obtain a solution to the proposed problem, when the process could in fact be automated.

Thus, a workflow-based system is very useful when the tasks that the user needs to solve are (as is usually the case) predefined and relationships between the tasks are well known. Our proposal provides an execution environment for related tasks, so that users can develop a workflow defining a set of related tasks in order to solve a specific problem, and subsequently execute the workflow with specific inputs. This process can be repeated several times with different inputs in order to derive biological conclusions.

In this paper we focus on the description of a workflow management environment that provides a set of applications for storing, executing and monitoring workflows defined by means of XML files in the Scufl [7] representation language. Our proposal extends this approach with the capability of using authentication-based systems, in which data confidentiality is ensured. Besides the use of a scheduler, which has statistics about the services published in the system, our proposal offers an optimized execution process, optimized error handling and standardized view of data.

Our proposal is presented in Section 2, each element being described in detail. Section 3 illustrates a workflow example performing a homology search and a phylogenetic study. It will also be shown how by using this workflow we can obtain biological knowledge about a family of proteins. Finally we round off with some conclusions and future work.

2 System and Methods

The National Institute for Bioinformatics (INB) in Spain has addressed the integration problem in bioinformatics through the design of a simple, dynamic and extensible platform in order to represent, recover, process, integrate and discover knowledge. The description of biological input/output objects is coordinated and standardized by means of an object-ontology in such a way that services can communicate with each other, wiring natural bioinformatics workflows. Automatic interfaces and help system builders have been incorporated into the architecture to make it more cohesive and to facilitate user communication. Beyond traditional bioinformatics platforms, data persistence system, user management and scheduling abilities have created a new generation of bioinformatics platforms.

The INB system architecture is organized at three main levels: (a) a web-interface at the top of the architecture facilitates communication between the user and the platform (b) the architecture core including the services' interface though bioMOBY API; and (c) at the bottom of the scheme the services' providers.

A web interface manages user sessions with an authentication mechanism. An automatic web interface builder is able to dynamically build on interfaces for browsing data objects, services and namespaces (associated with data containers). The list of services is displayed as a browsable tree from which the user gains access to procedures. In the same way automatic interfaces are built foequested, the system provides notification about the progress status of services, -including historical records of executed tasks-, together with the relationships between input data, service applied and output data. Frequently output data become the input for new services. The GUI provides a specific list of suitable services that can be applied.

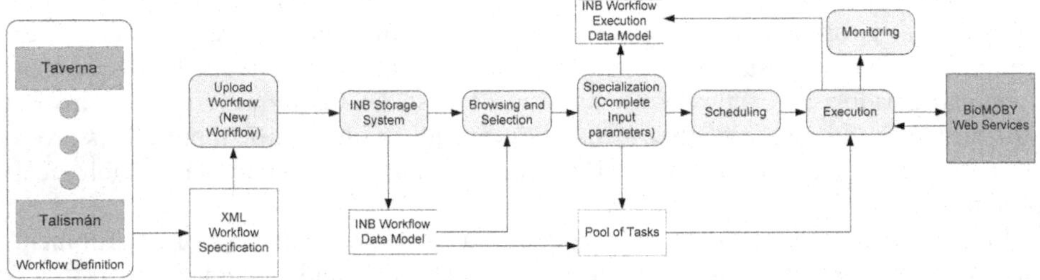

Fig. 1. Data and Process Flow for using the INB workflow management platform. Dark nodes are external applications or services, and the other ones are internal processes.

In this environment, the process to define and execute workflows that we propose is composed of the following steps (see Figure 1):

- To define workflows manually (writing an XML Scufl file) or using a graphical tool for this language, like Taverna Workbench [7], which allows users to graphically describe the services to be executed and how they are related (Section 2.1).
- To store user or generic workflows in the INB platform making use of the provided web interface (Section 2.2).
- To specialise workflows by defining their inputs, specific values for simple data objects or links to existent complex BioMOBY objects (Section 2.3).
- To execute workflows, taking the inputs and executing services as soon as their inputs are available (Section 2.3).
- To monitor the execution, by providing information about the changes in the status of the processors and data results (Section 2.3).

2.1 Defining Workflows

The execution of a set of related services can be done by selecting services and executing them making use of user data. However, an automated mode should be provided in the form of a workflow execution platform. These workflows have to be described in a well-known language for representing workflows in order to increase the utility of the proposed platform. Ultimately the workflow utility depends on the quality of the experiments designed by researchers. Thus, the main task of this kind of tool is to help users and to make the use of existing tools or services easier.

The proposal for defining workflows in our platform is to take advantage of a well-known representation language in the bioinformatics area. Taverna provides a graphical interface capable of dealing with BioMOBY services. Thus, after studying several proposals for building workflows, we have adopted Scufl [7] because it is a well-known workflow XML-based representation language. This language includes a complex structure for defining workflows which could make use of generic web services, BioMOBY services, scripts, etc.

A workflow in the Scufl language includes several entities: *Processors*, *Links*, *Coordination Constraints*, *Sources* and *Sinks*. A workflow also contains additional information such as the title, author and LSID (of the workflow), which can be used to differentiate workflows and help users in the correct selection of the most appropriate one.

Processors are the main elements of workflows because they represent the execution of the services. Processors can deal with generic web services, Soaplab services, Talisman services, sub-workflows, constant values or local functions and scripts, but a processor can also deal with a special type of web service, BioMOBY services. Links are the elements that connect outputs and inputs of processor executions. While sources are the inputs of workflows, sinks are the outputs. Finally, coordination constraints allow users to define conditions that must be fulfilled for executing certain processors.

Our goal is to provide support for loading and executing workflows composed of BioMOBY services published in the INB platform. However, we consider that it is important to offer a full solution that could make use of any kind of service, and the system should deal with the services of the INB and execute them, and delegate the execution of other types of services. An important problem here is that services of other MOBY centrals make use of other ontologies. Thus, in order to connect the services of two centrals, they have to be defined by means of the same concept.

2.2 Internal Representation and Storage

Once developed and represented by means of an XML file (using the Scufl language), a workflow can be executed. In order to provide a quality system it is necessary to offer a user authentication system and data persistency for all data retrieved for each processor. In addition, users must have available the use of services that require a long time to offer a result.

The INB platform offers an interface for loading, executing and monitoring the execution of workflows (see Section 2.3). This platform offers a persistent database to store workflows in order to promote their use and their analysis. This database contains a set of tables that can be queried to retrieve workflow information or for executing the workflow (Workflow, Processor, Link, Source and Sink). In the current version we offer the capability of executing BioMOBY services and scripts related with them, like those for getting the inputs and showing the outputs (*Create_moby_data* and *Parse_moby_data*).

The database tables designed include all the elements necessary to define a workflow (composed of BioMOBY services), so that there is no loss of information in the storage process. Furthermore, the use of constraints allows correct insertion of data in the database to be verified, and additionally it will be possible to rebuild a workflow (in XML format) from the database. Thus, if a workflow is published and shared between different users, all of them can retrieve the XML description of the workflow in order to make use of it. There are three types of workflows: *Generic*, *User* and *Specialized*.

User workflows are loaded taking advantage of the web interface, and are available only to this particular user. On the other hand, Generic workflows are loaded by the administrator by means of a similar interface. Since the "generic" workflows are shared by all users, an intermediate quality control is established to ensure the performance of the workflow. Workflows executed by users are only available to their owners. Thus, each user preserves the confidentiality of his/her data and experiments. On the other hand, the administrator can create Generic Workflows that will be shared by all users of the system. In this case, the administrator tests and verifies workflows submitted by the users, who have to provide a long description (such as documentation) to facilitate the use of a generic workflow by final users. This way of adding Generic Workflows provides a quality (and well documented) set of workflows.

Finally, specialized workflows are those that have been prepared to be executed by defining their inputs and parameters.

When a workflow is loaded, it is parsed, its elements (processors, links, sinks and sources) are stored in the database and a copy of the XML document (describing the workflow with Scufl language) is uploaded onto the INB web server. Additional information is added to workflows in order to improve the documentation of loaded workflows: Name, Short Description and an optional Long Description (as documentation of the workflow). This information is essential if we want to share our workflows with other people, because the title of a workflow (information included inside the XML description) is usually insufficient.

2.3 Executing and Monitoring Workflows

The INB provides a web interface for searching workflows (user and generic workflows), which shows a short description of each workflow (so users have a preliminary description that could be useful for selecting a service). Once a workflow is selected from the list, the next step is to define the inputs in order to execute it. Thus, if the input is a basic BioMOBY object, like String, Integer, Float or Datetime a text box is shown for a value to be introduced. However, if the input is a complex type, the user has three options: to select a stored BioMOBY object that is compatible with the input type, to upload an object stored in a local XML file or to create a new object of the required type. In order to create a bioMOBY object the platform offers a generic creation service, which analyses the structure of the object type and provides a web form for creating an object.

When a user inserts or selects the input/s, the system stores information about the workflow to be executed. Thus the inputs are stored for execution and future use. In addition, a set of internal tasks are created (one for each processor), and related processors imply that the corresponding tasks have the output of the predecessor as input. Relationships between the tables designed are quite similar to the relationships between the tables for storing the workflows, though the sources include a field for storing the value inserted by the user.

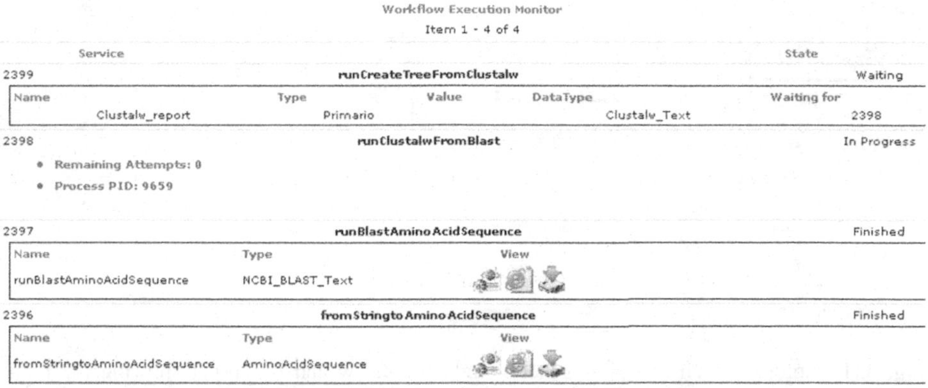

Workflow Execution Monitor
Item 1 - 4 of 4
Service State
2399 runCreateTreeFromClustalw Waiting
Name Type Value DataType Waiting for
Clustalw_report Primario Clustalw_Text 2398
2398 runClustalwFromBlast In Progress
• Remaining Attempts: 0
• Process PID: 9659
2397 runBlastAminoAcidSequence Finished
Name Type View
runBlastAminoAcidSequence NCBI_BLAST_Text
2396 fromStringtoAminoAcidSequence Finished
Name Type View
fromStringtoAminoAcidSequence AminoAcidSequence

Fig. 2. Monitoring the Execution of a Workflow. Each row contains a service (processor) in the workflow and its state. Finished services (2397 and 2396) include links to their results. Then, the services in execution can be executed several times if an error occurs. Finally, the other services indicate which service is being waited for. Thus, the execution process of a workflow shows a set of services that are changing their status right up until the end of the execution, when results can be analyzed.

The created tasks (that are related with the processors by means of the IdTask field) are executed by the system scheduler, which does not execute a task until the required input has been created (the objects have a state that indicates whether an object has been created or not). Thus, synchronization between processors is ensured.

The workflows, which a user has executed, can be monitored by means of a web interface that shows the workflow execution process on the web page. This monitoring tool also includes all the tasks created, their state: Waiting, Finished or Error, and their outputs. Thus, the partial and final results can be examined in order to extract biological conclusions to the workflows executed.

This first approach shows the execution process in a textual mode, but in future work we are planning to offer a graphical view of the execution, which will provide better comprehension of the data flow.

3 Workflow Results

The current INB interface presents a set of services, which carry out different analysis on biological data. These different tools can be automatically applied to a set of data to produce a complete analysis to solve biological problems using the same platform. To this end, we present a practical workflow to solve a phylogenetic study (this workflow is available in the platform as a generic workflow) using an amino acid sequence as the starting point (Figure 3). A homology search is conducted (Blast service) to obtain similar sequences with a common evolutionary history. Output from this service contains a set of putative homologous sequences to the query. A new service is linked (Clustalw from Blast) to build-up a multiple alignment with the most similar sequences reported (these sequences are previously extracted with getBestHits-FromBlast, including an e-value threshold). Finally a phylogenetic tree is obtained

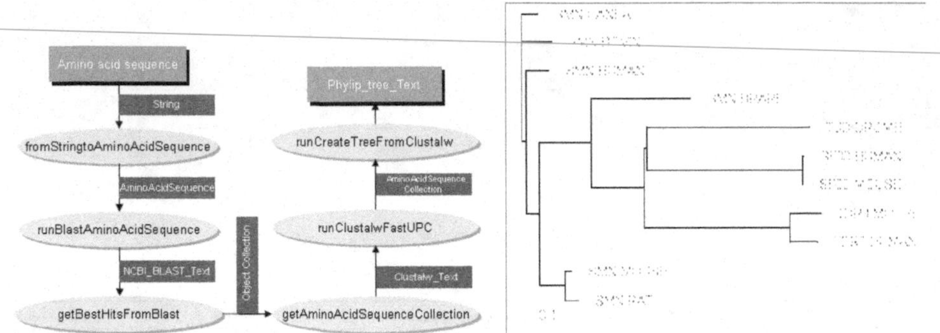

Fig. 3. Homology search and phylogenetic study workflow and its output showing the relatioships among the protein sequences related to SMN_HUMAN with the parameter 'threshold'from runCreateTreeFromClustalw service fixed to one. The SMN proteins from different organisms appear to be related to human ones, and other proteins containing tudor domain and old related SMN proteins appear more similar to fish SMN (SMN_BRARE). Interesting issues are observed in this result regarding maternal tudor protein from Drosophila melanogaster (see text).

using CreateTreeFromClustalw service highlighting the relationships among all the sequences.

This workflow has been tested with the human survival motor neuron protein (SMN; Accession Number: Q16637) running against a SWISS-PROT database and a relaxed e-value was used as a threshold to select distantly related homologous hits and to carry out the multiple alignment. Interesting results are reported such as the fact that several related sequences show a common domain identified as the 'Tudor domains', first identified as fragment repeats in Drosophila melanogaster [8]. As a result of our analysis this large Drosophila protein (TUD_DROME) appears separated from the remaining proteins in the phylogenetic tree (Figure 3) suggesting a relationship with splicing factors (SF30 proteins). In short, these results can conclude that the Drosophila protein, which is required during oogenesis for the formation of primordial germ cells and for normal abdominal segmentation, is a splicing factor assisting this process.

4 Conclusions

In this document we present a client engine based on semantic interconnection concepts. The platform is able to integrate into workflows various processing services developed by different users and groups through a web-based interface. This expands the functionality of current services, enabling the easy incorporation of new procedures to customize the system for specific concerns.

The support for loading, executing and monitoring workflows is based on a very common and well-defined representation language. Our proposal extends current tool with the capability of using authentication-based systems, in which the confidentiality of the data is ensured. In addition, the use of a scheduler based on statistics about the services stored in the system improves the efficiency in the use of computational resources. Another important advantage of this system is that data obtained from the execution of a workflow can be used to execute other services and even workflows.

Acknowledgements

This work has been partially supported by grant "GNV5-Bioinformática Integrada" from Genoma-España and the Spanish MEC Grant TIN 2005-09098-C05-01.

References

1. Stevens, R. D., Baker, P. G., Bechhofer, S., Ng, G., Jacoby, A., Paton, N., Goble, C. A., Brass, A. (2000) TAMBIS: Transparent Access to Multiple Bioinformatics Information Sources. Bioinformatics, 16:2 PP. 184-186.
2. Ludäscher, B., Gupta, A., Martone, M.E. (2003) A Model Based Mediator System for Scientific Data Management. In Z. Lacroix and T. Critchlow(eds.), Bioinformatics: Managing Scientific Data, pp. 335-370, 2003.
3. Lange, M., Freier, A., Scholz, U., Stephanik, A. (2001) A computational Support for Access to Integrated Molecular Biology Data.
4. Gracy, J., Chiche, L. (2005) PAT: a protein analysis toolkit for integrated biocomputing on the web. Nucl. Acids Res. 2005 33: W65-W71.
5. Aldana, J.F., Hidalgo-Conde, M., Navas, I., Roldán, M.M., Trelles, O. (2005) Bio-Broker: A biological data and services mediator system. IADIS International Conference, Applied Computing 2005, Algarve, Portugal, 22-25 February 2005. Pags. 527-534.
6. Wilkinson, M.D., Gessler, D., Farmer, A., Stein, L. (2003). The Bio-MOBY Project Explores Open-Source, Simple, Extensible Protocols for Enabling Biological Database Interoperability. Proceedings Virtual Conference Genomic and Bioinformatics (3):16-26. (ISSN 1547-383X).
7. Oinn, T., Addis, M., Ferris, J., Marvin, D., Senger, M., Greenwood, M. Carver, T., Glover, K., Pocock, M.R., Wipat, A., Li, P. (2004) Taverna: A tool for the composition and enactment of bioinformatics workflows Bioinformatics Journal 20(17) pp 3045-3054, 2004
8. Ponting C.P. (1997) Tudor domains in proteins that interact with RNA. Trends Biochem. Sci. 22: 51-52 (1997).

Knowledge Networks of Biological and Medical Data: An Exhaustive and Flexible Solution to Model Life Science Domains
(Systems Paper)

Sascha Losko, Karsten Wenger, Wenzel Kalus, Andrea Ramge, Jens Wiehler, and Klaus Heumann

Biomax Informatics AG, Lochhamer Str. 9, 82152 Martinsried, Germany

Abstract. The huge amount of unstructured information generated by academic and industrial research groups must be easily available to facilitate scientific projects. In particular, information that is conveyed by unstructured or semi-structured text represents a vast resource for the scientific community. Systems capable of mining these textual data sets are the only option to unveil the information hidden in free text on a large scale. The BioLT Literature Mining Tool allows exhaustive extraction of information from text resources. Using advanced tagger/parser mechanisms and topic-specific dictionaries, the BioLT tool delivers structured relationships. Beyond information hidden in free text, other resources in biological and medical research are relevant, including experimental data from "-omics" platforms, phenotype information and clinical data. The BioXM Knowledge Management Environment efficiently models such complex research environments. This platform enables scientists to create knowledge networks with flexible workflows for handling experimental information and metadata, including annotation or ontologies. Information from public databases can be incorporated using the embedded BioRS Integration and Retrieval System. Users can navigate and modify the information networks. Thus, research projects can be modeled and extended dynamically.

1 Introduction

Today, the life sciences generate an ever-increasing amount of information. This is mainly driven by two factors. First, the life sciences are highly complex fields of research. There are millions of enzymes, genes, chemical compounds, diseases, species, cell types and organs that interact and are related in many different ways. Second, new experimental methods are continuously developed; as their throughput increases, the amount of raw data generated increases with overwhelming speed.

For information technologies, the challenge remains to support scientists in the identification of relevant information, the integration of this information in specific "knowledge bases" and the formalization of this knowledge across multiple scientific domains to facilitate hypothesis generation and validation (and, therefore, the generation of new knowledge). Information technology (IT) solutions are needed to

U. Leser, F. Naumann, and B. Eckman (Eds.): DILS 2006, LNBI 4075, pp. 232–239, 2006.

support the knowledge generation cycle [1, 2] to ultimately gain an adequate understanding of whole biological systems. Systems Biology is a new field of research that has an intrinsic hierarchical nature, presenting a multiplicity of applicative fields that must be interconnected to give a complete description of the fundamental biological system (E-cell, virtual organs).

1.1 From Information to Knowledge

The most important source to collect a comprehensive set of relevant information available to the scientific community is the text body of published papers. Although this body of information is mostly unstructured, text-mining techniques have been developed to analyze text syntax and semantics. Text mining may be the most important answer to the mass production of scientific literature. It is, however, confronted with the same phenomena as Natural Language Processing (NLP): complexity and ambiguity. Natural language is diverse and can express the same thought in many syntactically different ways. Ambiguity arises on all levels of natural language: lexical ambiguities such as "bank", syntactical ambiguities such as "the man watches the girl with the telescope", and semantical ambiguities such as "every man loves a woman". Complexity and ambiguity often come together and make the resolution of the underlying meaning almost impossible, even for the human mind.

With respect to database integration, solutions to make all information accessible exist. A common approach is based on flat-file indexing, which emerged due to the flat-file origins of most biological databases. SRS [3] and the BioRS Integration and Retrieval System (http://www.biomax.com) are prominent examples of such technical solutions. Relational database management systems (RDBMS) are also widely used, with Oracle being one of the most popular RDBMS in the life science domain.

One issue in the integration of multiple databases is mapping the data semantics. A simple example is a case where a protein identifier is designated "`prot_id`" in one database, but is designated "`id`" in another. This problem is rather easy to solve. Both identifiers designate semantically identical entities (the protein) by semantically identical attributes. Common data access systems implement mechanisms to provide a "unified" search semantic across databases using this simple mapping technique. However, this technique is insufficient to describe, for example, the relationship between a protein and a protein complex in which the protein is likely to participate. Here, the semantic of the relationship has to be explicitly described: "Protein A *participates_in_complex* Complex B". In this way, diverse information, such as molecular processes, disease phenotypes or clinical information about patients, can be modeled as *complex semantic networks.*

The above Protein-Complex relationship example illustrates a simple approach to formalized knowledge. Though the actual definition of "knowledge" is indistinct, knowledge can be seen as the awareness of a validated interconnection of details, which, in isolation, are of lesser value. That "Protein A *participates_in_complex* Complex B" should therefore be supplemented by evidence *why* Protein A participates in Complex B. That evidence is annotation of the relationship. If it is possible to provide evidence for a defined relationship from different, independent sources (e.g., multiple scientific experiments based on various methods), the validity of the relationship is strengthened. For both Protein A and Complex B, further validated relationships with other "elements of a scientific domain" (such as compounds or diseases) may exist, which broaden the overall knowledge. With these elementary concepts, "elements", "relations" and "annotation", it becomes possible to formalize huge networks of knowledge.

1.2 This Study

In a proof-of-principle project implementing a simple knowledge-generation process, we have chosen to analyze PubMed abstracts, mining for co-occurrences of gene or protein names with cancer-related disease terms using text-mining techniques. Although cancer is a complex field of research with large numbers of published papers, a couple of comprehensive review articles summarize the more important genes responsible for the genesis of various cancers [4]. Qualitatively, we based the validation of our text-mining methods on these review articles.

All identified Gene-Cancer relationships were additionally analyzed for their association with certain compound or drug terms. The resulting data set was imported into the BioXM Knowledge Management Framework and a network of information was structured by, for example, classifying the disease term dictionary using the "NCI thesaurus" ontology [5]. Other ontologies such as the FunCat catalogue [6] and GO [7] were used to classify the molecular function of all identified genes or proteins. Additional information was mapped to the genes by integrating external public databases.

The result is a comprehensive knowledge base with a high-value core of information that can be extended with diverse proprietary information.

2 Methods

We have built a software environment composed of three core components. The BioRS Integration and Retrieval System forms an access layer for the integration of public databases such as MEDLINE. The BioLT Literature Mining Tool identifies information in the MEDLINE text body by applying text-mining techniques. The results of the text-mining process are imported into the BioXM Knowledge Management Environment, and their semantics are formalized and integrated into larger networks of knowledge.

2.1 Linguistic Text Analysis

The BioLT Literature Mining Tool is text-mining software that combines linguistic core functionalities with the power of a genuine retrieval component (BioRS Integration and Retrieval System) and controlled vocabularies. The user can take advantage of an augmented Boolean query language that allows exhaustive retrieval of entire text databases.

The BioLT tool can analyze any text corpus using several parsing mechanisms. A tagger records dictionary occurrences with a certain degree of fuzziness, such as dash elimination. The BioLT dictionary resource currently contains up to 20 different vocabularies and almost one million tokens (i.e., entries). A phrase parser recognizes chunks of tokens. Specialized mechanisms scan the text for gene name and polymorphism patterns. An acronym parser associates abbreviations of a certain kind with the likely expansions in the text. The parsing results, with indications of text positions, phrasal status, etc., go into a relational database. A separate index of the full text is created in the BioRS system. Queries to the BioLT tool are first issued to the BioRS system and the results are merged with the relational database. A number of standard weighting algorithms are available to rank the final results.

2.2 Integration of Public Databases

The BioRS Integration and Retrieval System, a data retrieval system, allows the integration of relational and flat-file databases into a common, homogeneous environment. The databases to be integrated, both public and proprietary, are organized differently according to storage (flat-file vs. relational) and format (EMBL format, XML format, etc.). The BioRS software allows the rapid retrieval of data (e.g., sequence, structure and literature) from multiple databases in parallel.

Using HTML-based query forms, searches can be as simple or complex as necessary using a sub-query option for search-result refinement. The BioRS system also supports queries for phrases and search-term synonyms. For example, a thesaurus of gene names may contain all corresponding alternative names for each gene. When searching a database for a specific gene name, entries containing synonyms of the gene name will also be retrieved. By mapping semantically equivalent attributes of different databases to a single BioRS attribute used for retrieval, the same information entities can be found in parallel in all integrated databases. Searchable cross-references between related information in different databases ensure complete information access.

The retrieval functionality can be incorporated into proprietary programs or scripts allowing for transparent access to entries in databases. The BioLT Literature Mining Tool and the BioXM Knowledge Management Environment use an embedded version of the BioRS system as middleware to access external databases.

2.3 Knowledge Management Software

The BioXM Knowledge Management Environment is designed for the aggregation of information and the semantic modeling of scientific processes. A particular area of scientific interest can be modeled as a network of related elements. The user can define different *element types* and *relation classes*. For example, elements of type "gene" or "protein" can be linked using a relation of type "Gene Regulation" or "Protein-Protein interaction". Sub-networks, called *contexts*, which allow biological pathways and processes to be organized as parts of the overall network of knowledge, can be defined. Relationships between contexts and other "*semantic objects*", such as elements, can be established. This allows efficient modularization and abstraction of knowledge. All "semantic objects" (such as *elements*, *relations*, *contexts* or *ontology concepts*) can be annotated. Annotations are form-based and support hierarchical organization of information (nested annotation forms). Multiple semantic objects can share annotation to imply relationships. The BioXM system supports the conceptualization of entire areas of interest using arbitrary ontologies. The taxonomy of "*is_a*" relationships, which formally structure the ontology, can be used to infer facts and abstract queries in the BioXM system. The software provides graphical browsing through the network and an advanced query builder for guided construction of complex queries with a natural-language-like syntax.

The BioXM Knowledge Management Environment allows access to all public databases integrated by the BioRS Integration and Retrieval System. External database entries can serve as either "virtual" semantic objects or "read-only" annotation of semantic objects. Although the information remains external, the database entries used as "virtual semantic objects" can be organized in the project tree, can become part of a network, and can be annotated by the user in the same way as any other semantic object in the BioXM system.

3 Workflow Results and Discussion

In the first step, PubMed abstracts were analyzed for co-occurrences of gene or protein names with cancer-related disease terms using the BioLT text-mining technology. We used a recent review article [4] to provide a preliminary verification of the quality of our predictions. All identified genes and proteins were then analyzed for their associations with compound or drug terms covered by the BioLT compound dictionary. The resulting data set was imported into the BioXM Knowledge Management Framework and the information network was structured by classifying the disease term dictionary using the "NCI thesaurus" ontology [5]. Other ontologies such as the FunCat catalogue [6] and GO [7] were used to classify the molecular function of all identified genes or proteins. Additional information was mapped to the genes by integrating external public databases.

3.1 Text-Mining Process

The next step in the knowledge-generation process was to find all relationships between genes or proteins and cancer-related disease terms, based on 30 years of MEDLINE abstracts starting in 1975. For all genes found with cancer associations, we also mined the MEDLINE abstract for compound- or drug-term associations.

Text mining provides a shortcut through the complexities of NLP and tries to guess the best results. Ambiguity, however, remains and there is no simple solution to deal with it. The following example shows the different spellings and meanings the gene symbol "psp" has in the MEDLINE corpus (table threshold is set to 10 occurrences).

Nine protein names (from the 33 acronym extensions of "psp") generate several hundreds of hits, a fraction of the total number of 7081 occurrences for "psp" in the complete text corpus. A substantial gap of 1245 occurrences remains unassigned with our approach. In addition, the clearly assigned protein names leave nine different meanings for the symbol "psp". One way to deal with this ambiguity is to detect and record the information for the user of the text-mining system.

3.2 Preliminary Quality Assessment of the Text-Mining Results

As a benchmark, the BioLT co-occurrence results were compared to a manually curated list of "all major pathways and hereditary cancer predisposition types" each

Table 1. The following example shows the different spellings and meanings the gene symbol "psp" has in the MEDLINE corpus (table threshold is set to 10 occurrences). The intended gene meanings are in bold font.

Acronym expansion	Occurrences	Acronym expansion	Occurrences
progressive supranuclear palsy	1704	plasmatocyte spreading peptide	23
- (no expansions found)	1245	phosphoserine phosphatase	23
paralytic shellfish poisoning	271	**prostate secretory protein**	**21**
pancreatic stone protein	**128**	paralytic shellfish poisons	20
primary spontaneous pneumothorax	122	period called pseudopregnancy	19
parotid secretory protein	**114**	**plasma protein**	**18**
Pancreatic spasmolytic polypeptide	107	**phage-shock protein**	**13**
polysaccharide peptide	58	photostimulable storage phosphor	12
paralytic shellfish poison	58	progressive supranuclear palsy	12
postsynaptic potential	55	posterior probability	11
perchloric acid-soluble protein	**43**	penicillin-susceptible pneumococci	11
phenyl saligenin phosphate	42	**phage-shock-protein**	**11**
Photostimulable phosphor	35	polysaccharide of spirulina platensis	10
polystyrene particles	33	premonitory sensory phenomena	10
phage shock protein	**33**	**parasitism-specific protein**	**10**
peak systolic pressure	28	post-suppression period	10
postsynaptic potentials	26		

related to one of 57 representative predisposition genes (Vogelstein and Kinzler, 2004). With 100% recall, all 57 genes and 57 cancer types were represented in the BioLT dictionaries for genes/proteins and diseases. Ninety-five percent (95%) of the relationships were ranked in the top three results of up to thousands of hits. For the remaining three genes, the corresponding diseases were found in the fourth and fifth positions. If, for example, one compares the relationships between the protein PDGFRA and 81 co-occurring disease terms from the BioLT disease dictionary with the disease relation "familiar gastrointestinal stromal tumors" noted in the publication above, the following four co-occurrences are presented at the top of the results in BioLT: hypereosinophilic syndrome (hes) (13 hits), gastrointestinal stromal tumors (19 hits), systemic mast cell disease (7 hits), and systemic mastocytosis (5 hits).

After the qualitative assessment of text-mining results, it is necessary to provide the possibility to manually validate and annotate results further. To base a decision-making process on this kind of automatic result, the information needs to be validated, not only by checking the original text source, but also by reviewing the findings in the context of an expanded knowledge base.

3.3 Information Integration

There were three main objectives in integrating the automatically derived text-mining results in the knowledge-generation process. A data model based on "Gene-Disease" and "Gene-Compound" relationships was established and the data was imported. The sentences were added to the appropriate relationship as annotation to provide evidence. The number of sentences found for each relationship gives a first, rough estimation of the validity of the particular relationship.

Further annotation forms were created to allow user-reviewed validation of the relationships. The annotation forms used several attributes to describe the nature of the evidence. An evidence ontology [8] was used as a controlled vocabulary to describe the evidence in a structured way and facilitate data mining.

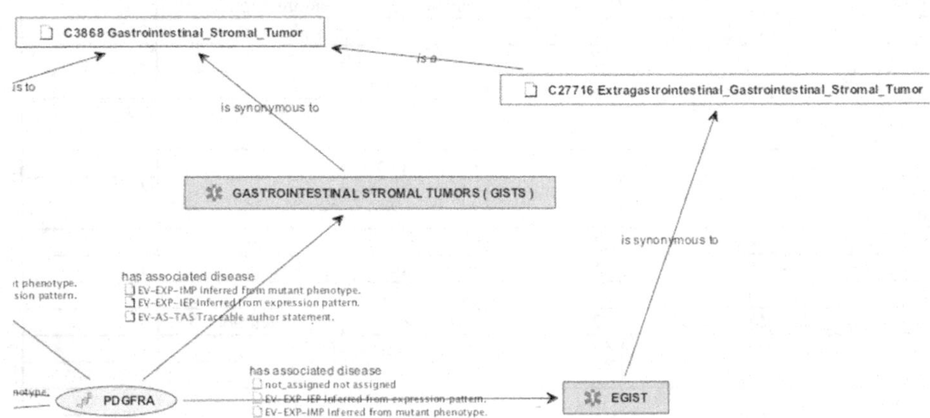

Fig. 1. Visualization of cancer terms and their classification using the "NCI thesaurus" ontology: PDGFRA was found to be associated with the disease "EGIST", which is a synonymous term for the ontology concept "Extragastrointestinal Gastrointestinal Stromal Tumor". That concept is derived from "Gastrointestinal Stromal Tumor", which itself has synonymous terms that were found to be associated with the gene PDGFRA. The manually annotated evidence codes [8] for the automatically generated relations are partly displayed.

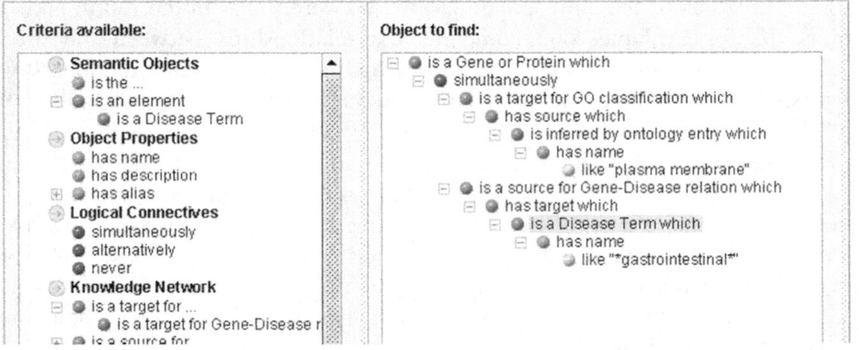

Fig. 2. Query Builder: This example query searches for all GO-inferred plasma membrane proteins associated with all "gastrointestinal" cancer-related diseases

The cancer terms were mapped to the "NCI thesaurus" ontology [5] (Fig. 1) to enable inferred searches for disease concepts. Gene-specific information was added by mapping the gene names dictionary to the EntrezGene database, integrated with the BioRS system. Knowledge provided by the KEGG and PubChem databases adds information about metabolic processes. The core data model of "Gene-Disease" and "Gene-Compound" relationships was extended with relations to functional ontologies such as GO [7] and the FunCat catalogue [6]. In this way, gene functionality in complex queries can be inferred, as shown in Fig. 2.

The concepts implemented by the BioXM Knowledge Management Environment are generic and flexible. Input from most scientific domains can be adapted to the requirements of specific projects easily. The BioXM system facilitates collaboration of experts on a project-specific data set and allows reviewing, modifying, editing and commenting on a core set of information. Multiple sources of information can be integrated to be mined for data, so that the resulting greater network of information provides value well beyond the originally intended scope of the project.

References

1. Lazebnik, Y., *Can a biologist fix a radio?--Or, what I learned while studying apoptosis.* Cancer Cell, 2002. **2**(3): p. 179-82.
2. Searls, D.B., *Data integration: challenges for drug discovery.* Nat. Rev. Drug Discov., 2005. **4**(1): p. 45-58.
3. Etzold, T., A. Ulyanov, and P. Argos, *SRS: information retrieval system for molecular biology data banks.* Methods. Enzymol., 1996. **266**: p. 114-28.
4. Vogelstein, B. and K.W. Kinzler, *Cancer genes and the pathways they control.* Nat. Med., 2004. **10**(8): p. 789-99.
5. Hartel, F.W., et al., *Modeling a description logic vocabulary for cancer research.* J. Biomed. Inform., 2005. **38**(2): p. 114-29.
6. Ruepp, A., et al., *The FunCat, a functional annotation scheme for systematic classification of proteins from whole genomes.* Nucleic Acids Res., 2004. **32**(18): p. 5539-45.
7. Ashburner, M., et al., *Gene ontology: tool for the unification of biology. The Gene Ontology Consortium.* Nat. Genet., 2000. **25**(1): p. 25-9.
8. Karp, P.D., et al., *An evidence ontology for use in pathway/genome databases.* Pac. Symp. Biocomput., 2004: p. 190-201.

On Characterising and Identifying Mismatches in Scientific Workflows

Khalid Belhajjame, Suzanne M. Embury, and Norman W. Paton

School of Computer Science
University of Manchester
Oxford Road, Manchester, UK
{khalidb, embury, paton}@cs.man.ac.uk

Abstract. Workflows are gaining importance as a means for modelling and enacting *in silico* scientific experiments. A major issue which arises when aggregating a collection of analysis operations within a workflow is the compatibility of their inputs and outputs: the analysis operations are supplied by independently developed web services which are likely to have incompatible inputs and outputs. We use the term mismatch to refer to such incompatibility. This paper characterises the mismatches a scientific workflow may suffer from and specifies mappings for their resolution.

1 Introduction

Scientific workflows are gaining considerable momentum as a mechanism for specifying and automating the execution of scientific experiments [1,10]. During the design of a scientific workflow, the designer's focus is on selecting and composing the analysis operations that will carry out the work of the experiment. Analysis operations are supplied by third parties and as such it is often the case that their inputs and outputs are incompatible with those of the other operations to which they must be connected. We use the term mismatch to refer to such incompatibility. In order to resolve a mismatch, the designer must expend some effort in discovering or implementing special operations that can be plugged into the workflow at the point of incompatibility, and can transform the data sets as necessary to resolve it.

Manual detection and correction of such mismatches is time-consuming and unreliable, and thus reduces the claimed benefits of scientific workflows in facilitating the rapid specification of experiments. In this paper, we propose a classification of the kinds of mismatches that can occur in data-driven workflows and derive the additional information that must be captured about workflow operations if potential mismatches are to be identified automatically. This additional information takes the form of annotations on web service inputs and outputs, based on three separate ontologies.

The remainder of the paper is organised as follows. First, in Section 2, we formally define scientific workflows. In Section 3, we describe the three additional ontologies used for annotating operation inputs and outputs, and use them (in Section 4) to present the mismatch classification and (in Section 5) to specify further annotations for transformation functions that characterise the kinds of mismatches they can address. Finally we close the paper by comparing our work against existing works, and drawing conclusions in Section 6.

U. Leser, F. Naumann, and B. Eckman (Eds.): DILS 2006, LNBI 4075, pp. 240–247, 2006.

2 Scientific Workflows

A scientific workflow is a set of operations connected together using data links. For the purposes of this paper, we define a scientific workflow *SWf* as *SWf* = ⟨*nameWf, OP, DL*⟩, where *nameWf* is a unique identifier for the workflow, *OP* is the set of operations from which the workflow is composed, and *DL* is the set of data links connecting the operations in *OP*.

Operation. An operation $op \in OP$ is defined as *op* = ⟨*nameOp, loc, in, out*⟩, where *nameOP* is the unique identifier for the operation, *loc* is the URL of the web service that implements the operation, and *in* and *out* are two sets representing the input and output parameters of the operation, respectively.

Parameter. A parameter provides information on the data type of a given operation input/output. It is defined by the pair ⟨*nameP, type*⟩, where *nameP* is the parameter's identifier (unique within the operation) and *type* is the parameter's data type. In our work, we assume an XML type system, so that parameter data types may be either simple types, such as *xs:string* and *xs:int*, or complex types, built from simple ones.

Data Links. Let $IN = \cup_{(op \in OP)} op.in$ be the set of inputs of all the operations comprising a scientific workflow, and $OUT = \cup_{(op \in OP)} op.out$ be the set of outputs of all its operations. The set of data links connecting the workflow operations must then satisfy the following: $DL \subseteq (OP \times OUT) \times (OP \times IN)$. A data link relating the output *o* of the operation *op1* to the input *i* of the operation *op2* is therefore denoted by the quadruple *(op1,o,op2,i)*.

3 Ontologies for Characterising Mismatches

Information on the types of operation parameters is usually easily available to scientific workflow systems. For example, where operations are actually web services, the data types can be extracted from the WSDL specification of the service. However, as we have seen, not all mismatches are visible in the types of the connected parameters. In order to automatically detect mismatches, the implicit information about the form and role of operation parameters must be made explicit, just as the information about the type of the parameter currently is. This additional information concerns the semantics of the parameter (i.e. the real world entity to which the parameter corresponds), the representation format used for the parameter over and above any data typing given to it and the extent of the parameter (i.e. the set of possible values which the parameter may take). For each of these, we must create an *ontology* of terms that can be used to annotate services with the information required to detect mismatches.

An ontology is commonly defined as an explicit specification of a conceptualisation [4]. Formally, an ontology θ can be defined as a set of concepts, $\theta = \{c1, \ldots, cn\}$. We use the following ontologies to annotate service parameters for mismatch detection.

Domain Ontology, θ_{domain}. This ontology captures information about the application domains covered by the operations, and enables us to describe the real world concepts to which each parameter corresponds. An example of such an ontology is that developed by the ^{my}Grid project describing the domain of molecular biology [9]. Typical concepts from this ontology are *ProteinSequence* and *MolecularWeight*.

For the identification of mismatches, we assume the existence of a function *domain()*, the signature of which is presented below. Given an operation and an input/output parameter, the function *domain()* returns the corresponding concept from the domain ontology.

$$domain: OP \times (IN \cup OUT) \rightarrow \theta_{domain}$$

Representation Ontology, $\theta_{represent}$. As in many other application areas, a variety of different formats have been defined for representing the same kind of biological data (i.e. data corresponding to the same domain concept). For example, a protein sequence can be represented using *Fasta* format or *Uniprot* format or any of several other similar formats. These formats can be represented using complex data types but at present it is much more common for workflow operations to treat them as simple string objects and to ignore their internal structuring. This is partly due to legacy design (since many of the more popular biological web services were implemented before the development of XML and its associated programming tools) and partly because current workflow systems do not always have very rich type systems.

In order to detect mismatches in representation format as well as data type mismatches, it is necessary for services to be annotated according to the formats expected and produced by their inputs and outputs. We therefore require an ontology of terms for describing data formats. Such an ontology has already been designed by the ^{my}Grid project [9].

We therefore assume the existence of a function *represent()*, with the signature presented below. Given an operation and an input or output parameter, the function *represent()* returns the corresponding concept from the representation ontology.

$$represent: OP \times (IN \cup OUT) \rightarrow \theta_{represent}$$

In order to compare two representation concepts for mismatch identification, we need an additional binary comparison operator for concepts in $\theta_{represent}$. We use the *contains()* function, the signature of which is presented below, to describe the relationship between these formats. Given two formats *x* and *y*, the function *contains(x,y)* is true if *x* contains all the data content needed for creating an instance of *y*, and false otherwise.

$$contains: (\theta_{represent} \times \theta_{represent}) \rightarrow Boolean$$

Extent Ontology, θ_{extent}. The concepts of this ontology define the scope of possible values of a given operation parameter. Although in general it is not possible to accurately describe the extent of a parameter, it is the case that the relationships between the extents of some web services is known in advance and can be used in detecting mismatches. For example, the TrEMBL database[1] is known to be a superset of SwissProt, whereas the various species specific gene databases are known not to overlap.

No ontologies currently exist for describing the extent of biological data sets, and we have therefore constructed one ourselves. We assume the existence of a function *extent()* for retrieving the extent of input/output parameters, with the signature:

[1] http://www.ebi.ac.uk/trembl/

$$extent: OP \times (IN \cup OUT) \rightarrow \theta_{extent}$$

In order to be able to compare extents, we use the function *coveredBy()*, the signature of which is presented below. Given two concepts from the extent ontology, *e1* and *e2*, *coveredBy(e1,e2)* is true if the space of values designated by *e1* is a subset of the space of values designated by *e2* and false otherwise.

$$coveredBy: (\theta_{extent} \times \theta_{extent}) \rightarrow Boolean$$

4 Characterising Mismatches

Using annotations of the form described in the previous section, we can automatically detect a variety of forms of mismatch that go beyond simple data type mismatches. We now present a classification of these mismatch types and define the criteria for identifying each one.

Type Mismatch. refers to incompatibility in terms of data type between connected parameters. In order to be compatible, the data type of the output parameter must be the same as or a subtype of the data type required by the input parameter. Formally, a data link $(op1,o,op2,i) \in DL$ suffers from a type mismatch iff[2]:

$$o.type \not\preceq i.type$$

Cardinality Mismatch. is a particular kind of type mismatch. For example, assuming a type system in which the only means of forming collection types is an array constructor, we can say that a data link $(op1,o,op2,i) \in DL$ suffers from a cardinality mismatch iff:

$$(o.type = ArrayOf\,(i.type)) \text{ or }$$
$$(i.type = ArrayOf\,(o.type))$$

ArrayOf(t) is a type. An instance of *ArrayOf(t)* is an array whose elements are *t* instances.

Domain Mismatch. refers to incompatibility in terms of semantic domain between connected output and input parameters. In order to be compatible, the domain of the output must be the same as or a sub-concept of the domain of the subsequent input. Specifically, a data link $(op1,o,op2,i) \in DL$ suffers from a domain mismatch iff[3]:

$$domain(op1,o) \not\subseteq domain(op2,i)$$

For example, consider a data link *(op1,o,op2,i)* such that *domain(op1,o)* = *DNA_sequence* and *domain(op2,i)* = *Protein_sequence*. According to the molecular biology ontology mentioned earlier [9], *DNA_sequence* $\not\subseteq$ *Protein_sequence*, therefore, *(op1,o,op2,i)* suffers from a domain mismatch.

[2] The symbol $\not\preceq$ stands for not a sub-type of.

[3] The symbol $\not\subseteq$ stands for not a sub-concept of.

Representation Mismatch. Two operation parameters, which belong to compatible semantic domains, can be represented using different data formats. Representation mismatch refers to the difference in terms of format between connected input and output parameters, which are domain compatible. Specifically, a data link $(op1,o,op2,i) \in DL$ suffers from a representation mismatch iff:

$$(domain(op1,o) \subseteq domain(op2,i)) \text{ and}$$
$$(represent(op1,o) \neq represent(op2,i))$$

For example, suppose that $domain(op1,o) = domain(op2,i) = Protein\_record$, $represent(op1,o) = Uniprot\_record$, and $represent(op2,i) = Fasta\_record$. The output and the input parameters have the same semantic domain. However, they adopt different representations. We conclude that the data link suffers from a representation mismatch.

Content Mismatch. is a particular kind of representation mismatch, in which the formats conflict in terms of data scope as well as in terms of pure representation—that is, the format of the output carries less data content than is required by the format of the subsequent input. Formally, a data link $(op1,o,op2,i) \in DL$ suffers from a content mismatch iff:

$$(domain(op1,o) \subseteq domain(op2,i)) \text{ and}$$
$$(represent(op1,o) \neq represent(op2,i)) \text{ and}$$
$$(contains(represent(op1,o),represent(op2,i)) = false)$$

This situation is distinguished because it represents a particularly serious form of representation mismatch. Even if we can find a web service that can perform the transformation between the two mismatched data formats, there may still be a problem with the workflow, since the transformed output may not contain all the information expected by succeeding operations. Note that we say there "may" be a problem with the workflow. It is possible that the succeeding operation will only access those parts of the transformed data structure that are contained within the initial data format, in which case there will be no problem.

As an example of this, consider the data link *(GetFasta,o,GetSequence,i)*. Here, *represent(GetFasta,o) = Fasta_record*, and *represent(GetSequence,i) = Uniprot_record*. This data link suffers from a content mismatch, since a *Fasta_record* does not contains all the elements required for creating a *Uniprot_record* instance: *contains(Fasta_record, Uniprot _record) = false*. However, in reality, the *GetSequence* operation will only read the protein sequence parts of the *Uniprot* record supplied as its input, and therefore a simple transformation between formats is sufficient to resolve the mismatch.

Extent Mismatch. refers to incompatibility in terms of the space of possible values between two connected output and input parameters. Specifically, a data link $(op1,o,op2,i)$ suffers from an extent mismatch if it does not suffer from a type mismatch, a domain mismatch or a representation mismatch, but the extent of the input i does not cover the extent of the output o. Formally, a data link $(op1,o,op2,i) \in DL$ suffers from an extent mismatch iff:

$$(o.type \preceq i.type) \text{ and}$$
$$(represent(op1,o) = represent(op2,i)) \text{ and}$$
$$(domain(op1,o) \subseteq domain(op2,i)) \text{ and}$$
$$(coveredBy(extent(op1,o),extent(op2,i)) = false)$$

For example, consider a data link *(op1,o,op2,i)* such that *domain(op1,o)* = *domain(op2,i)* = *ORF*. *ORF* stands for open reading frame. Suppose now that *extent(op1,o)* = *FlyBase* and *extent(op2,i)* = *SGD*. *FlyBase* is a database that stores information on the genetics and molecular biology of Drosophila. *SGD* is a scientific database of the molecular biology and genetics of the yeast *Saccharomyces cerevisiae*. The two databases are non-overlapping: none of the *ORFs* found in *FlyBase* are present in *SGD* (i.e. *coveredBy(FlyBase,SGD)* = *false*). Therefore, *(op1,o,op2,i)* suffers from an extent mismatch. Even though the parameters appear to match exactly in terms of domain, data type and representation format, the workflow will still not be able to produce a result.

5 Annotation of Parameter Mapping Operations

The same ontologies that allow us to create annotations for identifying mismatches can also support the annotation of transformation functions that can resolve them. In our context, we refer to such functions as *mappings*, since they map from one parameter type to another. In order to compare the available mappings with the identified mismatches, we annotate the mappings. Given a data link *(op1,o,op2,i)*, which suffers from a mismatch, a mapping is used for transforming the data produced by o to meet the requirements of the input i. Formally, a mapping is defined as follows:

$$\langle T1, T2, C1_{represent}, C2_{represent}, C1_{domain}, C2_{domain}, f_{map}\rangle$$

where $T1$ and $T2$ are data types, $C1_{represent}$ and $C2_{represent}$ are formats from the representation ontology, and $C1_{domain}$ and $C2_{domain}$ are concepts from the domain ontology. f_{map}: $T1 \rightarrow T2$ is a function. Given an instance $t1$ of $T1$ that follows the format $C1_{represent}$ and belongs to the domain $C1_{domain}$, $f_{map}(t1)$ returns an instance $t2$ of $T2$, which follows the format $C2_{represent}$ and belongs to the domain $C2_{domain}$. The extent of the mapping function $f_{map}()$ is specified by a pair (e_1,e_2), where $e_1, e_2 \in \theta_{extent}$; e_1 designates the extent of the domain of f_{map}, and e_2 designates the extent of its range.

The above annotation system can be used for locating the appropriate mappings for correcting the identified mismatches. It is possible that none of the existing mappings can be used for correcting a given mismatch. Instead of building a new mapping, there are cases in which the desired mapping can be obtained by composing in sequence two or more existing mappings.

6 Related Work and Concluding Remarks

Several problem solving environments (PSEs) have been proposed to support the design and enactment of scientific workflows [10]. Generally, they do not provide means for identifying and correcting mismatches. Taverna, for example, allows the designer to connect any two operations regardless of whether the connected outputs and inputs are compatible. Some PSEs are able to identify type mismatches. In Triana [8], data links are checked at design time and a warning message is displayed whenever two connected parameters have incompatible data types. In terms of parameter mapping, Kepler [3] is, to our knowledge, the only system which supports the mapping of operation parameters that have type mismatches. Note, however, that parameter mapping is a work in progress that is not supported by the current distribution of Kepler. The semantic domain of the operation inputs and outputs are described using an ontology. Whenever two connected parameters belong to compatible domains but have incompatible data types, a mapping is generated to transform the output parameter structure into the structure of the succeeding input parameter [2]. Incompatibilities due to differences in representation format, content and extent are not handled by this proposal.

In this paper, we have characterised a range of mismatches that can occur in scientific workflows. Our categorisation goes beyond existing work in this area by identifying the need for additional parameter annotations describing representation formats and extents, and in showing how they interact with the more familiar notions of domain and data type annotations. The categorisation can be used to implement mismatch detection and resolution services that allow workflow designers to concentrate their attention first on the core aspects of workflow semantics, and to consider the necessary data transformations afterwards. For complex workflows, this opens up the possibility for domain experts to rapidly specify abstract workflows, which they then pass to staff with technical expertise in managing data mapping and transformation to resolve any gaps in the workflow.

We have developed a prototype that implements the proposed framework as an extension of the Taverna workbench [7]. Using the prototype, we have conducted a preliminary evaluation of our mismatch categorisation. Real scale trials are not yet a practical possibility, due to the lack of rich service and annotation mappings. The Feta registry we are currently using, for example, contains the descriptions of around 30 service operations, though more extensive annotations are planned. However, we wished to gain some insight into the degree to which the mismatches we have identified occur in practice. To this end, we collected together a sample set of 14 bioinformatics workflows, which were designed in the context of e-science projects such as ISPIDER[4], myGrid[5] and Pegasys[6]. We then examined the operations contained within the workflows and made a judgement as to whether the operation was part of the core semantics of the workflow or whether its role in the workflow was to resolve incompatibilities between core operation parameters. We also attempted to classify the mismatches we found based on the categorisation presented in this paper.

The results of this small study showed that the most commonly occurring types of mismatches are the representation and domain mismatches. The majority of the workflows that we analysed suffered from these kind of incompatibilities, with the next most common kind being cardinality and extent mismatches. In fact, most of the type mismatches we identified were actually cardinality mismatches; non-cardinality-based type mismatches appear to be rare, if our small sample set is a reliable guide. This can be explained by the following two observations. First, the inputs and outputs of most bioinformatics analysis operations are weakly typed [5]. In most cases, parameters are either defined as strings or arrays of strings, regardless of the complexity of the data values actually being communicated. The second reason for the comparative rarity of non-cardinality-based type mismatch is that, for the time being, most of the

[4] http://www.ispider.man.ac.uk

[5] http://www.mygrid.org.uk

[6] http://bioinformatics.ubc.ca/pegasys/

available scientific workflow systems are not able to process complex types. Thus the work of parsing and constructing such data values is pushed down into the operations themselves.

Clearly, open questions remain regarding the best approaches to identify candidate mappings for identified mismatches. We are currently investigating the possibility of using mapping quality as a search criterion during mismatch resolution. This refers to the non-functional properties of mappings that may help the designer to select the best mapping for a given context [6]. Examples of such properties include the information and computational resources used for performing the mapping.

Acknowledgements

The work presented in this paper was funded by a grant from the BBSRC. We are also grateful to Duncan Hull and Robert Stevens, and our colleagues in the ISPIDER project, for useful discussions on mismatches in scientific workflows.

References

1. K. Belhajjame, S. M. Embury, H. Fan, C. A. Goble, H. Hermjakob, S. J. Hubbard, D. Jones, P. Jones, N. Martin, S. Oliver, C. Orengo, N. W. Paton, A. Poulovassilis, J. Siepen, R. Stevens, C. Taylor, N. Vinod, L. Zamboulis, and W. Zhu. Proteome data integration: Characteristics and challenges. In *UK All Hands Meeting*, 2005.
2. Sh. Bowers and B. Ludäscher. An ontology-driven framework for data transformation in scientific workflows. In *DILS*, pages 1–16, 2004.
3. Sh. Bowers and B. Ludäscher. Actor-oriented design of scientific workflows. In *ER*, pages 369–384, 2005.
4. T. Gruber. A translation approach to portable ontology specifications. *Knowledge Acquisition*, 5(2):199–220, 1993.
5. D. Hull, R. Stevens, P. Lord, C. Wroe, and C. Goble. Treating shimantic web syndrome with ontologies. In *First Advanced Knowledge Technologies workshop on Semantic Web Services (AKT-SWS04)*, 2004.
6. E. M. Maximilien and M. P. Singh. A framework and ontology for dynamic web services selection. *IEEE Internet Computing*, 8(5):84–93, 2004.
7. Th. M. Oinn, M. Addis, J. Ferris, D. Marvin, M. Senger, R. M. Greenwood, T. Carver, K. Glover, M. R. Pocock, A. Wipat, and P. Li. Taverna: a tool for the composition and enactment of bioinformatics workflows. *Bioinformatics*, 20(17):3045–3054, 2004.
8. I. J. Taylor, M. S. Shields, I. Wang, and O. F. Rana. Triana applications within grid computing and peer to peer environments. *J. Grid Comput.*, 1(2):199–217, 2003.
9. Ch. Wroe, R. Stevens, C. A. Goble, A. Roberts, and R. M. Greenwood. A suite of daml+oil ontologies to describe bioinformatics web services and data. *Int. J. Cooperative Inf. Syst.*, 12(2):197–224, 2003.
10. J. Yu and R. Buyya. A taxonomy of scientific workflow systems for grid computing. *SIGMOD Record*, 34(3):44–49, 2005.

Collection-Oriented Scientific Workflows for Integrating and Analyzing Biological Data*

Timothy McPhillips[1], Shawn Bowers[1], and Bertram Ludäscher[1,2]

[1] UC Davis Genome Center, University of California, Davis
[2] Department of Computer Science, University of California, Davis
{tmcphillips, sbowers, ludaesch}@ucdavis.edu

Abstract. Steps in scientific workflows often generate collections of results, causing the data flowing through workflows to become increasingly nested. Because conventional workflow components (or actors) typically operate on simple or application-specific data types, additional actors often are required to manage these nested data collections. As a result, conventional workflows become increasingly complex as data becomes more nested. This paper describes a new paradigm for developing scientific workflows that transparently manages nested data collections. Collection-oriented workflows have a number of advantages over conventional approaches including simpler workflow designs (*e.g.*, requiring fewer actors and control-flow constructs) that are invariant under changes in data nesting. Our implementation within the KEPLER scientific workflow system enables the explicit representation of collections and collection schemas, concurrent operation over collection contents via multi-level pipeline parallelism, and allows collection-aware actors to be composed readily from conventional actors.

1 Introduction

Scientists today require access to data from diverse sources. Nowhere is this need more pressing than in the life sciences, where multiplying databases and rapidly growing data repositories promise to provide researchers with a wealth of information relevant to the systems they study. Effectively exploiting diverse sources of data requires a spectrum of data integration approaches.

In the database community, data integration traditionally means resolving different data structures that represent fundamentally the *same* kind of information [11]. This information may be stored using heterogeneous schemas, and may use different representations for data values (*e.g.*, for identifying objects). In such cases, data integration involves determining mappings between source schemas, and then transforming these schemas into a common schema and corresponding integrated data set that can be used for some other purpose. These mappings and transformations typically represent logically necessary relationships between different data sources.

In contrast, data integration in the life sciences often entails applying fundamentally *different* kinds of information to answer scientific questions, make discoveries,

* Work supported in part by SciDAC/SDM (DE-FC02-01ER25486), NSF/SEEK (DBI-0533368), and NSF/GEON (EAR-0225673).

U. Leser, F. Naumann, and B. Eckman (Eds.): DILS 2006, LNBI 4075, pp. 248–263, 2006.

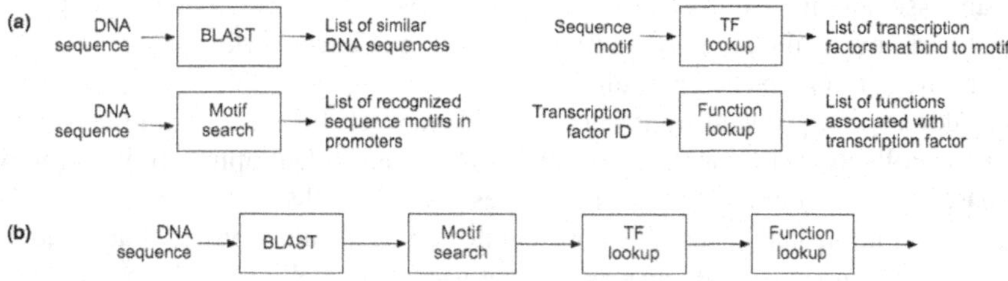

Fig. 1. Scientific workflow components frequently produce lists of results: (a) typical bioinformatics components; and (b) a hypothetical workflow composed from these components that leads to increasingly nested data collections

and test theories. Such scientific data integration procedures necessarily invoke scientific theories that cannot be inferred from schemas or data alone. For example, consider a systematist who wishes to use both genomic sequence data and morphological data in the process of inferring the evolutionary relationships among organisms. Instead of simply mapping DNA sequences and morphological data into a uniform data format, different processes may be applied to each data source to infer evolutionary (*i.e.*, phylogenetic) trees. The systematist then may use the assumption that the organisms have only one true set of evolutionary relationships, and that the phylogenetic trees inferred from genomic and morphological data approximate the true relationships. By employing this theory, the researcher may "integrate" these distinct data sources by computing a consensus tree that reflects commonalities in the distinct phylogenetic trees inferred from the different data sources. These consensus trees (*i.e.*, the resulting data product of integration) can then be analyzed further or applied in other studies.

The challenge of integrating life-science data from multiple sources becomes even more daunting as disciplines become increasingly specialized and as more diverse types of scientific data are desired. Scientific workflow systems [12,13,15,20,22,4] aim at facilitating these types of integration and analysis.[1] However, current scientific workflow systems still offer little or no support for effectively managing (and hiding) the inherent complexity of life-science data, leading to overly complex workflows that are hard to create, reuse, and optimize.

As shown in Figure 1, scientific workflow components (or actors) frequently generate lists of results. When carried out one after the other, such operations naturally yield increasingly nested collections of data that must be managed during workflow execution. This situation is further complicated by the fact that the steps in such workflows in general operate on different nesting levels. For example, a query of a database mapping sequence motifs to known transcription factors might take a single motif as an input, while the operation upstream of this step in the workflow might generate a list of motifs to operate upon. Similarly, the collection of all transcription factors associated with a number of different sequence motifs might be required as input to a downstream component. As these examples demonstrate, scientific workflows must be able to

[1] Figure 4 shows an implementation of a workflow for inferring and analyzing phylogenetic trees using the KEPLER system.

maintain associations within and between nested lists of intermediate results throughout the workflow, while at the same time presenting to each workflow component data inputs of the correct type and granularity.

We address this problem by proposing a framework for representing and managing nested collections in scientific workflows (Section 2). Our approach is inspired by flow-based programming [18] and techniques used in collection-based [3] and functional programming languages. We represent nested data collections as "flat" sequences of data tokens embedded with special control tokens for delimiting the beginning and end of each collection. We previously have described [17] how our implementation of this approach within the KEPLER scientific workflow system provides convenient high-level operations for managing nested collections; facilitates highly pipelined execution of actors operating at different levels of collection nesting; simplifies workflow design; enables context-dependent, dynamic configuration of actors; and supports robust workflow exception handling.

In this paper we define an abstract data model for collection-oriented workflows (Section 3). Using this abstract data model, we then define a lightweight schema language for restricting collection-oriented structures. Collection schemas can be used for a number of purposes. They allow developers to "publish" reusable collection definitions. Schemas are also used in defining *scope* parameters (Section 4), which declare the type of data an actor operates over, and how the actor should be invoked over that data. In general, scope parameters are declarative expressions used to configure collection-aware actors and to simplify actor development. Finally, we introduce an approach that allows collection-aware actors to be composed readily from conventional KEPLER actors, and show how this approach can simplify the development of new collection-aware actors and further facilitate reusability in scientific workflows.

2 The Collection-Oriented Workflow Approach

2.1 The KEPLER Scientific Workflow System

The KEPLER scientific workflow system [1,12] is being developed jointly by a collaboration of application-oriented scientific research projects.[2] KEPLER extends the PTOLEMY II[3] system (hereafter, PTOLEMY) with new features and components for scientific workflow design and for efficient workflow execution using distributed computational and experimental resources. PTOLEMY was originally developed as a modeling and simulation environment, *e.g.* to study complex computation models and embedded system applications.

In KEPLER, users develop workflows by selecting appropriate components (called *actors*) and placing them on the design canvas. Once on the canvas, components can be "wired" together to form the desired dataflow graph, *e.g.*, as shown in Figure 4. Actors have *input ports* and *output ports* that provide the communication interface to other actors. Workflows can be hierarchically defined, using *composite actors* to contain subworkflows. Control-flow elements such as branches and loops are also supported.

[2] `http://www.kepler-project.org/`

[3] `http://ptolemy.eecs.berkeley.edu/`

In KEPLER, actors can be written directly in Java or can wrap external components. For example, KEPLER provides mechanisms to create actors from web services, C/C++ applications, scripting languages, R[4] and Matlab, database queries, SRB[5] commands, and so on.

In KEPLER, data is represented as a sequence of *tokens*, which are passed from one actor to another via actor connections. KEPLER differs from other scientific workflow systems in that the overall execution and component interaction semantics of a workflow is not determined by actors, but instead is defined by a separate component called a *director*. This separation allows actors to be reused in workflows having different models of computation. KEPLER (via PTOLEMY) includes directors that specify, *e.g.*, process network (PN), synchronous dataflow (SDF), continuous time (CT), discrete event (DE), and finite state machine (FSM) computation models.

Most scientific workflows defined using KEPLER use the PN director (based on [9]), or SDF, a restricted version of PN. The PN director executes each actor in a workflow as a separate process (or thread). Actors communicate asynchronously in process networks through buffered channels implemented as queues of effectively unbounded size. The PN director can be used to pipeline data tokens through scientific workflows, enabling highly concurrent execution. In SDF, actors *a priori* define fixed token consumption and production rates. This allows the SDF director to statically schedule actors [10], while guaranteeing certain properties of workflows. PTOLEMY's support for composite actors allow multiple computing models to be used within a single workflow by optionally specifying distinct directors for particular subworkflows, *e.g.*, the PaupHSearch composite actor employs the SDF director (Figure 8), but may be used within a workflow based on the PN director (Figure 4).

2.2 Managing Nested Data Collections in Kepler

KEPLER currently does not provide explicit support for managing nested collections, and workflow authors use a variety of approaches to add this support to KEPLER workflows. The general approach used to support nested collections in KEPLER is shown in Figure 2. Figure 2 (a) shows two conventional KEPLER actors A and B, where the output of A is connected to the input of B. Here, A produces singleton data items of type β (where individual items are denoted β_1, β_2, etc.), which are directly consumed by B. Figure 2 (b) shows a similar workflow, but where actor A has been replaced by actor A′, which produces lists of items of type β instead of only singleton β values. The block labeled *CF* indicates where special control-flow actors are used to unpack and repack list elements.[6] Figure 2 (c) uses the same underlying workflow; in this case however, actor A′ receives a list of input values, introducing additional control-flow blocks. Figure 2 (d) shows the case where an actor A″ produces pairs of items of type (δ,β), the β items are routed using a control-flow bock to the B actor (which expects only β items), and the δ items are routed downstream where they are paired with B's output (again, using a control-flow block) and passed as input to the C actor.

---

[4] `http://www.r-project.org/`

[5] Storage Resource Broker, `http://www.sdsc.edu/srb/`

[6] Control-flow blocks are implemented in a number of ways in practice, but are typically modeled using multiple low-level actors possibly placed within a composite.

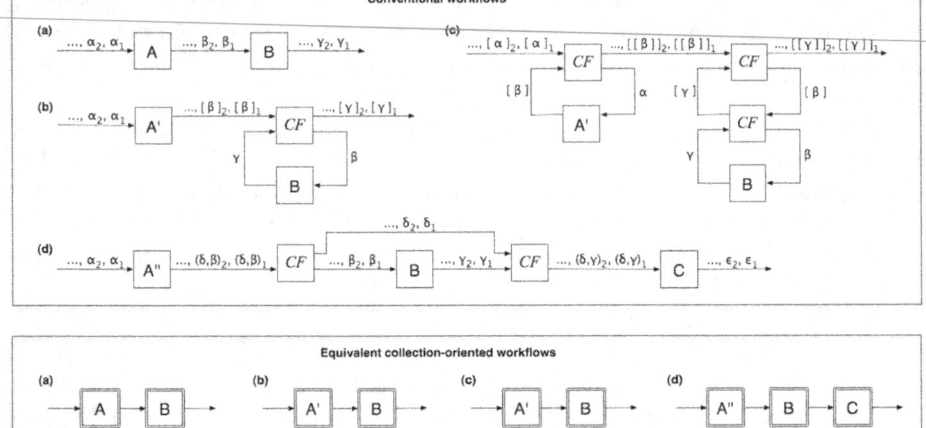

Fig. 2. Conventional scientific workflows with control-flow constructs for handling complex data (top), and corresponding collection-oriented workflows (bottom) in which the control-flow is managed explicitly by the framework

As Figure 2 demonstrates, a significant weakness of using special actors to manage collections is that the resulting workflows must be modified to handle changes in (upstream) data nesting. In principle, one could tailor variants of actors A–C to support particular collection structures, *e.g.*, by embedding the logic represented by the *CF* blocks within custom code in each actor. This approach, however, limits the ability to reuse these actors in other workflows and contexts. In general, *ad hoc* approaches for managing nested collections in scientific workflows leads to code duplication and tightly couples actor implementations with workflow designs; hampers rapid prototyping of workflows and associated data structures; makes comprehension, reuse, and refactoring of existing workflows difficult; and limits reuse of actors designed for these workflows.

Our solution is to provide explicit support for developing "collection-aware" actors. These actors employ a common framework for managing nested collections efficiently and transparently. Moreover, workflows composed from collection-aware actors do not suffer from the reuse limitations inherent in *ad hoc* approaches to managing nested collections. The lower panel of Figure 2 shows collection-oriented workflows equivalent to the conventional workflows in the upper panel. Note that introducing additional levels of data nesting does not change the collection-oriented workflow definitions. Collection-oriented workflows and actors are by design immune to such changes and thus far more reusable. In this example, each collection-aware actor defines their input of interest using a *scope expression* (*e.g.*, α for A and β for B). The framework automatically performs the necessary control-flow functions for providing each actor with their data of interest. In addition, input data outside of an actor's scope is automatically forwarded downstream.

Figure 3 illustrates our approach for streaming nested collections through workflows. Data streams are "flattened" into a sequence of tokens by denoting nested collections via pairs of explicit opening and closing delimiter tokens. Delimited collections may

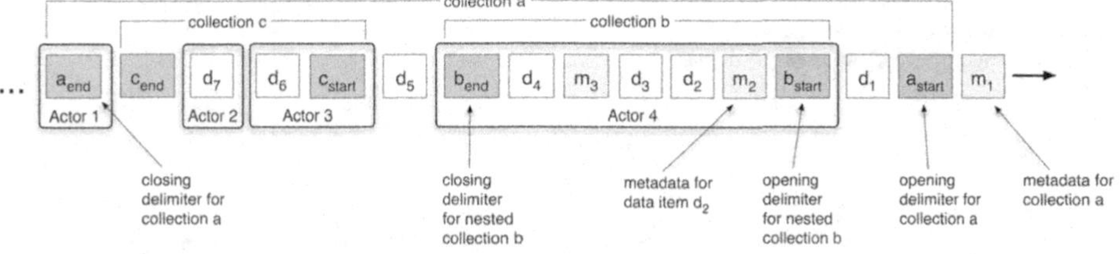

Fig. 3. Collection-oriented workflows represent nested data collections as flat token streams, where collection-aware actors can concurrently process collections

contain data tokens (labeled d_i in Figure 3), explicit metadata tokens (labeled m_j in Figure 3), and other sub-collections (denoted using embedded control tokens, *e.g.*, b_{start} and b_{end}). Metadata tokens are used to carry information that applies to the collections or data items that follow them in the stream. As shown in Figure 3, a series of actors may operate concurrently on the contents of collections. For example, in Figure 3, Actors 1-4 all simultaneously process parts of collection *a*, Actors 2 and 3 each simultaneously process a part of collection *c*, and so on.

Figure 4 shows a collection-oriented workflow implemented within KEPLER for inferring phylogenetic trees. The AddData actor is used to specify a list of files containing input data in the Nexus file format [14]. The ReadFile actor reads these Nexus files from disk and outputs a generic TextFile collection for each; ParseNexus transforms these text collections into corresponding Nexus collections. The PaupHSearch actor executes the PAUP* [21] external application (as a separate system process) on each Nexus collection it receives, adding the phylogenetic trees it infers to the collection. The PhylipConsense actor applies the CONSENSE[7] external application to the trees inferred by the PaupHSearch actor, adding a consensus tree (reflecting commonalities in the trees inferred by PAUP*) to each Nexus collection. The ExceptionCatcher actor discards Nexus collections that triggered exceptions in upstream actors. The TreeReporter actor displays each tree and associated statistics for each tree in a web-browser interface. Finally, the ComposeNexus and WriteFile actors save the results of analyzing each Nexus collection back to disk in the Nexus file format.

Note that each Nexus collection created by the ParseNexus actor pass through five downstream actors. These actors operate on the Nexus collections in turn, assembly-line style, reading data from the collections, and adding new information back to the collections. In particular, PaupHSearch expects to find data representing a character matrix in each Nexus collection, and PhylipConsense expects to find the phylogenetic trees inferred by PaupHSearch. The TreeReporter actor requires access to both the character matrix and the trees. As described in detail in the next section, each actor in a collection-oriented workflow declares what collection types (*e.g.*, *Nexus*) and data types (*e.g.*, *CharacterMatrix*) it operates on using a scope expression. As previously mentioned, the framework transparently passes any data not required by an actor to downstream actors, *i.e.*, an actor is never made aware of data it does not declare interest in. The result is that composing collection-oriented workflows simply entails stringing

[7] `http://evolution.gs.washington.edu/phylip.html`

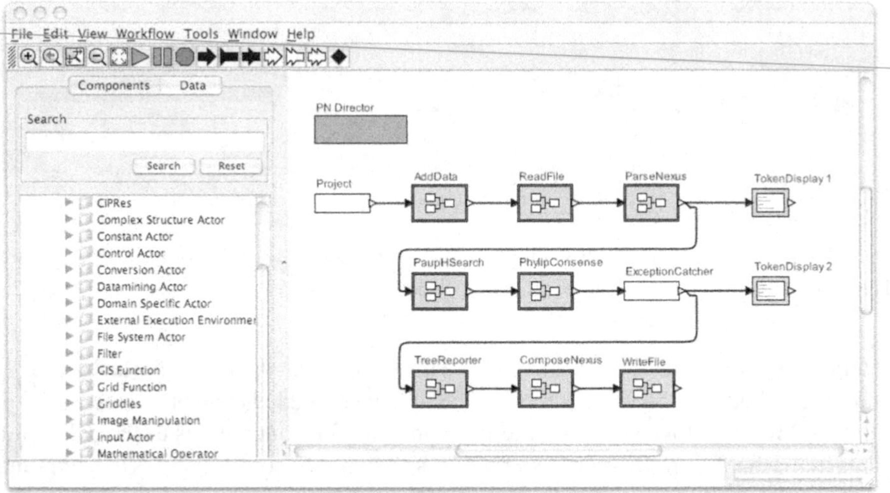

Fig. 4. A KEPLER collection-oriented workflow for inferring phylogenetic trees

together actors in an intuitive order (*e.g.*, it makes sense to run TreeReporter after PaupHSearch and PhylipConsense), without regard to the details of the data structures flowing between actors at runtime.

3 Abstract Data Model for Collection-Oriented Workflows

In this section we describe an abstract data model and syntax for representing collection-oriented structures (instances and schemas). Our model represents nested data collections as node-labeled ordered trees that are "flattened" into sequences of underlying data tokens.

3.1 Collection Instances

A *collection instance* in our abstract data model denotes a node-labeled ordered tree (similar to XML). Tree order represents the serialization order of a collection. In general, the order of items within a collection may or may not be "scientifically" meaningful. Node labels are applied to collections, metadata, and data values. Syntactically, a collection is denoted $l[\ldots]$, a metadata value is denoted $@l{:}d$, and a data value is denoted $l{:}d$, where l is a label and d a data value. A data (or metadata) value is either an atomic value such as a string or int, or a complex value represented by an object identifier.

A *collection-oriented sequence* can consist of labeled collections, labeled metadata values, and labeled data values. We require each label within a particular metadata sequence to be unique. The abstract syntax for sequences is defined by the following grammar. Note that in the abstract syntax, a collection defines a tree by encapsulating a collection-oriented sequence, where each item represents a child of the collection.

$$s ::= v \mid v, s \qquad (Sequence)$$
$$v ::= l{:}d \mid @l{:}d \mid l[s] \qquad (Data, Metadata, or Collection Value)$$

We convert collection-oriented sequences into KEPLER *token sequences* as follows. Each nested data collection is represented as a flat sequence of tokens within KEPLER(see Figure 3), such that each collection instance is enclosed by special opening and closing delimiter tokens (representing the '[' and ']' collection symbols). Delimiter tokens carry the label of the associated collection. Tokens are also used to store metadata and data items, and to provide actors with explicit access to item labels and to atomic and object-based values.

Nested data collections are often used to model the physical structure of a system under study. The following example, taken from structural biology, represents a portion of a protein structure described in a Protein Data Bank[8] (PDB) file. The PDB collection contains a protein-chain collection that in turn contains two atom objects o_1 and o_2.

PDBCollection[ProteinChain[Atom:o_1, Atom:o_2]].

The next example defines a Nexus collection nested within a project collection, along with associated metadata.

Project[@FilePath:'/myproject/aquatic/turtles.nex',
Nexus[CharacterMatrix:o_1, @CI:0.88, Tree:o_2, @CI:0.82, Tree:o_3]]

This Nexus collection has a file-path metadata value, and each tree within the collection has a CI (consistency index) metadata value. Note that the character matrix and trees inherit the file-path metadata value of the Nexus collection.

In the abstract model, we require metadata values for a given data or collection item to directly precede the item in a sequence. This restriction guarantees that as a data item is received by an actor, the actor has seen the item's associated metadata values. Metadata values are automatically cached for an actor in the KEPLER implementation of collection-oriented workflows. In general, this approach simplifies the processing of metadata, and for many cases limits the amount of data that must be cached, maximizing the performance of pipelining.

The function *descendents*(c) returns the contents of a collection c as a sequence of items, given by a top-down, left-to-right traversal of c. The function *metadata*(v) returns, as a sequence, the metadata values directly associated with a data or collection item v. Metadata values "cascade" to the descendents of a collection, unless otherwise overridden by an item. Thus, the function *metadata**(v) returns all metadata values, as a sequence, for data and collection items v.

The abstract data model for nested data collections is similar to XML. In particular, data and collection items correspond to XML elements, where data "elements" contain only simple content, collection "elements" contain complex content (*i.e.*, subelements), and metadata items correspond to attributes. Our model is simpler in that it does not have constructs corresponding to XML documents, identifiers (IDs), references (IDREFs), or mixed content. Also, we treat nesting explicitly as denoting "part-of" relationships, with the result that metadata is inherited by contained "parts." Because of

[8] `http://www.rcsb.org`

the similarity to XML, we can use standard XML languages over nested collections such as XPath expressions, *e.g.*, to retrieve portions of collection-oriented sequences.[9]

3.2 Collection Schemas

Collection schemas are similar to regular tree grammars [19]. However, our approach is tailored to collection-oriented workflows, in that: (1) we do not assume a "closed" schema model by default, and instead allow conforming instances to contain additional information; (2) we do not restrict the particular nesting levels of sub-collections, and allow conforming instances to contain unspecified intermediate collections; and (3) we do not restrict the ordering of sub-items (collections or data items).

A simple example of a collection schema and conforming instance are given in Figure 5. The schema, shown on the left, defines a PDB collection of interest as containing an optional header collection and one or more protein chain collections, where each protein chain contains one or more atoms having a *name* metadata value. A conforming instance of the schema is shown on the right of Figure 5. The PDB collection instance does not directly contain a protein chain, and instead contains multiple "molecule" collections. Similarly, each protein chain does not directly contain an atom data item, and instead the atoms are nested within residue collections. Thus, unlike with XML Schema or XML DTDs, collection schemas allow instances to have additional items including intermediate collections (*e.g.*, matching PDBCollection//ProteinChain//Atom instead of PDBCollection/ProteinChain/Atom).

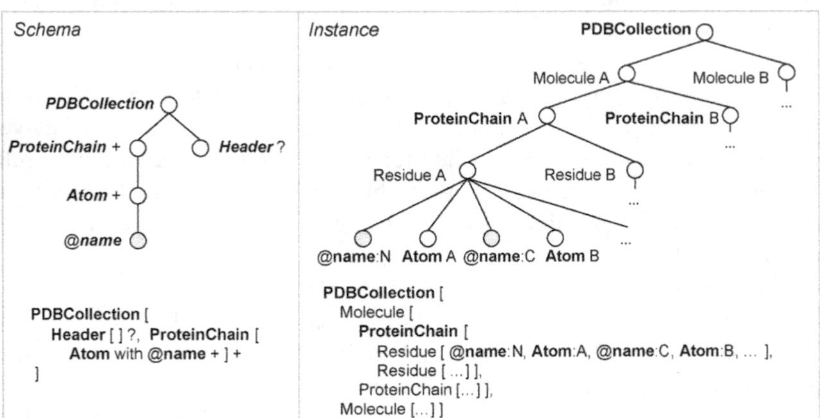

Fig. 5. A collection schema (left) shown as both a tree pattern and using the abstract schema language, and a conforming instance (right) shown both as a nested collection and using the abstract collection language

A *sequence type* specifies the kinds of items expected within a given sequence. In addition to expected item types, one can also specify item types that are not permissible in conforming sequences (via the not expression as shown below). An item type is either

[9] With the caveat that metadata values, treated as attributes, "cascade" to nested items.

$$\frac{v :: \tau_v,\ (\neg\exists v' \in s)\ v' :: \tau_v}{v,\ s :: \tau_v} \qquad \frac{v :: \tau_v,\ (\neg\exists v' \in s)\ v' :: \tau_v}{v,\ s :: \tau_v?,\ s :: \tau_v?} \qquad \frac{v :: \tau_v}{v,\ s :: \tau_v+}$$

$$\frac{}{s :: \tau_v*} \qquad \frac{\neg\,(s :: \tau_v)}{s :: \mathsf{not}\ \tau_v} \qquad \frac{(\forall \tau \in \tau_s)\ s :: \tau}{s :: \tau_s}$$

Fig. 6. Typing rules for occurrence definitions and sequences

$$\frac{}{l{:}d :: \mathsf{data}} \qquad \frac{}{l{:}d :: l} \qquad \frac{d :: \omega}{l{:}d :: {:}\omega} \qquad \frac{d :: \omega}{l{:}d :: l{:}\omega}$$

$$\frac{}{l[s] :: [\,]} \qquad \frac{}{l[s] :: l[\,]} \qquad \frac{\mathrm{decendents}(l[s]) :: \tau_s}{l[s] :: [\tau_s]} \qquad \frac{\mathrm{decendents}(l[s]) :: \tau_s}{l[s] :: l[\tau_s]}$$

$$\frac{}{@l{:}d :: @} \qquad \frac{}{@l{:}d :: @l} \qquad \frac{d :: \omega}{@l{:}d :: @\ {:}\omega} \qquad \frac{d :: \omega}{@l{:}d :: @l{:}\omega}$$

$$\frac{v :: \tau_q,\ @l{:}d' \notin \mathrm{metadata}(v)}{@l{:}d,\ v :: \tau_q} \qquad \frac{v :: \tau_q,\ v' \in \mathrm{metadata}^*(v),\ v' :: \tau_m}{v',\ v :: \tau_v\ \mathsf{with}\ \tau_m}$$

Fig. 7. Typing rules for data, collection, and metadata items

a data type or a collection type (itself a sequence type). Data and collection types can have occurrence qualifiers restricting the number of times an item may occur within a sequence. The occurrence qualifiers are zero or one (?), one or more (+), zero or more (*), or exactly one (the default). Data and collection types also can have associated metadata types.

A collection type can specify a label and a sequence type. A data or metadata type can specify a label and a value type. Value types (denoted ω) are given by their type names. We do not further specify value structures for complex objects. As for collection instances, metadata types given for a data or collection type "cascade" to nested items.

$$\begin{array}{lcll}
\tau_s & ::= & \tau_q \mid \mathsf{not}\ \tau_v \mid \tau_s, \tau_s & (\mathit{SequenceType}) \\
\tau_q & ::= & \tau_v \{ + \mid {}^* \mid ? \} & (\mathit{QualifiedType}) \\
\tau_v & ::= & \tau_d \{ \mathsf{with}\ m \} \mid \tau_c \{ \mathsf{with}\ m \} & (\mathit{ItemType}) \\
\tau_d & ::= & \mathsf{data} \mid l \mid {:}\omega \mid l{:}\omega & (\mathit{DataType}) \\
\tau_c & ::= & \{ l \} [\{ \tau_s \}] & (\mathit{CollectionType}) \\
m & ::= & \tau_m \mid m, m & (\mathit{MetadataSet}) \\
\tau_m & ::= & @ \{ l \} \{ {:}\omega \} & (\mathit{MetadataType})
\end{array}$$

Given a sequence s and a sequence type τ_s, we write $s :: \tau_s$ if s conforms to the type τ_s. Figure 6 defines the typing rules for occurrence definitions and general sequences. Note that the zero-or-many occurrence qualifier, as shown, does not restrict collection contents. However, this qualifier is useful for defining collection-oriented actors, which we discuss in more detail in the next section. The last rule of Figure 6 defines the general case for matching entire sequences. Figure 7 gives the typing rules for data items, collections, and metadata items.

Using schema expressions, it is possible to define standard representations for use in collection-oriented workflows. In particular, a given schema description can be "published," allowing it to be reused by actors. These published schemas also can enable certain forms of static type checking, *i.e.*, to ensure that a given collection instance satisfies the target schema within a workflow. Schema expressions also form the basis for scope expressions, as described in the next section.

4 Scope Expressions for Collection-Oriented Actors

Collection-oriented actors are typically designed to process data within a particular *scope*, as opposed to entire streams of heterogeneous data collections. Here we introduce scope parameters for explicitly defining the portion of an incoming data stream that is relevant to a collection-aware actor. Scope parameters can significantly reduce the effort of developing collection-oriented actors. For example, all data that falls outside of an actor's scope specification can be automatically "passed through" the actor unchanged. The use of scope parameters in this way also facilitates actor reusability, allowing actors to be used on selected portions of complex data streams, and without the actors needing to understand the structure or contents of the entire stream. Workflow designers also can more readily configure a collection-aware actor to work over particular subsets of data by specializing scope parameters, allowing actors to be flexibly reused in distinct workflows. We have found the following types of scope parameters to be useful in practice.

- **Read Scope.** A read scope specifies the portion of an incoming data sequence that is relevant to an actor. Typically, the read scope is used to identify the items generally required for an actor to execute. For example, consider an actor *A* whose read scope is given as a Nexus collection. Here, each particular Nexus collection within an input stream "triggers" *A* to execute.
- **Write Scope.** A write scope specifies where output data is placed within a given stream. For example, actor *A* may add new data items within each input Nexus collection. Alternatively, the actor may add a new collection as a sibling of the Nexus collection, or even replace the Nexus collection with an altogether new type of collection.
- **Iteration Scope.** An iteration scope extends a read scope and describes in more detail (1) what specific data items within the read scope are used by an actor for processing, and (2) how the actor should be invoked over those data items. For example, using an iteration-scope parameter, actor *A* may state that it should be invoked once for each phylogenetic tree in a collection. Alternatively, the actor may state that it should be invoked once over all trees within a collection.
- **Scope Filter.** A scope filter further specializes a read scope. Scope filters are typically used by workflow developers to control processing within a scientific workflow. For example, one might specialize the read scope of actor *A* by adding a metadata restriction (*i.e.*, that a particular metadata value is required) or by requiring the Nexus collection to be nested within another type of collection (*e.g.*, a particular kind of sub-project collection).

Here we focus on read and iteration scope parameters. Our approach is to use collection schemas for expressing read scopes (*i.e.*, for stating the type of incoming data of interest), and to model iteration scopes as queries over schema instances. The result obtained from applying an iteration-scope query to a read-scope instance is then used to control the iteration of the actor (for the particular read-scope instance). We give a simple query language for specifying iteration scopes, where parts of the read scope of an actor are embedded with variable bindings. Both read and iteration scopes are used to facilitate the construction of collection-aware composite actors that wrap traditional actors and subworkflows, as we discuss further in the following section.

The following is an example of a read scope for the PaupHSearch actor of Figure 4.

```
PaupHSearch.read-scope := Nexus[ CharacterMatrix, WeightVector ? ]
```

The PaupHSearch works over Nexus collections that contain exactly one character matrix data item and zero-or-one weight vector. The iteration scope of the PaupHSearch actor is straightforward. For each Nexus collection, the actor consumes the character matrix and weight vector (if it exists), and produces a set of phylogenetic trees. The PaupHSearch iteration scope is given by the following expression.

```
PaupHSearch.iteration-scope ($c, $v) :=
    Nexus[ CharacterMatrix {$c}, WeightVector {$v} ].
```

This iteration expression is shorthand for the following Datalog query.

R(c, v) :- Collection(n), Label(n, Nexus), Descendents(n, c),
Label(c, CharacterMatrix), Descendents(n, v), Label(v, WeightVector).

The relations used in the body of the query access portions of a given instance of the read-scope schema. For example, the Label relation associates a collection, data, or metadata item with its label, the Collection relation contains the collection items within the instance, and the Descendents relation relates collection items with their (transitively) contained items.

The read scope of the TreeReporter actor of Figure 4 is given by the following expression.

```
TreeReporter.read-scope := Nexus[ CharacterMatrix, Tree + ].
```

In this case, the TreeReporter actor displays a report for each tree in the nexus collection using the given character matrix. Thus, for a given nexus collection, the actor is *repeatedly* invoked, once for each tree. This invocation pattern is expressed by the following scope iteration.

```
TreeReporter.iteration-scope ($c, $t) := Nexus[ CharacterMatrix {$c}, Tree {$t} ].
```

Finally, the read scope of the ComposeNexus actor of Figure 4 is given by the following expression.

```
ComposeNexus.read-scope := Nexus[ CharacterMatrix ?, WeightVector ?, Tree *]
```

The ComposeNexus actor converts an optional character matrix, weight vector, and a list of zero-or-more trees into a Nexus file. Note here that the actor is invoked exactly once for each input Nexus collection, unlike the TreeReporter actor, which is invoked once per tree. This invocation pattern is described by the following iteration scope.

```
ComposeNexus.iteration-scope ($c, $v, collect($t in $n)) :=
    Nexus{$n}[ CharacterMatrix {$c}, WeightVector {$v}, Tree {$t} ]
```

The collect expression constructs a list of trees, where each tree is contained in the given Nexus collection. Every collect expression in an iteration scope consists of a data or metadata variable (in this case $t) combined with a collection variable (in this case $n).

In general, an iteration scope defines a mapping from instances I of a collection schema S to a relation $R(x_1, \ldots, x_n)$, for $n \geq 1$. We call each x_i of R an *attribute* of the iteration scope. Let I be an instance of the read-scope S. We write $R(I)$ to denote the result of applying the iteration scope to I, where each x_i attribute value for a tuple in $R(I)$ consists of either a metadata value, a data value, or a list of values resulting from a collect expression. Further, the actor is invoked once for each tuple in $R(I)$. We note that $R(I)$ can be "lazily" constructed (similar to a standard database iterator) such that the actor is invoked immediately as each new tuple is obtained.

5 Developing Collection-Aware Actors

We provide two approaches for developing collection-aware actors in KEPLER. The first, which we discuss in more detail in [17], is to directly implement collection-aware actors *natively* using a Java API. This API simplifies the implementation of collection-aware actors by providing comprehensive support for streaming, managing, and operating on nested data collections. However, we do not expect all actors to be developed in this way. A large number of "legacy" conventional actors already exist and are in use, including web-services and application components that are not designed to be collection aware. Furthermore, it is often easier and more intuitive to implement conventional KEPLER actors, especially those actors that do not explicitly operate on collections. Examples include straightforward data-transformation actors that take a single input and produce a single output, and actors that provide low-level functions for reading and writing files.

Thus, the second approach for developing collection-aware actors, which we introduce here, involves wrapping traditional actors, or entire subworkflows, within *collection-aware composite actors*. This approach facilitates the use of KEPLER itself for specifying collection-aware actors, allows conventional actors to be reused within multiple collection-aware actors, and reduces the need for writing *ad hoc*, single-purpose collection-aware actors from scratch.

To demonstrate the approach, Figure 8 shows how the PaupHSearch, TreeReporter, and ComposeNexus composite actors of Figure 4 are defined. Each composite actor contains a subworkflow employing an SDF director and one or more conventional actors. The ComposeNexus subworkflow illustrates how a single conventional actor may be wrapped in a composite to yield a collection-aware version of the actor. The subworkflow input ports labeled CharacterMatrix?, WeightVector? and Tree* map to attributes of the iteration scope parameter of the enclosing collection-aware composite actor. Like any other collection-oriented actor, the ComposeNexus subworkflow is invoked each time a match is found for the iteration scope of the actor. On each

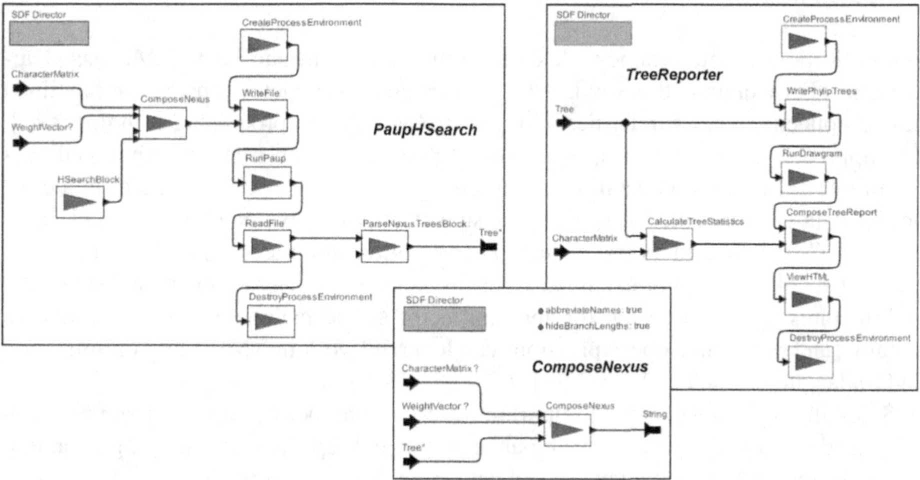

Fig. 8. The PaupHSearch, TreeReporter, and ComposeNexus collection-aware actors defined in terms of conventional actors

invocation, data values of the iteration-scope attributes are passed to the corresponding subworkflow input ports, the actor within the subworkflow operates on these data, and the outputs written by the enclosed actor are accumulated by the subworkflow output port labeled String. The enclosing composite actor then inserts the output of the subworkflow back into the data stream. The labels on the ports specify the types and quantity of data consumed or produced by the subworkflow, and provide anchors for mapping the iteration scope attributes to the ports.

The PaupHSearch and TreeReporter composite actors are more sophisticated. Both create and destroy temporary file system directories for running external processes, on each invocation, by employing the CreateProcessEnvironment and DestroyProcessEnvironment conventional actors. Both run external applications (PAUP* and DRAWGRAM[10]), write temporary files for these external programs to read, and parse output files created by these programs. Note that the ComposeNexus conventional actor is used both in PaupHSearch and in the ComposeNexus composite actor.

Employing the SDF director in collection-aware composite actors, rather than the PN director used to control the overall collection-oriented workflow, offers a number of practical advantages. The use of SDF simplifies the specification of these subworkflows, requiring each actor to have a well-defined token consumption and production rate. In addition, these lower-level actors can benefit from the optimized static schedule computed by the SDF director, since they typically perform a single function or are meant to be executed only once per composite invocation. The ability to use multiple models of computation in a single overall workflow is one of the main strengths of KEPLER, and is essential for supporting these SDF-based subworkflows in our collection-oriented workflow framework.

[10] `http://evolution.gs.washington.edu/phylip.html`

6 Conclusion

Our collection-oriented framework shares a number of similarities to XML-based approaches. For example, the way in which collection-aware actors operate on pipelined nested collections has similarities with some XML stream processing techniques [6]. The approach is also similar in spirit to list processing constructs in functional programming [2] as well as dataflow programming [18]. Because the abstract model of nested data collections is (essentially) a subset of XML, we can leverage and adapt existing XML-based query processing [7] and optimization techniques [8] for managing nested collections. For example, algorithms for XML-based publish and subscribe architectures [23] are relevant for applying actor scope parameters to incoming data streams, and iteration-scope expressions can leverage work in XML query optimization and on languages such as XPathLog [16].

Scientific workflows play an important role in a number of ongoing large research projects dealing with scientific data management, and represent an emerging paradigm for analyzing and integrating biological data from diverse sources. The development of "rigid" workflow modeling and design frameworks has recently been identified as a major bottleneck for scientific workflow reuse and repurposing [5]. We have found that this lack of flexibility is often due to the use of control-flow within workflows for managing, integrating, and analyzing inherently complex life-science data. The collection-oriented framework extends the capabilities of existing systems by facilitating the management of scientific data within scientific workflows. In particular, collection-oriented workflows are often significantly simpler and more intuitive than their conventional counterparts, can support higher-levels of concurrency and pipelining, and allow flexible actor configuration enabling greater levels of actor reuse. By additionally allowing collection-aware actors to be composed from conventional actors and KEPLER sub-workflows, our approach can support the reuse and repurposing of a wide variety of actors and workflows.

References

1. I. Altintas, C. Berkley, E. Jaeger, M. Jones, B. Ludäscher, and S. Mock. Kepler: An Extensible System for Design and Execution of Scientific Workflows. In *SSDBM*, 2004.
2. P. Buneman, S. A. Naqvi, V. Tannen, and L. Wong. Principles of Programming with Complex Objects and Collection Types. *Theoretical Computer Science*, 149(1), 1995.
3. S. Davidson, C. Hara, and L. Popa. Querying an Object-Oriented Database using CPL. In *Brazilian Symposium on Databases (SBBD)*, 1997.
4. E. Deelman, J. Blythe, Y. Gil, C. Kesselman, G. Mehta, S. Patil, M.-H. Su, K. Vahi, and M. Livny. Pegasus: Mapping Scientific Workflows onto the Grid. In *European Across Grids Conference*, 2004.
5. A. Goderis, C. Goble, U. Sattler, and P. Lord. Seven Bottlenecks to Workflow Reuse and Repurposing. In *ISWC*, 2005.
6. L. Golab and M. T. Özsu. Issues in Data Stream Management. *ACM SIGMOD Record*, 2003.
7. A. K. Gupta and D. Suciu. Stream Processing of XPath Queries with Predicates. In *ACM SIGMOD*, pages 419–430, 2003.
8. Z. G. Ives, A. Y. Halevy, and D. S. Weld. An XML Query Engine for Network-Bound Data. *VLDB Journal*, 11(4):380–402, 2002.

9. G. Kahn and D. B. MacQueen. Coroutines and Networks of Parallel Processes. In *IFIP Congress*, 1977.
10. E. A. Lee and D. G. Messerschmitt. Static Scheduling of Synchronous Data Flow Programs for Digital Signal Processing. *IEEE Trans. Comput.*, C-36, 1987.
11. U. Leser and F. Naumann. (Almost) Hands-Off Information Integration for the Life Sciences. In *Conference on Innovative Data Systems Research (CIDR)*, 2005.
12. B. Ludäscher, I. Altintas, C. Berkley, D. Higgins, E. Jaeger, M. Jones, E. A. Lee, J. Tao, and Y. Zhao. Scientific Workflow Management and the Kepler System. *Concurrency and Computation: Practice & Experience*, 2005.
13. R. S. MacLeod, D. M. Weinstein, J. Davison de St. Germain, C. R. Johnson, S. G. Parker, and D. Brooks. SCIRun/BioPSE: Integrated Problem Solving Environment for Bioelectric Field Problems and Visualization. In *Symposium on Biomedical Imaging (ISBI): From Nano to Macro*, 2004.
14. D. Maddison, D. Swofford, and W. Maddison. NEXUS: An Extensible File Format for Systematic Information. *Systematic Biology*, 46(4):590–621, 1997.
15. S. Majithia, M. S. Shields, I. J. Taylor, and I. Wang. Triana: A Graphical Web Service Composition and Execution Toolkit. In *ICWS*, 2004.
16. W. May. XPath-Logic and XPathLog: A Logic-Programming-Style XML Data Manipulation Language. *Theory and Practice of Logic Programming*, 4(3):239–287, 2004.
17. T. McPhillips and S. Bowers. An Approach for Pipelining Nested Collections in Scientific Workflows. *ACM SIGMOD Record*, 34(3):12–17, 2005.
18. J. Morrison. *Flow-Based Programming*. Van Nostrand Reinhold, 1994.
19. M. Murata, D. Lee, and M. Mani. Taxonomy of XML Schema Languages using Formal Language Theory. In *Extreme Markup Languages Conferences*, 2001.
20. T. M. Oinn, M. Addis, J. Ferris, D. Marvin, M. Senger, R. M. Greenwood, T. Carver, K. Glover, M. R. Pocock, A. Wipat, and P. Li. Taverna: A Tool for the Composition and Enactment of Bioinformatics Workflows. *Bioinformatics*, 20(17), 2004.
21. D. Swofford. PAUP*: Phylogenetic Analysis Under Parsimony (*and Other Methods). Version 4. Sinauer Associates, Sunderland, Massachusetts.
22. D. Thain, T. Tannenbaum, and M. Livny. Distributed Computing in Practice: The Condor Experience. *Concurrency – Practice and Experience*, 17(2-4), 2005.
23. F. Tian, B. Reinwald, H. Pirahesh, T. Mayr, and J. Myllymaki. Implementing a Scalable XML Publish/Subscribe System Using a Relational Database System. In *ACM SIGMOD*, pages 479–490, 2004.

Towards a Model of Provenance and User Views in Scientific Workflows

Shirley Cohen, Sarah Cohen-Boulakia, and Susan Davidson

Department of Computer and Information Science
University of Pennsylvania, USA
{shirleyc, sarahcb, susan}@seas.upenn.edu

Abstract. Scientific experiments are becoming increasingly large and complex, with a commensurate increase in the amount and complexity of data generated. Data, both intermediate and final results, is derived by chaining and nesting together multiple database searches and analytical tools. In many cases, the means by which the data are produced is not known, making the data difficult to interpret and the experiment impossible to reproduce. Provenance in scientific workflows is thus of paramount importance.

In this paper, we provide a formal model of provenance for scientific workflows which is general (i.e. can be used with existing workflow systems, such as Kepler, myGrid and Chimera) and sufficiently expressive to answer the provenance queries we encountered in a number of case studies. Interestingly, our model not only takes into account the chained and nested structure of scientific workflows, but allows asks for provenance at different levels of abstraction (*user views*).

1 Introduction

Fueled by technologies capable of producing massive amounts of data, scientists are faced with an explosion of information which must be rapidly analyzed and combined with other data to form hypotheses and create knowledge. Scientific analyses are thus becoming increasingly large and complex, with a commensurate increase in the amount and complexity of data generated.

To address this problem, over the past several years a number of scientific workflow systems have been developed to support scientists in the analysis of their data. Such systems differ from business-oriented workflow systems in the focus on data – e.g. sequences, phylogenetic trees, proteins – and its transformation into hypotheses and knowledge [23]. Examples of scientific workflow systems include myGrid/Taverna [19], Kepler [5], Chimera [12] and DiscoveryNet [22] (see [30]). Still other interesting examples of workflow systems include MHOLline [25], HKIS-Amadea [9], and AdaptFlow [14]. Some integration solutions also include workflows to add value to warehoused data. For example, the GUS [11] system allows users to import data of interest, run bioinformatics tools over that data, and store the results obtained; pipelines are expressed using Perl.

Scientific workflows are specified using a variety of graph-based models. Nodes in the workflow specification represent *step classes* (alternatively called tasks,

U. Leser, F. Naumann, and B. Eckman (Eds.): DILS 2006, LNBI 4075, pp. 264–279, 2006.

actors, processes, boxes) and edges capture the flow of data between step classes. In many workflow systems (e.g. Kepler and myGrid), a step class may itself be a workflow. An execution of a workflow generates a partial order of *steps*, each of which has a set of *input* and *output* data objects. Each step is an instance of a step class, and the input-output flow of data and class associated with each step must conform to the workflow specification (see for example [16]).

In workflow systems, data, both intermediate and final results, is thus derived by chaining and nesting together multiple database searches and analytical tools. In many cases, the means by which the data are produced is not known, making the data difficult to interpret and the experiment impossible to reproduce. Provenance in scientific workflows is thus of paramount and increasing importance, as evidenced by recent specialized workshops [2] and surveys [23] dedicated to the subject of provenance of scientific information.

Many systems using scientific workflows provide a way to keep track of the origins of data. For example, the GUS schema contains about twenty tables dedicated to provenance information. Some scientific workflow systems, such as myGrid [28], record various kinds of metadata related to provenance. Recently, Kepler has developed a logging mechanism for tracking information and dependencies between components of the data flow [4]. Nevertheless, no formal model of provenance for workflow systems has to our knowledge been developed which precisely defines the meaning of provenance taking into account the nested structure of step classes and the data produced.

Formal models of provenance do exist within the database community (see for example [6,3,27]). However, these models reason over restricted forms of algebraic queries and give very fine-grained reasoning; for example, a tuple in a result gets its value from a particular set of tuples in the input (*where* provenance) and is there because of a (possibly bigger) set of input tuples (*why* provenance). More recently, [7] considers the problem of copying data between databases, and describes an approach in which these actions can be automatically recorded in a convenient, queryable form with acceptable overhead. However, the problem of tracking provenance in scientific workflow systems raises new challenges. First, since the operators in workflows are black boxes (step classes), fine grained reasoning cannot be performed. The most that can be assumed is that steps are *deterministic*, i.e. that given the same set of input the output will be the same. This input must include not only data but also user input (e.g. the selection of results based on visual inspection), parameter settings (e.g. the kind of matrix used in a Blast tool), and any other input used by the step (e.g. a randomize number used in a bootstrap). Second, scientific workflow systems frequently provide a notion of *user views* which determines whether or not a user can zoom into a step class to see a sub-workflow. User views therefore affect the granularity at which provenance is reasoned about.

The aim of this paper is to present a formal model of provenance in workflow systems which takes into account the chained and nested structure of scientific workflows as well as user views. The model has been formulated by interviewing numerous scientists in several domains (e.g. genomic research, and phylogenetic tree construction and analysis) and analyzing what several important scientific workflow systems are currently doing. The model is abstract, *i.e.* it details the minimum information that must be provided by a workflow system in order to perform the types of reasoning about provenance that scientists wish to perform. It is generic in the sense that it can be used by any workflow systems providing this minimum information.

This paper is organized as follows. We first present one of the use cases collected (Section 2) from our interviews of scientists, whose data provenance requirements are representative of those of other studies. We then introduce our model of provenance (Section 3) and in Section 4 show how it can be used to express the provenance queries of Section 2. In Section 5 we show the connection to nested transactions, and discuss whether or not the required provenance information is provided by the logging mechanisms of Kepler, myGrid and Chimera. Finally, Section 6 concludes the paper.

2 Tree Inference Use Case

Systematic biologists are attempting to develop a comprehensive history of life's origins by studying the phylogenetic relationships of the millions of earth species. Assembling these species and placing them on the "tree of life" requires increased amounts of information about each one as well as sophisticated analytical tools to build an understanding of the relationships among species. At present, the infrastructure used to manage the flow of phylogenetic data lacks the querying capabilities needed to address many important scientific challenges.

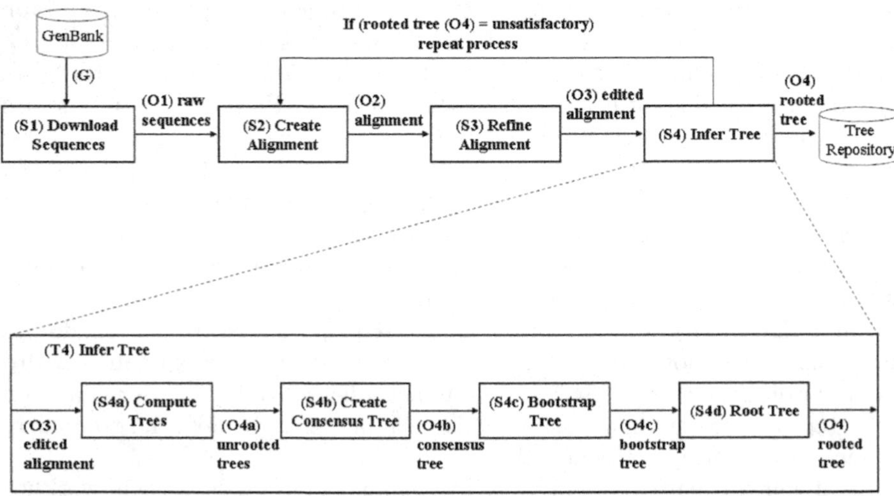

Fig. 1. Tree inference use case

As an example, consider a typical tree inference workflow depicted in Figure 1. This workflow is composed of four main step classes (S1 to S4); the last step class is nested and composed of four step classes (S4a to S4d).

The Download Sequence step class (S1) is responsible for obtaining a set of chosen DNA sequences from GenBank. Note that the input to this step class is a user-driven event. The second step class, Create Alignment (S2), takes in the raw sequences and runs an alignment program, such as ClustalW [15], to generate a multiple sequence alignment. The third step class, Refine Alignment (S3), is where the biologist verifies and improves the quality of the multiple sequence alignment by manually adjusting gaps inserted by the alignment program.

The fourth step class, Infer Tree (S4), takes the edited alignment and produces from it a phylogenetic tree. Note that this step class contains multiple substeps within it. The first substep class, Compute Trees (S4a), runs a tree inference program like PAUP [24] or Phylip [21] and generates a set of unrooted trees from the alignment. The second substep class, Create Consensus Tree (S4b), computes a consensus tree from the set of unrooted trees. The third substep class, Bootstrap Tree (S4c), calculates a confidence score for each node of the consensus tree. The last substep class, Root Tree (S4d), consists of rooting the consensus tree by selecting a site as an outgroup. The output from this substep class, a rooted tree, is saved if it is considered biologically meaningful. Otherwise, the alignment and inference step classes are repeated until a suitable rooted tree is derived.

A typical lab executes this scenario several times a year, resulting in vast amounts of intermediate and final data products. However, with current workflow technology this scenario is carried out without the ability to ask questions about how a phylogenetic tree came to be and what alignment and sequences it originated from. A biologist wishes to not only review the current state of a phylogenetic analysis that is in progress, but also guide it to some desired future state; such as refining the parameters to Clustal to produce a more precise alignment or foreseeing (based on historical results) that Phylip may produce fewer trees than PAUP*. The biologist also wishes to know which sequences were dropped by the alignment program and consequently were not used to infer the rooted tree. A related goal is to be able to assess the quality and impact of a data product such as a rooted tree by reviewing both the DNA sequences and alignment used to produce it, and understanding which subsequent workflow executions used the same alignment as input to a tree inference step. In related studies, increased knowledge of data provenance will allow the biologist to reuse intermediate products, such as the many unrooted trees which can be quite time-consuming to generate. To address these needs we collected some data provenance queries that describe in words the semantics of the queries we are interested in:

1. What direct data products did this tree originate from?
2. What are all the data products which have been used to produce this tree?
3. What step produced this tree?
4. What sequence of steps produced this tree?

5. What parameters and steps produced this tree?
6. What alignments in the space of stored data objects were used as inputs to steps in subsequent workflows?
7. What trees in the data space were inferred using the same sequence of steps, parameters, and input data?
8. What steps require user input data?

It should be noted that there is a strong connection between questions about data provenance and general questions about workflow execution, and the biologist is interested in discovering useful facts about both. Some general workflow queries are shown below:

- What steps in this workflow did not complete or execute?
- What steps ran concurrently in the same workflow instance?

In this paper, we will concentrate on the first set of queries, the *data provenance queries.* However, a longer-term goal is to allow biologists to interactively explore other aspects of a workflow execution without needing to become an expert in the logging mechanisms of the system.

3 Model of Provenance

Provenance is defined over a workflow execution as a function which takes as input the identifier of a data object and returns the sequence of steps and input data objects on which it depends. All data that is produced by some step is called *calculated* data. In contrast, *user* or *parameter* data is injected into the data space of the workflow execution by a user; its provenance is whatever information is recorded about how it was input, e.g. the user who input the data and the time at which the input occurred. We call this *Info*(d).

Definition 1. *The provenance of a data object d (Prov(d)) is given as:*

$$Prov(d) = \begin{cases} (sid, \{d_1 : Prov(d_1), ..., d_n : Prov(d_n)\}) & d \text{ is calculated data} \\ Info(d) & d \text{ otherwise} \end{cases}$$

where sid is the id of the step that produced d as output, and d_i is the id of data input to the step.

As an example, consider the simple toy workflow in Fig. 2, in which S_1 takes as input $\{I_1, I_2\}$, produces as output $\{D\}$, which is taken as input to S_2, which produces as output $\{O_1\}$. Then

$$Prov(O_1) = (S_2,\ \{D : (S_1,\ \{I_1 : Info(I_1), I_2 : Info(I_2)\})\})$$

Note that *Prov*(d) gives complete information about how the data object came to be (*deep* provenance). We could also have defined *Prov*(d) to consider just the immediate provenance or n-deep provenance of a data product.

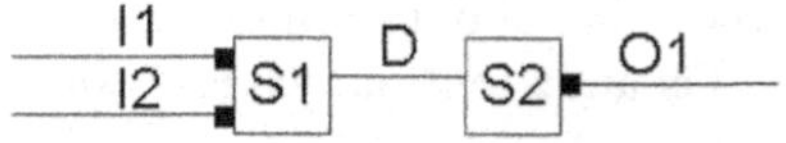

Fig. 2. Example of workflow

Throughout this section, we will use Datalog to define the minimal information needed from the workflow system (base predicates) as well as the reasoning we can perform over that information. We choose Datalog since it is a simple, declarative language which easily expresses recursive queries; it is thus a natural model for graph data and queries which entail finding paths. Using known translations to the relational model, the provenance system described in this paper could then be implemented using a relational database system which provides support for transitive closure (e.g. Oracle or DB2). (See [26] for a description of Datalog and its translation to the relational model and [7] for a discussion of how to optimize performance.)

Another option would have been to use an object-oriented data model as suggested by the definition of *Prov*(*d*). While this model avoids the problem of having to flatten nested sets of data, it does not naturally capture transitive closure. Furthermore, our model of provenance does not need any of the object features it supports (such as inheritance or polymorphism). We therefore opt for a simpler and more declarative model.

3.1 Minimal Information to Reason About Provenance

To be able to reason about provenance, we make a number of assumptions about the information provided by the workflow system:

- **Provenance information for user or parameter data is provided.** That is, *Info*(d) is available.
- **Each output data object has a unique id.** The notion of unique ids for output objects is ubiquitous in proposals for scientific workflow. For example, in [27], data is never overwritten or updated in place, and each version of data has a different id; in [5], each token has a unique id although two tokens may correspond to the same data object.
- **The system maintains information about steps and the ordering of input/output operations to steps.** In order to reason about provenance, some sort of logging must be performed by the system. We will discuss how to achieve this in Section 5.

We therefore model the minimal information that must be provided to a provenance reasoning system as the following base predicates, where *did* is the id of a data object, *annot* is the provenance information of user or parameter data (*Info*(did)), *sid* is the id of a step, and *ts* is an integer that captures the partial order of input and output events to a step:

$info(did, annot)$
$input(sid, did, ts)$
$output(sid, did, ts)$

To allow users to see the value of data objects and obtain information about the step class of which *sid* is an execution, we use *csid* as the id of a step class and add:

$value(did, v)$
$instanceOf(sid, csid)$
$infoClass(csid, info)$

Using these base predicates, we can express *Prov*(d) for calculated data using the following Datalog rule:

$$prov(did, sid, iid) : -input(sid, iid, tsi) \wedge output(sid, did, tso) \wedge tsi \leq tso$$

Note that our definition of provenance includes both the step and input data to that step. However, it will also be useful to talk about the set of data objects on which calculated data depends (*dProv*), either directly or indirectly, as well as the set of steps on which were used in calculating the data (*sProv*):

$dProv(did, iid) : -prov(did, \_, iid)$
$dProv(did, iid) : -prov(did, \_, x) \wedge dProv(x, iid)$

$sProv(did, sid) : -prov(did, sid, \_)$
$sProv(did, sid) : -prov(did, \_, x) \wedge sProv(x, sid)$

Returning to the toy example of Fig. 2, since $prov(D, S1, I1)$, $prov(D, S1, I2)$, and $prov(O1, S2, D)$ are true, we can infer $dProv(O1, D)$, $dProv(O1, I1)$, $dProv(O1, I2)$. We can also infer $sProv(O1, S2)$, $sProv(D, S1)$ and $sProv(O1, S1)$.

3.2 Composite Steps

In many workflow systems, a step class may itself be a workflow. We call such step classes *composite*, and their executions *composite steps*; step classes that do not contain workflows will be called *base*, and their executions *base steps*. Typically, each input to a composite step class is input to one or more of its substep classes, and the output of a substep class is either input to another substep class or becomes the output of the composite step class.

There are several reasons why composite step classes are used in workflows. First, users may wish to focus on a certain level of abstraction and ignore lower levels of detail. Second, they may represent levels of "authorization"; users without the appropriate clearance level would not be allowed to see the lower level executions of a step class.

Definition 2. *Given a workflow specification, the* user view *of a user (or class of users) U, $UserView(U)$, is the set of lowest level step classes that U is entitled to see.*

Note that a user view cannot contain two step classes such that one is contained in the other. We assume that the user view is *valid*, i.e. that each of the highest level step classes in the workflow specification is either in the view, or that at some lower level all of its contained substeps are in the user view. For example, consider Fig. 3. In this workflow, S_C directly contains S_{C1} and transitively contains step classes S_1 and S_2. The composite step class at the highest level, S_C, has input set $\{I_1, I_2\}$ and output set $\{O_1, O_2\}$. Within S_C there is a composite step class S_{C1} which takes $\{I_1\}$ as input and produces $\{O_1\}$ as output; S_C also contains step class S_3 which takes $\{I_2\}$ as input and produces $\{O_2\}$ as output. Within S_{C1} there is a step class S_1 which takes $\{I_1\}$ as input and produces $\{D\}$ as output; $\{D\}$ is then input to step class S_2, which produces $\{O_1\}$ as output.

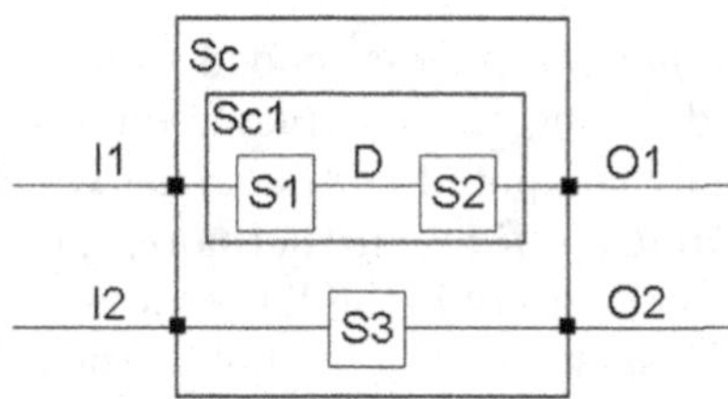

Fig. 3. Example of composite Step

Three examples of user classes for this workflow are:

- $UserView(U_1) = \{S_C\}$ (the "black box" user class)
- $UserView(U_2) = \{S_{C1}, S_3\}$
- $UserView(U_3) = \{S_1, S_2, S_3\}$ (the "admin" user class)

However, the user view $\{S_{C1}\}$ is not valid since S_3 is missing.

A partial ordering $<_u$ on user views can be defined using the containment of step classes.

Definition 3. *Given two user views U_1 and U_2, we say that U_2 is a* finer level *than U_1 (or U_1 is a* higher level *than U_2), $U_1 <_u U_2$, iff* $\forall s_2 \in UserView(U_2)\ \exists s_1 \in UserView(U_1)$ *such that* $s_1 = s_2$ *or* s_1 *contains* s_2 *either directly or transitively.*

For example, $U_1 <_u U_2$, $U_2 <_u U_3$ and $U_1 <_u U_3$.

To answer questions of provenance, we must take the user view into account and reason about the input and output to steps which are instances of step classes that are in the user view. That is, we must know the connection between the specification and the execution of a workflow, as well as the containment relationship between step classes. We therefore assume that the workflow system provides the following information:

- **The user view of each class of users.** A variety of techniques could be used to capture this information. For example, the GUI in Kepler allows users to zoom in on steps. We can imagine capturing this information by taking each composite class, zooming in to the appropriate level, and taking the union of the resulting classes.
- **The input and output to each step, whether composite or base.**

Thus we use the following as our base predicates, where *sid* is the id of a step (either base or composite), *did* is the id of a data object, *ts* captures the partial order of input and output events to a step, *cid* is the id of a step class (either base or composite), and *ccid* is the id of a composite step class.

$Cinput(sid, did, ts)$
$Coutput(sid, did, ts)$
$immContains(ccid, cid)$
$userView(u, cid)$

Cinput (*Coutput*) is *input* (*output*) extended to composite steps. The relation *contains*(*ccid*, *cid*), denoting the complete containment relation between step classes, can be trivially computed as the transitive closure of the immediately contains relation, $immContains$. Furthermore, the following constraint on $userView$ expresses the fact that *cid* is the lowest level that u is entitled to see: If *contains*(*ccid*, *cid*) and $userView(u, ccid)$ holds, then $userView(u, cid)$ does not hold.

It will also be convenient to talk about steps (whether base or composite) that are allowed to be seen by a particular user:

$userInstance(u, sid) : -instanceOf(sid, cid) \wedge userView(u, cid)$

Using these predicates, we calculate provenance as a function of the user view as follows:

$userProv(u, did, sid, idid) : -Cinput(sid, idid, tsi) \wedge Coutput(sid, did, tso) \wedge$
$tsi \leq tso \wedge userInstance(u, sid)$

We can also redefine the data (step) provenance with respect to a user view, $userDProv(u, did, iid)$ ($userSProv(u, did, sid)$) using $userProv$ instead of *prov*. (Details are omitted.)

3.3 Reasoning with User Views

We now explore properties of provenance as a function of user view. In particular, when a user views the execution at a finer level he may see data objects that are not visible at a higher level which are the output of hidden substeps. Reasoning about provenance at a finer level will also allow a more precise view of the provenance of a data object.

For example, in the workflow of Figure 3, from user views U_1 and U_2 the data object D is not visible as a data object on which O_1 or O_2 depends. Furthermore, at user view U_1 both I_1 and I_2 are seen as data objects on which O_1 depends, while at user views U_2 and U_3 only I_1 is included.
The observation about what data objects d are visible within a user view u can be formalized as follows:

$invisible(d, u) : -output(\_, d, \_) \wedge \neg visible(d, u)$
$visible(d, u) : -userProv(u, d, \_, \_)$
$visible(d, u) : -userProv(u, \_, \_, d)$

For example, consider the workflow of Figure 3 and the user view U_2. Then $userProv(U_2, O1, SC1, I1)$ and $userProv(U_2, O2, S3, I2)$ hold, meaning that we can infer $visible(O1, U_2)$, $visible(I1, U_2)$, $visible(O2, U_2)$, and $visible(I2, U_2)$. Furthermore, since $output(S1, D, \_)$ holds but not $visible(D, U_2)$, $invisible(D, U2)$ holds. Similarly, we could show that $invisible(D, U_1)$ holds.

To formalize the second observation, given data object d and two user views u_1 and u_2, let $DProv(u_1, u_2, d)$ be the set of all data objects that d depends on either directly or indirectly as seen in user view u_2 that are visible in u_1. More precisely, it is the set of data objects X in $ans(X)$ below (where parameter $\$U1$ is set to u_1, $\$U2$ is set to u_2 and $\$D$ is set to d):

$$ans(X) : -userDProv(\$U2, \$D, X) \wedge visible(X, \$U1)$$

As an illustration, consider the workflow of Fig. 3 with $\$U1{=}U_1$, $\$U2{=}U_3$ and $\$D{=}O1$. Then $userDProv(U_3, O1, D)$, $userDProv(U_3, O1, I1)$ and $visible(O1, U_1)$ hold, but $visible(D, U_1)$ does not hold. Thus $DProv(U_1, U_3, O1){=}\{I1\}$.

The observation about the refinement of data provenance as a function of user view can now be stated as follows:

Lemma 1. *Given a data object did and two user views u_1 and u_2, such that $u_1 <_u u_2$ and did is visible in u_1. Then*

$$DProv(u_1, u_1, did) \supseteq DProv(u_1, u_2, did).$$

Returning to our example, recall that $U_1 <_u U_3$. It can be easily checked that $DProv(U_1, U_1, O1){=}\{I1, I2\}$ and thus $DProv(U_1, U_1, O1) \supseteq DProv(U_1, U_3, O1)$.

3.4 Discussion

Much of the information (base predicates) that we are assuming are easily obtainable from either the workflow specification (*immContains*, *userView*, *info*, *infoClass*), or from low-level logging/execution knowledge (*input*, *output* and *instanceOf*). However, many workflow systems do not keep intermediate data products, that is $value(did, v)$ may not be available for all *did*. In this case, the workflow system may be able to provide only partial information about provenance, i.e. the *did* of data objects.

The remaining predicates, *Cinput* and *Coutput*, are the topic of Section 5.

Is it reasonable to require that the value of all intermediate data objects be kept? An increasing number of optimization and compression techniques to efficiently record provenance information have been proposed in the database community. In particular, [7] exploits the hierarchical structure of data to optimize provenance storage, and gives experimental results to show that provenance can be tracked and managed efficiently. In the context of scientific workflows, which are run many times and generate a large number of intermediate results, the nesting of composite steps and use of user views also gives the ability to limit the results. However, the results are kept around only if they are visible in some user view. By specifying appropriate user views, the system can therefore limit the promises made to users about provenance information.

4 Querying Provenance

We now turn to the queries about provenance introduced in Section 2, and show that they can be answered using the predicates developed in Section 3. Note that these queries concern data (1,2,5) and step (3,4) provenance and use immediate (1,3) as well as deep (2,4,5) provenance information.

In what follows, we assume that the user view is input as parameter $\$U$ and the data object as parameter $\$D$. Examples are given in terms of data object O4 in the *Tree inference* workflow of Figure 1.

1. **Which data objects have been directly used to produce this result?**
   $ans(X) : -userProv(\$U, \$D, \_, X)$

   If the input user view contains step S4, then the immediate provenance of O4 given by $ans(X)$ above is {O3}. However, if the input user view contains steps S4a-d, then $ans(X)$ is {O4c}.
2. **What are all the data objects which have been used to produce this result?**
   $ans(X) : -userDProv(\$U, \$D, X)$

   ans returns {O1,O2,O3} if the input user view contains step S4, and {O1,O2,O3,O4a,O4b,O4c} if it contains steps S4a-d.
3. **What step class produced this data product?**
   $ans(X) : -userProv(\$U, \$D, X, \_)$

   If the input user view contains step S4, then $ans(X)$ is {S4}. However, if the input user view contains steps S4a-d, then $ans(X)$ is {S4d}.
4. **What sequence of steps produced this data product?**
   $ans(X) : -userSProv(\$U, \$D, X)$ ans returns {S1, S2, S3, S4} if the input

   user view contains step S4, and {S1, S2, S3, S4a,S4b,S4c,S4d} if it contains steps S4a-d.

5. **What parameters and steps produced this data product?**
   The intent of this query is to know the input to each step that led to the data product. Note that to distinguish parameters from other input data we need additional information from the workflow system, e.g. the predicates $parameter(d)$, $userInput(d)$ and $calculated(d)$, which could then be used by our system in a straightforward way. $ans(X, Y) : -userProv(\$U, \$D, X, Y)$,

   parameter(Y)
   $ans(X, Y) \ : \ -userProv(\$U, Z, X, Y), ans(\_, Z)$ Assuming the input user

   view contains step S4, ans returns {(S1,G), (S2,O1), (S3,O2), (S4,O3)}. To answer the original query, this set would be filtered for the second component to be a parameter resulting in the empty set (all inputs are calculated data in this example).

Details of queries 6-8 can be found in [10].

5 Obtaining *Cinput* and *Coutput* from Logs

Up to this point, we have assumed that *Cinput* and *Coutput* are available to define the provenance of a data object. We now argue that this information is achievable using standard nested transaction logging mechanisms, and discuss how to obtain this information in Kepler, MyGrid and Chimera.

Logging of nested transactions. Using ideas from nested transactions [18], the log of the workflow system would contain the events – start (s), read (r), write (w), and commit (c) – not just of base transactions but of transactions within which they are nested. For example, the following could be the log of the (composite) transaction T_1 which contains subtransaction T_2, which in turn contains the (base) transaction T_3:

$$s(T_1), s(T_2), r(d_1), w(d_2), s(T_3), r(d_2), r(d_3), w(d_3), c(T_3), w(o_1), c(T_2), c(T_1)$$

In this case, $input(sid, did, ts)$ is computed as the data read and the order in which it was read. For example, $input(T_3, d_2, 5)$ could be true. With composite transactions, the output would be calculated as all the data that is output by some subtransaction and not input to another subtransaction; the input of a composite transaction is defined analogously. For example, $input(T_1, d_1, 2)$ and $input(T_1, d_3, 5)$ would be true but $input(T_1, d_2, 6)$ would not be true.

Note that we can compute *Cinput* and *Coutput* from the log events of nested transactions since it contains the notion of execution of composite steps as well as base steps.

Kepler. In Kepler, a workflow consists of a collection of nodes called *Actors* (corresponding to step classes) which communicate through input and output *ports*. Communication occurs through the passing of *tokens* (corresponding to data input and output) which are globally unique; tokens are read and written, and each token is written only once. The model of computation of a workflow is defined by a *Director* who mediates communication between actors.

The log associated with this model records the reading and writing of tokens on ports, which are uniquely associated with Actors [4]. Each execution of an actor corresponds to a step in the terminology of this paper.

Conceptually, the first read event on a port associated with an Actor begins the execution (transaction) of that Actor. Subsequent writes by that Actor on this port depend on all its previous reads, where "previous" is captured by an integer called a firing. Since this implies that the state (read tokens) of the Actor gets bigger and bigger as time goes on, the notion of a *clear* event is introduced and recorded in the log, the effect of which is to clear the state of the Actor. Thus any write after the clear event will depend only on the read events since the state was cleared. In terms of transactions, this can be thought of as committing a transaction and beginning a new transaction.[1]

[1] This is a simplification of the model, which also uses a notion of "firings" to capture the set of read tokens on which a write depends rather than an ordering of events.

Using the Kepler log, it is certainly possible to capture *input* and *output*. Moreover, Kepler supports composition of Actors, and enables users to zoom in and view finer levels of detail of an Actor. However, since the log records only events of base steps, there is currently no notion of the execution of a composite step. Thus it is not clear how to calculate *Cinput* and *Coutput* for composite steps. The Kepler group is exploring a variety of approaches to work around this problem [17].

myGrid. In myGrid [19],[2] a workflow is a network of processors and links. A *processor* (corresponding to a step class) is a transformation that accepts a set of input data and produces a set of output data. Several types of processors exists, one of which is the nested processor. Two kinds of links are considered: *data links*, which mediate the flow of data between a data source and sink; and *coordination constraint links*, which control the execution of two processors (roughly speaking, playing the role of a director in Kepler). The log file in myGrid is an XML file which records global execution information: the user of the workflow, the start time, the end time and the set of services invocations performed (each invocation corresponds to a step). Exploiting the nested structure of XML, information is also provided for each service invocation: start time, end time, parameters of the service invocations, input data, and output data. Life Sciences Identifiers (LSIDs) [8] are used to uniquely and persistently identify data resources and their associated metadata.

An interesting aspect of myGrid is the automatic annotation of provenance logs with concepts drawn from the myGrid ontology. The COHSE[3] system performs this task by augmenting documents with links based on the semantic content of those documents. This process allows users to dynamically generate a hypertext collection of provenance documents, data, services, and workflows based on their associated concepts, and to perform reasoning over the ontology (see [28], [29], and [1] for more details).

Using the myGrid log, it is indeed possible to capture *input* and *output*. While the current literature does not focus on the provenance of nested processors, the intrinsically nested structure of the myGrid log file seems naturally suitable for capturing nested transactions, thus allowing the calculation of *Cinput* and *Coutput*.

Chimera. In Chimera [12],[4] a *transformation* is a program (script file) and an execution of a transformation is a *derivation*, corresponding to a step class and a step, respectively. Data products are represented as abstract typed datasets (virtual data) and as materialized replicas. *Derivations* can be connected to form workflows that consume and produce replicas (input and output data).

---

[2] We omit here the internal relationships between myGrid, the Scufl language, Taverna and freefluo tools.

[3] Conceptual Open Hypermedia Services Environment.

[4] We omit here the internal relationships between GriPhyN, Chimera, Pegasus and Condor.

The Chimera virtual data schema defines a set of relations used to represent and capture descriptions of how a program can be invoked, and to record its potential and/or actual invocations. Upon execution, workflows automatically create *invocation objects* for each derivation in the workflow, annotated with the information of the runtime process. Invocation objects are an *annotation scheme* for representing provenance information and thereby providing a mechanism for linking input and output data products.

In Chimera, provenance information can be retrieved from the Virtual Data Catalog (VDC) [13] expressed in the Virtual Data Language (VDL). VDL supports both recursive searches and can output all the derivations in the system that produced a particular dataset. VDL interacts with an end-user query system, the Virtual Data Browser (VDB), to interactively access the catalog.

In the current implementation of Chimera, nested transformations are allowed since each transformation can call other transformations. As each derivation has its own provenance information, it should be possible to populate *Cinput* and *Coutput*.

6 Conclusion

This paper examines data provenance through the prism of large-scale scientific applications. Motivated by phylogenetic analyses which produce volumes of data, our research extends existing ideas of data provenance to scientific workflows. In this context, we formulate a model for provenance and define notions of *data provenance*, *step provenance*, and *user views* for computing user-oriented queries over workflow executions.

User views are especially helpful for reasoning about data provenance through nested executions. They are essential for defining the level of detail of a provenance query and determining what data must be kept by the system. As such, we devise ways in which a user can effectively query a workflow execution in an intuitive fashion without needing to become an expert in the system's logging facility. We demonstrate the expressiveness of our model by answering a collection of queries supplied by systematic biologists.

Our model is simple and generic enough to capture information that is (or soon will be) available in existing scientific workflow system, and we demonstrate this with Kepler, myGrid, and Chimera. From this, we show that a scientific workflow system which provides basic execution logging could implement our model and benefit from our approach.

We are currently exploring new ways to improve the expressiveness of our model. First, we will consider the general workflow queries in Section 2 related to partial and concurrent executions. Second, we will augment our model with additional semantics such as object typing to allow finer-grained queries, and explore the use of an object-oriented data model augmented with transitive closure. Third, we wish to experiment with storage models such as that proposed by the Pasoa project [20] to improve query performance.

Acknowledgment

The authors wish to thank the members of the Kepler group, particularly Bertram Ludäscher, Timothy McPhillips, and Shawn Bowers, for the many fruitful discussions about scientific workflows.

References

1. Alpdemir, N., Mukherjee, A., Paton, N. W., Fernandes, A., Watson, P., Glover, K., Greenhalgh, C. Oinn,T.,Tipney,H: Contextualised Workflow Execution in myGrid, *Proc of European Grid Conference*, Springer-Verlag, LNCS **3470**, 444-453, 2005.
2. Berry, D., Buneman, P., Wilde, M., and Ioannidis, Y. editors. e-Science Workshop on Data Provenance and Annotation, *National e-Science Centre, Edinburgh*, 2003.
3. Bhagwat, D., Chiticariu, L., Tan W. C., Vijayvargiya,G.: An Annotation Management System for Relational Databases, *Proc. Conference on Very Large Data Bases (VLDB)*, 900–911, 2004.
4. Bowers,S., McPhillips, T., Ludäscher, B., Cohen, S., Davidson, S.B.: A Model for User-Oriented Data Provenance in Pipelined Scientific Workflows. *To appear in Proc. of IPAW'06 International Provenance and Annotation Workshop*, 2006.
5. Bowers, S., Ludäscher, B.: Actor-Oriented Design of Scientific Workflows, *Proc of ER'05, International Conference on Conceptual Modeling*, 369–384, 2005.
6. Buneman, P., Khanna, S., Tan, W.: Why and Where: A Characterization of Data Provenance, *Proc. of Int. Conf. on Database Theory (ICDT)*, 316–330, 2001.
7. Buneman, P., Chapman, A., Cheney, J.: Provenance Management in Curated Databases, *To appear in Proc. of SIGMOD International Conference on Management of Data*, 2006.
8. Clark, T., Martin, S., Liefeld, T.: Globally distributed object identification for biological knowledgebases. *Briefings in Bioinformatics*, **5(1)** 59–70, 2004.
9. Cohen-Boulakia, S., Lair, S., Stransky, N., Graziani, S., Radvanyi, F., Barillot, E., Froidevaux, C.: Selecting biomedical data sources according to user preferences, *Bioinformatics, Proc. ISMB/ECCB04*, **20**, i86-i93, 2004.
10. Cohen, S., Cohen-Boulakia, S., Davidson, S.: Towards a Model of Provenance in Scientific Workflows, *University of Pennsylvania, Internal Report*, #MS-CIS-06-03, 2006.
11. Davidson, S., Crabtree, J., Brunk, B., Schug, J., Tannen, V., Overton, C., Stoeckert, C.: K2/Kleisli and GUS: Experiments in integrated access to genomic data sources *IBM Systems Journal*, 2001.
12. Foster, I., Vockler, J., Woilde, M., Zhao, Y.: Chimera: A Virtual Data System for Representing, Querying, and Automating Data Derivation, Proc. of the 14th *Intl. Conf. on Scientific and Statistical Database Management (SSDBM)*, 2002.
13. Foster, I., Voeckler, J., Wilde, M., Zhao, Y.: The Virtual Data Grid: A New Model and Architecture for Data-Intensive Collaboration Proc of Conference on Innovative Data System Research (CIDR), 2003.
14. Greiner, U., Mller, R., Rahm, E., Ramsch, J., Heller, B., Lffler, M.: AdaptFlow: Protocol-based Medical Treatment Using Adaptive Workflows. *Methods of Information in Medicine*, **44**, 80–88, 2005.
15. Higgins, D. G., Sharp, P. M.: Clustal: A package for performing multiple sequence alignment on a microcomputer. Gene 73: 237-244, 1998.

16. Kiepuszewski, B., ter Hofstede, A. H. M., van der Aalst, W. M. P. : Fundamentals of control flow in workflows. Acta Inf., **39(3)**, 143–209, 2003.
17. McPhillips, T., Bowers,S.: An approach for pipelining nested collections in scientific workflows. *SIGMOD Record*, **34(3)**, 12–17, 2005.
18. Moss, J.E.B.: Nested Transactions: An Approach to Reliable Distributed Computing, *Ph.D. dissertation, Dept. of Electrical Engineering and Computer Science, MIT*, April 1981.
19. Oinn, T.M., Addis, M., Ferris, J., Marvin, D., Senger, M., Greenwood, R.T., Carver, K., Glover, Pocock, M.R., Wipat, A., Li, P. : Taverna: a tool for the composition and enactment of bioinformatics workflows, *Bioinformatics, Proc. ISMB/ECCB03*, **20(1)**, 3045–3054, 2003.
20. The Pasoa Project Luc Moreau et al. http://www.pasoa.org/
21. Phylip Programs and Documentation: http://evolution.genetics.washington.edu/phylip/phylip.html.Swofford
22. Rowe, A., Kalaitzopoulos, D., Osmond, M., Ghanem, M., Guo Y.: The discovery net system for high throughput bioinformatics *Bioinformatics*, **19(1)**, i225–i231, 2004.
23. Simmhan, Y., Plale, B., Gannon, D.: A survey of data provenance in e-science. *SIGMOD Record*, **34(3)**, 31–36, 2005.
24. Swofford D. L: PAUP*: Phylogenetic Analysis Using Parsimony (*and other methods). Sinauer Associates, Sunderland, MA, 2000.
25. Targino, R., Cavalcanti, M.C., Mattoso M.: An Environment to Define and Execute In-Silico Workflows Using Web Services. Proc. of *DILS 2005, Data Integration in the Life Sciences*, Springer-Verlag, LNBI **3615**, 288–291, 2005.
26. Ullman, J.D., Widom, J.: A First Course in Database Systems. Prentice-Hall, 1997.
27. Widom, J.: Trio: A System for Integrated Management of Data, Accuracy, and Lineage. *CIDR'05, Conference on Innovative Data Systems Research*, 262–276, 2005.
28. Zhao, J., Wroe, C., Goble, C., Stevens, R., Quan, D. and Greenwood, M.: Using Semantic Web Technologies for Representing e-Science Provenance, *Proc of Semantic Web Conference (ISWC)*, 92-106, 2004.
29. Zhao, J., Goble, C., Stevens, R., Bechhofer, S.: Semantically Linking and Browsing Provenance Logs for e-Science. *Proc of International Conference on Semantics of a Networked World (IC-SNW)*, Springer-Verlag, LNCS **3226**, 157–174, 2004.
30. http://www.extreme.indiana.edu/swf-survey/

An Extensible Light-Weight XML-Based Monitoring System for Sequence Databases

Dieter Van de Craen*, Frank Neven, and Kerstin Koch

Hasselt University and Transnational University of Limburg
School for Information Technology
firstname.lastname@uhasselt.be

Abstract. Life science researchers want biological information in their interest to become available to them as soon as possible. A monitoring system is a solution that relieves biologists from periodic exploration of databases. In particular, it allows them to express their interest in certain data by means of queries/constraints; they are then notified when new data arrives satisfying these queries/constraints. We describe a sequence monitoring system XSeqM where users can combine metadata queries on sequence records with constraints on an alignment against a given source sequence. The system is an XML-based solution where constraints are specified through search fields in a user-friendly web interface and which are then translated to corresponding XPath-expressions. The system is easily extensible as addition of new databases to the system then only amounts to the specification of new mappings from search fields to XPath-expressions. To protect private source sequences obtained in labs, it is imperative that researchers do not have to upload their sequences to a general untrusted system, but that they can run XSeqM locally. To keep the system light-weight, we therefore introduce an optimization technique based on query containment to reduce the number of XPath-evaluations which constitutes the bottleneck of the system. We experimentally validate this technique and show that it can drastically improve the running time.

1 Introduction

Motivation. Due to the increase in the speed of sequencing of genes and proteins, sequence databases, such as Genbank, double in size every two years [26]. This rapid expansion of data motivates researchers to repeat search queries over time. Indeed, a BLAST-search [13] that does not produce any useful result today might do so tomorrow. In this paper, we therefore propose a user-friendly sequence monitoring system XSeqM (*eXtensible Sequence Monitor*) that relieves researchers from repeating such searches over time.

We provide two motivating examples:

1. Researchers in a lab have obtained one or a few sequences of genes or proteins for which a BLAST-search only gives similarities for small regions of

* Corresponding author.

U. Leser, F. Naumann, and B. Eckman (Eds.): DILS 2006, LNBI 4075, pp. 280–296, 2006.

the sequence. No highly similar, annotated sequences are available in any database which might give hints for the function of the gene or protein. Therefore, the researchers regularly repeat BLAST-searches against several databases to find genes or proteins with a higher similarity.
2. A researcher has obtained a gene g expressed in the central nervous system (CNS) of the rainbow trout and is interested to learn about genes similar to g which are expressed in the peripheral nervous system (PNS) in any fish organism or mammal. She therefore repeats a BLAST-search with the gene g on a weekly basis.

The two tasks described above are tedious and time consuming when executed manually: not only the BLAST-searches themselves, but also the post-processing of the BLAST-reports (if any) to sort out relevant matches from irrelevant ones. Indeed, in situation (1), a match could be irrelevant as the matched part of the sequence is too small or the likelihood of the match expressed by the E-value is too large. In situation (2), all BLAST-hits from non-fish and non-mammal species should be discarded together with those that are not mRNA and that do not refer to the PNS.

A Solution: The XSeqM-System. In the XSeqM-system users can register BLAST-requests combined with constraints on the metadata of a sequence record. All requests are checked locally by the system after retrieval of the daily updates from the respective databases and users are informed, for instance through email, when relevant results are found. Figure 2 shows part of the monitor request related to situation (2). In brief, every such request specifies the following information:

- a database of interest (e.g., Genbank, SwissProt, ...),
- a sequence of interest (e.g., the gene g),
- constraints on the metadata (e.g., classification should contain the string 'fish' and molecular type should equal 'mRNA')
- an alignment program and its parameters (e.g., BLAST with word size 11 and matrix PAM30)
- relevance constraints (e.g., size of match should be greater than 20 and E-value should be smaller than e^{-10}).

The XSeqM-system has the following characteristics:

1. XSeqM is light-weight. It can be installed locally in a lab on a computer with average system requirements. This is important, as, referring to situation (1) above, research labs can be hesitant to upload their newly found sequences in a public system as some of them might be candidates for a patent application.
2. XSeqM is user-friendly as it hides all use of XML: users interact with the system through a Web-interface where search fields can be combined using the logical operators, much like other query and monitoring systems such as SRS and PubCrawler [22].
3. XSeqM is a flexible XML-based solution to which any sequence database can be added that makes updates available and whose format can be transformed

into XML. Almost all sources nowadays allow to export information in XML-format or there are third party tools available to convert existing formats to XML. The administrator determines for every sequence database a number of search fields. For every search field f, an XPath-expression P_f is created that selects the corresponding value in every XML-file in the update. Table 1, for instance, lists the interesting search fields for a GenBank record and the corresponding XPath-expressions. Every user request is then translated under the hood to a Boolean combination of XPath-expressions. Similarly, relevance constraints on BLAST-reports are translated into XPath-expressions over the XML-representation. Therefore, in principle, any XPath-expressible constraints can be used.

Efficient Evaluation. The main technical part of the paper deals with efficient execution of all monitoring requests. In brief, the system executes the following steps. Let $m_1, \ldots, m_k$ be all monitoring requests with corresponding constraints $p_1, \ldots, p_k$ on the metadata, i.e. Boolean combinations of XPath-expressions. For every sequence record s in the update, we need to check which expressions p_i match s. When p_i is successfully matched, we BLAST the sequence in s against the sequence in m_i. When all relevance constraints of m_i on the BLAST-report are satisfied, the owner of request m_i is alerted. As an alignment of sequences through BLAST is expensive, it is imperative to first check the metadata constraints and only start BLAST for those sequences which are selected.

As every local lab is considered to have its own system, we consider systems of moderate size (say, a few thousands of monitoring requests). Daily updates to Genbank vary in size from 50 to 200 Megabytes (zipped): these contain between 30000 and 150000 sequences. The bottleneck of the system is in the evaluation of the constraints $p_1, \ldots, p_k$ for every sequence record s in the update. A direct approach using a standard XPath-evaluator like Xalan[1] takes more than 24 hours and is therefore not an option. Powerful fast streaming XPath-engines have been proposed over the past years [21,12] which can handle millions of XPath-expressions. Unfortunately, we cannot use these engines directly: to ensure high throughput streaming engines do not consider full XPath. In particular, they do not consider arbitrary Boolean combinations of XPath-expression or allow to test whether a certain given string occurs as a substring of a text element. We therefore make use of the state-of-the-art evaluator YFilter [18,19] as a first pre-processing step to extract string-values from sequence records. More precisely, by evaluating for every search field the corresponding expression P_f on the update, we get for every sequence record a complex value representation on which the metadata constraints can be checked. E.g., Table 2 contains such a representation for the GenBank record of Figure 1 through the XPath expressions in Table 1. In a second step, we then evaluate every pattern p_i on this representation. An additional advantage of this method is that more advanced pattern matching on string values can be used than is available in XPath. For instance, one could require that the string value matches a given regular expression.

We consider an optimization based on containment of constraints. As the system runs at a local lab, chances are high that many constraints on the metadata

Table 1. Search fields for a GenBank record and corresponding XPath-expressions

f	P_f
organism	/p/e[@class="source"]/Qualifier[@value-type="organism"]/@value
accession	/p/@ic-acckey
gi	/p/Attribute[@name="primary_id"]/@content
author name	/p/q[@title="Sequence References"]/Reference/RefAuthors/text()
title	/p/q[@title="Sequence References"]/Reference/RefTitle/text()
keyword	/p/Attribute[@name="keyword"]/@content
comment	/p/Attribute[@name="comment"]/@content
classification	/p/Attribute[@name="classification"]/@content
Feature key	/p/e/@class
Gene name	/p/e[@class="gene"]/Qualifier[@value-type="gene"]/@value
Protein name	/p/e[@class="cds"]/Qualifier[@value-type="product"]/@value
chromosome	/p/e[@class="source"]/Qualifier[@value-type="chromosome"]/@value
molecular type	/p/e[@class="source"]/Qualifier[@value-type="mol_type"]/@value
tissue type	/p/e[@class="source"]/Qualifier[@value-type="tissue_type"]/@value
tissue library	/p/e[@class="source"]/Qualifier[@value-type="tissue_lib"]/@value
cell line	/p/e[@class="source"]/Qualifier[@value-type="cell_line"]/@value
development stage	/p/e[@class="source"]/Qualifier[@value-type="dev_stage"]/@value
EC Number	/p/e[@class="cds"]/Qualifier[@value-type="EC_number"]/@value
p	/Bsml/Definitions/Sequences/Sequence
e	Feature-tables/Feature-table[@title="Features"]/Feature
q	Feature-tables/Feature-table

are related. For instance, a constraint could require that the organism should contain the string 'Oncorhynchus' while another query could require that the organism should equal 'Oncorhynchus mykiss' and the tissue type equals 'brain'. Clearly, the second constraint implies the first. So, we know that the first constraint is true when the second is, and the second is false when the first is. Our optimization technique exploits these ideas to reduce the number of evaluations. More precisely, we define a graph structure that captures the relationships between the constraints and consider two forms of propagation: false propagation and true propagation. We experimentally show that false propagation outperforms true propagation and the pure streaming approach.

Finally, we discuss how to incrementally maintain the containment graph. It never has to be computed from scratch. The insertion operation is time consuming as in the worst case it involves a linear number of containment checks (a coNP-hard problem [20]). Luckily only a limited number of insertions are expected on a daily basis, say at most hundred, which for a system already containing 5000 requests can be done in less than 100 minutes. In case a larger number of insertions is required, we discuss a technique that accelerates the containment check at the expense of introducing more requests: constraints are transformed into disjunctive normal form, testing containment of conjuncts can then be done in linear time. For instance, adding 100 request to a containment graph with 5000 nodes then only takes 12 seconds.

Table 2. Complex value representation of the GenBank record in Figure 1

f	values
organism	{"Oncorhynchus mykiss"}
accession	{ "AM181351" }
gi	{ "84993308" }
author name	{ "Zarkadis,I.K. and Marioli,D.", "Zarkadis,I.K." }
title	{ "Cloning of the vitronectin gene in rainbow trout", "Direct Submission" }
keyword	{ "vitronectin protein 1", "vtn1 gene" }
comment	{ }
classification	{ "mykiss Oncorhynchus Salmonidae Salmoniformes Protacanthopterygii Euteleostei Teleostei Neopterygii Actinopterygii Euteleostomi Vertebrata Craniata Chordata Metazoa Eukaryota" }
Feature key	{ "source", "gene", "cds" }
Gene name	{ "vtn1" }
Protein name	{ "vitronectin protein 1" }
chromosome	{ }
molecular type	{"mRNA"}
tissue type	{ "liver" }
tissue library	{ }
cell line	{ }
development stage	{ }
EC Number	{ }

Outline. This paper is organized as follows. In Section 2, we survey other monitoring approaches. Section 3 introduces XML and XPath. Section 4 gives an overview of XSeqM. In Section 5, we outline several evaluation strategies. Section 6 reports on our experiments. In Section 7, we discuss the incremental maintenance of the containment graph. We conclude in Section 8.

2 Related Work

Existing alerting systems like BioMail, JADE or Science Direct are used for literature alerts [3,4,9]. They search the PubMed database in given intervals and alert users via email if new publications matching special keywords are available [25]. The only system integrating query possibilities for Genbank in addition to literature alerts is PubCrawler [22,23]. PubCrawler provides a user with the possiblity to define two types of queries. The first type is a keyword search and the second is a neighborhood query. With a neighborhood query a user can express his interest in articles or sequences that are similar to given articles or sequences already present in the database. A limitation of this approach is that the user can not enter an unpublished sequence which has no identifier assigned yet. Also, advanced options in the comparison with other sequences are not provided, e.g., the minimal length of a match or the E-value. XSeqM does provide these possibities and allows for the combination of a keyword search and an alignment with any given sequence.

XML filtering systems evaluate a set of queries against a stream of documents. The XMLTK system [21] combines all path expressions into a single deterministic finite automaton. YFilter [18,19], the successor of XFilter [12], combines all expressions in one nondeterministic finite automaton. These systems thus employ a finite state automaton for all the XPath-expressions. The XML stream is parsed by a SAX parser and the SAX events are streamed through the finite state automaton. A query matches a document if during parsing an accepting state for that query is reached. The main limitation of these systems compared to XSeqM is that they do not support full XPath. As the translation of user queries in XSeqM can result in complex XPath-expressions, these systems can not be applied directly in our situation.

In [14,15] and [24] optimization of navigational queries on life science sources is investigated. In this setting alternate paths are possible to evaluate a query. The focus in [14,15] is on finding a set of paths that maximizes the number of results while satisfying a constraint on the evaluation cost. Minimizing the total number of accesses to sources when evaluating multiple queries in batch mode is discussed in [24]. The goal of XSeqM differs from these as we want to monitor multiple sources seperately rather then answering queries over multiple sources.

3 XML and XPath

The eXtensible Markup Language (XML) is a standard for data exchange on the Web [10]. Most bioinformatics data formats can be converted into an XML representation. Numerous XML formats for a wide range of biological data are available. Some examples are BSML, SPTr-XML, GO-XML,... [16].

XPath is an XML pattern language for locating information in XML documents [17]. In particular, XPath can retrieve the value of elements or attributes and can test whether that value satisfies a certain condition. We give an example of both. The expression `//Attribute[@name="classification"]/@content`, for instance retrieves the classification of an entry as the actual classification is the value of the `content` attribute of an `Attribute` element that has a `name` attribute with value 'classification'. The expression

```
boolean(//Attribute[@name="classification" and contains(@content,"Mammalia"])
```

checks whether the classification contains the string 'Mammalia'. XPath can also be used to query the XML-representation of a BLAST-report. For instance, the expression `//Hit[Hit_num/text()="1"]/Hit_hsps/Hsp/Hsp_evalue/text()` selects the E-value of the first hit.

4 Monitoring System

We detail the three different components of XSeqM which are graphically illustrated in Figure 3.

```
LOCUS       AM181351               674 bp    mRNA    linear   VRT 16-JAN-2006
DEFINITION  Oncorhynchus mykiss partial mRNA for vitronectin protein 1 (vtn1
            gene), isolated from liver.
ACCESSION   AM181351
VERSION     AM181351.1  GI:84993308
KEYWORDS    vitronectin protein 1; vtn1 gene.
SOURCE      Oncorhynchus mykiss (rainbow trout)
  ORGANISM  Oncorhynchus mykiss
            Eukaryota; Metazoa; Chordata; Craniata; Vertebrata; Euteleostomi;
            Actinopterygii; Neopterygii; Teleostei; Euteleostei;
            Protacanthopterygii; Salmoniformes; Salmonidae; Oncorhynchus.
REFERENCE   1
  AUTHORS   Zarkadis,I.K. and Marioli,D.
  TITLE     Cloning of the vitronectin gene in rainbow trout
  JOURNAL   Unpublished
REFERENCE   2  (bases 1 to 674)
  AUTHORS   Zarkadis,I.K.
  TITLE     Direct Submission
  JOURNAL   Submitted (11-JAN-2006) Zarkadis I.K., Dept. of Biology, School of
            Medicine, University of Patras, Rion, Panepistimioupolis, ...
FEATURES             Location/Qualifiers
     source          1..674
                     /organism="Oncorhynchus mykiss"
                     /mol_type="mRNA"
                     /db_xref="taxon:8022"
                     /tissue_type="liver"
     gene            <1..>674
                     /gene="vtn1"
     CDS             <1..>674
                     /gene="vtn1"
                     /codon_start=1
                     /product="vitronectin protein 1"
                     /protein_id="CAJ57657.1"
                     /db_xref="GI:84993309"
                     /translation="SCCMDF..."
ORIGIN
        1 agctgctgca tggacttcga cagtgcctgc cctaggaaga tttcccgcgg tgacacattt
          ...
      661 tgtgtgcgct tgac
```

Fig. 1. Example GenBank entry

1. The Input Module consists of the WWW Interface and the Query Translation Module. As illustrated in Figure 2(top), a query is created in the WWW Interface by uploading a sequence and specifying search terms in search fields. These search fields are then linked together by selecting the appropriate logical connectors: AND, OR and NOT, and parentheses. This method of operation is similar to the one used in other query and monitoring systems such as SRS and PubCrawler. Queries entered by users are then translated by the Query Translation Module to a Boolean formula and a mapping from which the corresponding XPath expression can be constructed. An example of part of this translation is given in Figure 2(bottom). The user therefore does not have to be aware of the underlying technology used. The monitor request is stored in the local repository.
2. The Evaluation Module is responsible for the actual evaluation of the monitor requests. It consist of the Request Evaluator, the Alignment Module and the Report Generator. When the Evaluation Module receives an update of a database, say GenBank, the monitor requests concerning GenBank are

fetched from the local repository. The evaluation then proceeds as follows. First, the Request Evaluator evaluates the metadata constraints of the monitoring requests on all sequence records in the update. The Alignment Module then aligns every selected sequence with the corresponding source sequences. Finally, the Report Generator constructs a report from every BLAST-report that satisfies the relevance constraints and notifies the owner of the monitor request.

3. The Update Module consist of the BioDBInterface and the XML Converter Module. The BioDBInterface Module checks at regular timepoints whether updates to some of the monitored databases are available. If such an update is available, then the BioDBInterface Module fetches this update and passes it on to the Evaluation Module. Despite the fact that more and more biological data is available as XML, not all data is. In such a case, the XML Converter Module will convert the update into an XML-format.

5 Evaluation Strategies

We provide an abstract view of the different parts of the evaluation algorithm. Let $m_1, \ldots, m_k$ be an enumeration of all monitoring requests. Every request $m_i = (p_i, s_i, r_i)$ consists of a metadata constraint p_i, a sequence s_i, and a relevance constraint r_i. Let $u_1, \ldots, u_\ell$ be an enumeration of sequence records constituting an update. The system provides the following steps:

1. Compute the set of pairs $N = \{(j, i) \mid u_j \text{ matches } p_i\}$.
2. For every $(j, i) \in N$, align the sequences u_j and s_i through BLAST resulting in a BLAST-report $R_{j,i}$.
3. When $R_{j,i}$ matches r_i, warn the owner of request m_i.

The bottleneck of the system is located in step (1) above: testing the metadata constraints. When M and N are the number of monitoring requests and the number of sequence records in an update, respectively, then $M \times U$ constraints need to be checked. We consider systems where M can be 5000 and U can be 10^5. Step (2) can be evaluated quite fast (on average 10^4 pairs of sequences can be aligned with BLAST on a local system in less then half a minute). For step (3), the same techniques as for step (1) can be used, although in general this step can be done quite fast as the number of BLAST-reports will be much smaller than $M \times U$. In the rest of this section, we outline several evaluation strategies for step (1) which are experimentally evaluated in the next section.

5.1 Naive Brute Force Evaluation

The first evaluation strategy is a simple brute force method which tests every constraint p_i for every entry in the update. To evaluate the XPath-expressions, we use Xalan [1].

Insert query concerning GenBank

Fill in this form to insert a new query.

☑ Add Sequence?

Sequence name : MyTestSequence Program : Blast

E value : 10 Word Size : 11

Size of Match : 20 Other Advanced :

Sequence

gcagtgccctgcagcgggaagcccttcgacgccttcctccagctcaagaacggatccatc
tacgccttcagaggtgattatttctttgagtcgatgagagggcggttgtccaccggttac
cccaaactgatctacgacaagtggggcatcagaggacctatagatgctgcctttactcgc

Pattern :

Fill in your query:

		classification	contains	fish
AND		tissue type	contains	brain
AND		molecular type	equals	mRNA

Add Query Reset

Blast

ID	sequence	Evalue	wordsize	MatchSize
...				
51	gcagtgcc...	10	11	20
...				

Query

ID	userID	database	formula
...			
51	8	genbank	v_51_1 & v_51_2 & v_51_3
...			

Mapping

ID	variable	querytype	keyword	value
...				
51	v_51_1	contains	classification	fish
51	v_51_2	contains	tissue_type	brain
51	v_51_3	equals	molecular_type	mRNA
...				

Fig. 2. Example of a monitoring request and its translation

5.2 XML Streaming Approach

An XML stream query processing system takes as input a stream of XML documents on which it evaluates queries simultaneously. Filtering systems such as XFilter, YFilter, XMLTK, ... are freely available and provide efficient evaluation of large numbers of XPath-expressions. The problem with these systems is that the XPath fragment they consider is not powerful enough to express our

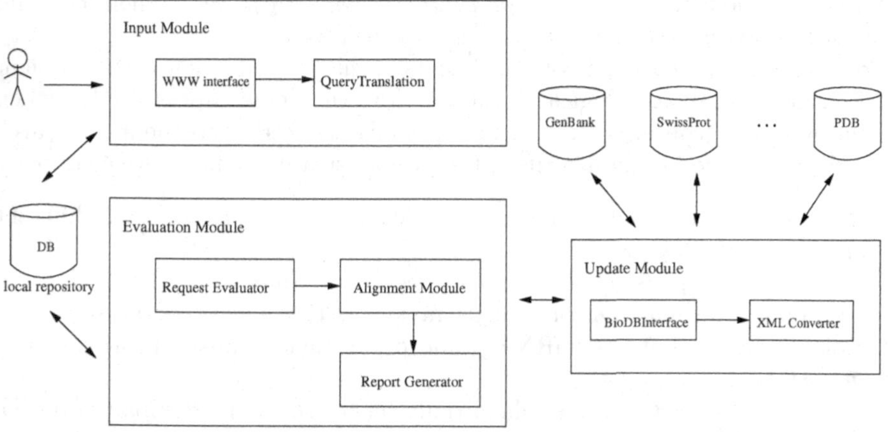

Fig. 3. Overview of the modules

user constraints. However, if we look at the number of search fields that can be queried in our setting, we observe that this number is small (typically 10 to 20) and fixed in advance. So, instead of evaluating the XPath-expressions generated from the user constraints directly on the updates, we proceed in two steps:

1. Using YFilter, we retrieve all the values for the search fields for a sequence record of an update and create for each record a complex value representation. E.g., Table 2 contains a complex value representation of the GenBank record of Figure 1 obtained through the XPath expressions in Table 1.
2. In a second step, we evaluate the metadata constraints on this complex value representation. For instance, the constraint

   classification.contains('Teleostei') AND tissue_type.contains('brain') AND molecular_type.contains('mRNA')

   is not satisfied on the record in Table 2 as the tissue_type does not contain brain. The semantics of the contains('s') operator is that at least one of the strings in the set should contain the string 's' as a substring.

5.3 Query Containment

The evaluation of expressions in step two above is still naive: all expressions are matched against all entries in the update. As the XSeqM-system runs at a local lab where researchers are working on related topics, chances are high that some constraints on the metadata are related. A useful notion in this context is the following: a constraint p is contained in a constraint p', denoted $p \subseteq p'$, if whenever a sequence record satisfies p it also satisfies p'. For instance, let p be the constraint

organism.equals('Oncorhynchus mykiss') AND tissue_type.contains('brain') AND molecular_type.contains('mRNA') and let p' be the constraint *organism.contains('Oncorhynchus')*. Then it should be clear that every record which satisfies p also satisfies p'.

So, containment checking of constraints basically reduces to containment checking of propositional logical formulas. However, some care is needed when dealing with the 'contains' and 'equals' predicate referring to the same search field. We use the following algorithm that we illustrate on the above example:

- Rewrite the constraints p and p' to logical formulas q and q' over different propositional symbols.

  That is, q equals $a \wedge b \wedge c$ and q' equals d. Here, a, b, c and d stand for organism.equals('Oncorhynchus mykiss'), tissue_type.contains('brain'), molecular_type.contains('mRNA'), and organism.contains('Oncorhynchus'), respectively.
- Let γ be a propositional formula initially set to true. For every pair of propositional variables x and y referring to the same search field, test whether the constraint corresponding to x is contained in the constraint corresponding to y. If so, add $\neg x \vee y$ to γ. The intuition is that γ restricts the set of possible truth assignments to those that correspond to the semantics of the 'contains' and 'equals' predicates. In particular, the formula $\neg x \vee y$ only accepts truth assignments that assign true to y when x is also true, which encodes that x implies y.

  The only variables referring to the same search field are a and d. Clearly, a is contained in d as every record satisfying organism.equals('Oncorhynchus mykiss') also satisfies organism.contains('Oncorhynchus'). So, γ is the formula $\neg a \vee d$.
- Now, $p \subseteq p'$ iff $q \wedge \neg q' \wedge \gamma$ is unsatisfiable.

  So for our example, we need to test that $a \wedge b \wedge c \wedge \neg d \wedge (\neg a \vee d)$ is not satisfiable, which is the case. Indeed, the only way to satisfy the first four conjuncts is to set a, b, and c true and d false, but this is prohibited by the last conjunct.

In general testing unsatisfiability is coNP-complete [20]. Fortunately, the expressions we consider are very small. We make use of the state-of-the-art SAT-solver Limmat [6]. As our formulas are in general not in CNF, we use Limboole [5], a front end to Limmat that allows to check unsatisfiability of arbitrary formulas and not just formulas in CNF.

The **containment DAG** of a set of constraints is a directed acyclic graph (DAG) without any transitive edges where every node represents a set of equivalent constraints and there is an edge from node n to node n' if every expression in n is contained in every expression in n'. Note that it is sufficient to test if one expression from n is contained in one expression from n'. A **source** is a node without incoming edges; a **sink** is a node without outgoing edges. Note that a DAG can have multiple sources and sinks.

We make the following observations:

- to check whether a sequence record matches the expressions in a node n, it suffices to test this for one expression in n;

– when an expression in n is true for a sequence record, then all expressions in descendant nodes of n are true for that record; and
– when an expression in n is false for a sequence record, then all expressions in ancestor nodes of n are false for that record.

In the following, the evaluation of a node against a sequence record corresponds to selecting one of the equivalent expressions the node represents and matching this expression against the record. The above observations lead to two related optimization techniques allowing to discard nodes in the containment DAG:

1. **false propagation:** start evaluation in the sinks, when a node evaluates to false all ancestors can be discarded as they evaluate to false, when the node evaluates to true all parents have to be addressed;
2. **true propagation:** start evaluation in the sources, when a node evaluates to true all descendants can be discarded as they evaluate to true as well, when the node evaluates to false all its children have to be addressed.

Note that a node can be reached by multiple paths. So, to avoid multiple evaluations of nodes every node carries a bit indicating whether the node is already evaluated or not. It is clear that if expressions seldom match entries in the update then false propagation will result in a strong decrease in the number of actual evaluations. In the case that expressions frequently match entries, the use of true propagation can be advantageous.

6 Experiments

In this section, we experimentally validate our optimization techniques. We created monitoring requests resulting in three types of containment DAGs T1, T2, and R (cf. Figure 4). We repeated our experiments for different numbers of monitoring request (from 1000 till 5000). We only report on the case with 5000 requests as all experiments produced similar results. The experiments were performed on a Pentium IV (3.0 GHz) architecture with 1 GB of internal memory running under Linux 2.6. All programs are written in Java.

The metadata constraints were created by extracting possible values out of available updates. The first type of containment DAG (T1) is specially tailored for false propagation. Part of a DAG of type T1 is given in Figure 4(top). It is a reversed tree consisting of a small number of sinks. It is constructed by only making use of AND-operators. The idea is that every sink represents the most general constraint which is subsequently refined by additional constraints when progressing upwards. For instance, a sink may state that the organism in the update matches 'Oncorhynchus mykiss', its parent may refine this by adding another constraint, namely that the molecular type must be 'mRNA'. Trees can be disjoint, for instance, when each of them corresponds to an organism.

The shape of the second type of containment DAG (T2) is the reverse of the first one and is ideal for true propagation. Part of a DAG of this type is given in

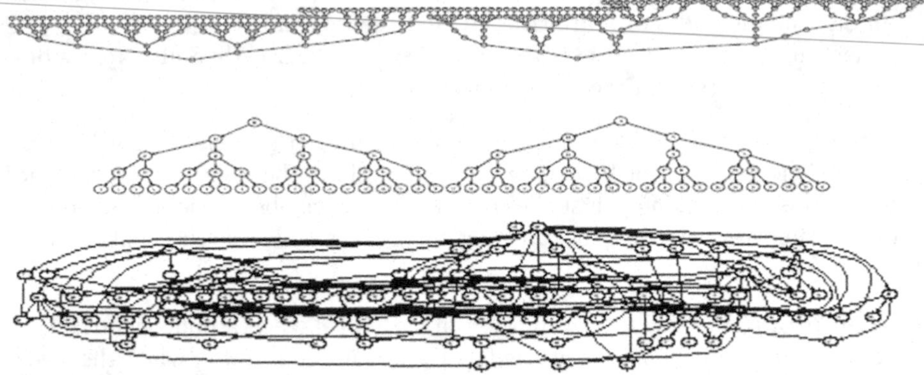

Fig. 4. Fragment of example containment graphs T1 (top), T2 (middle), and R (bottom). Edges point downwards.

Figure 4(middle). The idea is that each source is the most restrictive constraint which gets relaxed by every descendant.

The last type of containment DAG (R) was created by generating random constraints (using AND, OR, and NOT-operators) and creating the containment DAG. Figure 4(bottom) shows the typical shape of such a containment DAG. To keep the comparison of the different evaluation strategies fair, we have eliminated all equivalent constraints but one from every node.

Figure 5(left) shows the average time in seconds to evaluate the constraints of 5000 monitoring requests for an update containing 10^5 sequence records. Note that the scale is logarithmic. The figure clearly indicates that the naive brute

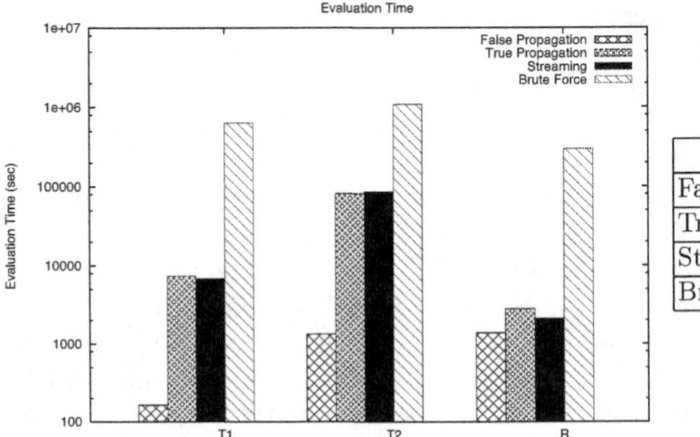

	T1	T2	R
False prop	0.5%	52%	38%
True prop	99%	99%	93%
Streaming	100%	100%	100%
Brute force	100%	100%	100%

Fig. 5. (left) Average evaluation time in seconds for 5000 monitoring requests on an update consisting of 10^5 sequence records; (right) Average percentage of nodes in the DAG that are evaluated

force method is unsatisfactory and that the pure streaming method presents a definite improvement. Further, false propagation outperforms every other method where the obtained acceleration ranges from twice the speed (on R) to several orders of magnitude (on T1 and T2). True propagation does not result in any improvement as only very few constraints evaluate to true, thereby severely limiting the effect of true propagation. That only a very small part of the input data is selected, is inherent to the situation of a monitoring system where on the one hand users are interested in very specific sequences and on the other hand research labs usually upload new sequences in bulk giving rise to many related sequences, for instance, of the same organism. So, when no request related to those organisms are specified a lot of updates are already discarded.

Figure 5(right) shows the average percentage of nodes in the DAG that every method evaluates. For the pure streaming and the brute force approach this is of course 100%. On graphs of type T2, false propagation only needs to evaluate on average 50% of the nodes. So, one would expect that false propagation is twice as fast as the pure streaming approach. However, the experiments show that the latter is in fact 100 times faster. The reason is that false propagation starts at the sinks which contain the simplest and fast to evaluate constraints. So, the 50% of the nodes the method allows to discard contain the largest and most time consuming to evaluate expressions. Actually, the reason that the pure streaming method takes much more time on graphs of type T2 than on the other graphs is that due to the construction of the graph (starting from the most specific constraint which gets relaxed by every descendant) constraints are on average more involved. Further, for random graphs, false propagation needs only to evaluate 40% of the nodes on average, but only a speed up of a factor two is obtained w.r.t. pure streaming. The reason is that due to the random DAG structure of the graph, a lot of time is spent keeping track of which nodes still need to be evaluated.

So, in all cases false propagation is the best method. The amount of improvement depends on the shape of the containment graph.

7 Incremental Maintenance of the Containment DAG

The timings in the experiments of the previous section only concern the evaluation of the constraints and not the time needed to compute the containment DAG. Indeed, the containment DAG is independent of the updates and can therefore be computed beforehand. Actually, the DAG never needs to be computed from scratch but can be maintained incrementally when new monitor requests arrive or are removed.

Removing a request with constraint p is easy: locate the corresponding node in the DAG and remove p from it; when no more constraints are present in the node remove it and add edges from all its parents to all its children. Adding a request with constraint p is more time consuming as in the worst case a linear number of containment checks need to be done. In particular, we need to compute the set of nodes U and L of upper and lower border nodes, respectively, such that the following holds:

- for every constraint p_1 in a node in U, $p_1 \subseteq p$; there is no descendant node of a node in U that contains a constraint p_1 with $p_1 \subseteq p$;
- for every constraint p_2 in L, $p \subseteq p_2$; there is no ancestor node of a node in L that contains a constraint p_2 with $p \subseteq p_2$.

When there is a node $n \in U \cap L$, then p is equivalent to all constraints in n: add p to n. Otherwise, add a new node n with constraint p, add edges to n from all nodes in U, add edges from n to all nodes in L.

Instead of using a naive brute-force approach which checks for every node in the DAG whether it is in U or L, we compute an initial upper and lower border, and gradually refine them:

1. Initially, let U be the set of all sources, and let L be the set of all sinks;
2. repeat until no more changes
   (a) if there is a child (with constraint p_1) of a node n in U such that $p_1 \subseteq p$ then replace n in U by all such children of n; and,
   (b) if there is a parent (with constraint p_2) of a node n in L such that $p \subseteq p_2$ then replace n in L by all such parents of n.

Although our incremental algorithm reduces the number of containment tests, the time spend performing containment tests is not negligible. For instance, inserting a constraint in a containment DAG already containing 1250, 2500 and 5000 nodes takes on average 15, 30, and 60 seconds, respectively. In general, this is not a problem as only a limited number of monitor requests will be added/removed (say at most hundred).

At present, the construction of the containment graph is not completely satisfactory. The bottleneck is the high complexity of the containment test. One possibility to speed up the containment test is to transform constraints into disjunctive normal form giving rise to a containment test of quadratic time complexity (as opposed to exponential). Preliminary experiments on graphs of type R show that this increases the total size of constraints by 25%. Constructing a containment graph with 5000 nodes from scratch then only takes 60 seconds. It remains to investigate in further detail the trade-of between fast containment graph construction and the increase in size of the containment graph. Another option, of course, is to require that users enter metadata constraints in disjunctive normal form.

8 Conclusion

We have shown that the combination of state-of-the-art tools together with an optimization technique suffices to implement a monitoring system for sequence databases. A prototype will be made available soon for download at [11]. Currently, only Genbank is supported and as relevance constraints the system only allows to specify constraints on E-value and the size of the match. At present, some of the biologists of our university are testing the system and we are incorporating their feedback. For instance, one feature that needs to be added is shredding of input sequences. Rather than blasting a complete sequence they want to blast every subsequence of a certain size. The latter will pose new computational challenges. In particular, we want to improve the evaluation of the false propagation method on highly linked DAGs (of type R). Further, all tests in this paper are performed on generated data. We hope that the cooperation with the biologists will give us enough real world data to test and improve our algorithms on.

References

1. The Apache Xalan Project. http://xalan.apache.org.
2. Bioinformatic Sequence Markup Language (BSML). http://www.bsml.org.
3. Biomail. http://biomail.sourceforge.net/biomail.
4. Jade. http://www.biodigital.org/jade.
5. Limboole. http://fmv.jku.at/limboole/.
6. Limmat. http://fmv.jku.at/limmat/.
7. Pubcrawler. http://www.pubcrawler.ie.
8. PubMed Cubby. http://www.pubmed.gov.
9. Sciencedirect. http://www.sciencedirect.com.
10. World Wide Web Consortium. Extensible Markup Language (XML). http://www.w3.org/XML.
11. The XSeqM website. http://alpha.uhasselt.be/dieter.vandecraen/XSeqM/.
12. M. Altinel and M. J. Franklin. Efficient filtering of XML documents for selective dissemination of information. In *Proceedings of the 26th International Conference on Very Large Data Bases (VLDB 2000)*, pages 53–64. Morgan Kaufmann Publishers Inc., 2000.
13. S.F. Altschul, W. Gish, W. Miller, E.W. Myers, and Lipman D.J. Basic local alignment search tool. *Journal of Molecular Biology*, 215(3):403–410, 1990.
14. J. Bleiholder, S. Khuller, F. Naumann, L. Raschid, and Y. Wu. Query planning in the presence of overlapping sources. In *Proceedings of the 10th International Conference on Extending Database Technology (EDBT 2006)*, pages 811–828. Springer, 2006.
15. J. Bleiholder, Z. Naumann, F.and Lacroix, L Raschid, H. Murthy, and M.-E. Vidal. Biofast: challenges in exploring linked life sciences sources. *SIGMOD Record*, 33(2):72–77, 2004.
16. E. Cerami. *XML for Bioinformatics.* Springer-Verlag, 2004.
17. J. Clark. XML Path Language (XPath). http://www.w3.org/TR/xpath.
18. Y. Diao, P. Fischer, M. Franklin, and R. To. YFilter: Efficient and Scalable Filtering of XML Documents. In *Proceedings of the 18th International Conference on Data Engineering (ICDE'02)*, page 341. IEEE Computer Society, 2002.
19. Y. Diao and M.J. Franklin. High-Performance XML Filtering: An Overview of YFilter. *IEEE Data Engineering Bulletin*, 26(1):41–48, 2003.
20. M.R. Garey and D.S. Johnson. *Computers and Intractability: A Guide to the Theory of NP-Completeness.* Freeman, 1979.
21. T.J. Green, G. Miklau, M. Onizuka, and D. Suciu. Processing XML streams with deterministic automata. In *Proc. 9th International Conference on Database Theory (ICDT 2003)*, pages 173–189, 2003.
22. K. Hokamp and K. Wolfe. What's new in the library? What's new in GenBank? Let PubCrawler tell you. *Trends in Genetics*, 15(11):471–472, 1999.

23. K. Hokamp and K.H. Wolfe. PubCrawler: keeping up comfortably with PubMed and GenBank. *Nucleic Acids Research*, 32(Web Server Issue):W16–W19, 2004.
24. F. Neven and D. Van de Craen. Optimizing monitoring queries over distributed data. In *Proceedings of the 10th International Conference on Extending Database Technology (EDBT 2006)*, pages 829–846. Springer, 2006.
25. M. Shultz and S.L. De Groote. MEDLINE SDI services: how do they compare? *Journal of the Medical Library Association*, 91(4):460–467, 2003.
26. J. F. Wilson. The rise of biological databases. *The Scientist*, 16(6):34, 2002.

Author Index

Lecture Notes in Bioinformatics

Vol. 4075: U. Leser, F. Naumann, B. Eckman (Eds.), Data Integration in the Life Sciences. XI, 298 pages. 2006.

Vol. 3939: C. Priami, L. Cardelli, S. Emmott (Eds.), Transactions on Computational Systems Biology IV. VII, 141 pages. 2006.

Vol. 3916: J. Li, Q. Yang, A.-H. Tan (Eds.), Data Mining for Biomedical Applications. VIII, 155 pages. 2006.

Vol. 3909: A. Apostolico, C. Guerra, S. Istrail, P. Pevzner, M. Waterman (Eds.), Research in Computational Molecular Biology. XVII, 612 pages. 2006.

Vol. 3886: E.G. Bremer, J. Hakenberg, E.-H.(S.) Han, D. Berrar, W. Dubitzky (Eds.), Knowledge Discovery in Life Science Literature. XIV, 147 pages. 2006.

Vol. 3745: J.L. Oliveira, V. Maojo, F. Martín-Sánchez, A.S. Pereira (Eds.), Biological and Medical Data Analysis. XII, 422 pages. 2005.

Vol. 3737: C. Priami, E. Merelli, P. Gonzalez, A. Omicini (Eds.), Transactions on Computational Systems Biology III. VII, 169 pages. 2005.

Vol. 3695: M.R. Berthold, R.C. Glen, K. Diederichs, O. Kohlbacher, I. Fischer (Eds.), Computational Life Sciences. XI, 277 pages. 2005.

Vol. 3692: R. Casadio, G. Myers (Eds.), Algorithms in Bioinformatics. X, 436 pages. 2005.

Vol. 3680: C. Priami, A. Zelikovsky (Eds.), Transactions on Computational Systems Biology II. IX, 153 pages. 2005.

Vol. 3678: A. McLysaght, D.H. Huson (Eds.), Comparative Genomics. VIII, 167 pages. 2005.

Vol. 3615: B. Ludäscher, L. Raschid (Eds.), Data Integration in the Life Sciences. XII, 344 pages. 2005.

Vol. 3594: J.C. Setubal, S. Verjovski-Almeida (Eds.), Advances in Bioinformatics and Computational Biology. XIV, 258 pages. 2005.

Vol. 3500: S. Miyano, J. Mesirov, S. Kasif, S. Istrail, P. Pevzner, M. Waterman (Eds.), Research in Computational Molecular Biology. XVII, 632 pages. 2005.

Vol. 3388: J. Lagergren (Ed.), Comparative Genomics. VII, 133 pages. 2005.

Vol. 3380: C. Priami (Ed.), Transactions on Computational Systems Biology I. IX, 111 pages. 2005.

Vol. 3370: A. Konagaya, K. Satou (Eds.), Grid Computing in Life Science. X, 188 pages. 2005.

Vol. 3318: E. Eskin, C. Workman (Eds.), Regulatory Genomics. VII, 115 pages. 2005.

Vol. 3240: I. Jonassen, J. Kim (Eds.), Algorithms in Bioinformatics. IX, 476 pages. 2004.

Vol. 3082: V. Danos, V. Schachter (Eds.), Computational Methods in Systems Biology. IX, 280 pages. 2005.

Vol. 2994: E. Rahm (Ed.), Data Integration in the Life Sciences. X, 221 pages. 2004.

Vol. 2983: S. Istrail, M.S. Waterman, A. Clark (Eds.), Computational Methods for SNPs and Haplotype Inference. IX, 153 pages. 2004.

Vol. 2812: G. Benson, R.D. M. Page (Eds.), Algorithms in Bioinformatics. X, 528 pages. 2003.

Vol. 2666: C. Guerra, S. Istrail (Eds.), Mathematical Methods for Protein Structure Analysis and Design. XI, 157 pages. 2003.